INDONESIA

ENERGY POLICY, LAWS AND REGULATIONS HANDBOOK

VOLUME 1

STRATEGIC INFORMATION AND BASIC LAWS

International Business Publications, USA
Washington DC, USA - Indonesia

INDONESIA

ENERGY POLICY, LAWS AND REGULATIONS HANDBOOK

VOLUME 1 STRATEGIC INFORMATION AND BASIC LAWS

UPDATED ANNUALLY

Cover Design: International Business Publications, USA

We express our sincere appreciation to all government agencies and international organizations which provided information and other materials for this handbook

2015 Edition Updated Reprint International Business Publications, USA
ISBN 1-5145-1247-5

This handbook provides basic information for starting or/and conducting business in the country. The extraordinary volume of materials covering the topic, prevents us from placing all these materials in this handbook. For more detailed information on issues related to any specific investment and business activity in the country, please contact Global Investment Center, USA

Please acquire the list of our business intelligence and marketing reports and other business publications. We constantly update and expand our business intelligence and marketing materials. Please contact the center for the updated list of reports on over 200 countries.

in the USA: **Global Investment Center, USA.**
P.O.Box 15343, Washington, DC 20003
Phone: (202) 546-2103, Fax: (202) 546-3275, E-mail: **ibpusa3@gmail.com**

For additional analytical, marketing and other information please contact Global Investment Center, USA

Printed in the USA

INDONESIA

ENERGY POLICY, LAWS AND REGULATIONS HANDBOOK

VOLUME 1

STRATEGIC INFORMATION AND BASIC LAWS

TABLE OF CONTENTS

STRATEGIC AND DEVELOPMENT PROFILES

STRATEGIC PROFILE

Capital (and largest city)	Jakarta 6°10.5′S 106°49.7′E6.175°S 106.8283°E
Official language(s)	Indonesian
Demonym	Indonesian
Government	Unitary presidential republic
- President	Susilo Bambang Yudhoyono
- Vice President	Boediono
Independence	following Dutch colonial rule
- Declared	17 August 1945
- Acknowledged	27 December 1949
	Area
- Land	1,919,440 km^2 (16th) 735,355 sq mi
- Water (%)	4.85
	Population
- 2009 estimate	238,000,000 (4th)
- 2000 census	206,264,595
- Density	119.8/km^2 (84th) 312.7/sq mi
GDP (PPP)	2010 estimate
- Total	$1,027.279 billion
- Per capita	$4,379
GDP (nominal)	2010 estimate
- Total	$670.421 billion
- Per capita	$2,858
Gini (2002)	34.3
HDI (2007)	▲0.734 (medium) (111th)
Currency	Rupiah (IDR)
Time zone	various (UTC+7 to +9)
- Summer (DST)	not observed (UTC)
Drives on the	Left
ISO 3166 code	ID
Internet TLD	.id
Calling code	+62

Indonesia officially the **Republic of Indonesia** is a sovereign state in Southeast Asia and Oceania. Indonesia is an archipelago comprising 13,466 islands. It encompasses 33 provinces and 1 Special Administrative Region (for being governed by a pre-colonial monarchy) with over 238 million people, making it the world's fourth most populous country. Indonesia's republic form of government comprises an elected legislature and president. The nation's capital city is Jakarta. The country shares land borders with Papua New Guinea, East Timor, and Malaysia. Other neighboring countries include Singapore, the Philippines, Australia, Palau, and the Indian territory of the Andaman and Nicobar Islands. Indonesia is a founding member of ASEAN and a member of the G-20 major economies. The Indonesian economy is the world's 10th largest by nominal GDP.

The Indonesian archipelago has been an important trade region since at least the 7th century, when Srivijaya and then later Majapahit traded with China and India. Local rulers gradually absorbed foreign cultural, religious and political models from the early centuries CE, and Hindu and Buddhist kingdoms flourished. Indonesian history has been influenced by foreign powers drawn to its natural resources. Muslim traders brought the now-dominant Islam, while European powers brought Christianity and fought one another to monopolize trade in the Spice Islands of Maluku during the Age of Discovery. Following three and a half centuries of Dutch colonialism, Indonesia secured its independence after World War II. Indonesia's history has since been turbulent, with challenges posed by natural disasters, corruption, separatism, a democratization process, and periods of rapid economic change.

Indonesia consists of hundreds of distinct native ethnic and linguistic groups. The largest – and politically dominant – ethnic group are the Javanese. A shared identity has developed, defined by a national language, ethnic diversity, religious pluralism within a majority Muslim population, and a history of colonialism and rebellion against it. Indonesia's national motto, *"Bhinneka Tunggal Ika"* ("Unity in Diversity" *literally,* "many, yet one"), articulates the diversity that shapes the country. Despite its large population and densely populated regions, Indonesia has vast areas of wilderness that support the world's second highest level of biodiversity. The country has abundant natural resources, yet poverty remains widespread

Background: The world's largest archipelago, Indonesia achieved independence from the Netherlands in 1949. Current issues include: implementing IMF-mandated reforms of the banking sector, effecting a transition to a popularly elected government after years of rule by dictators, addressing charges of cronyism and corruption among the Chinese-dominated business class, dealing with alleged human rights violations by the military, and resolving growing pressures for some form of autonomy or independence in certain regions such as Aceh and Irian Jaya. On 30 August 2009 a provincial referendum for independence was overwhelmingly approved by the people of Timor Timur. Concurrence followed by Indonesia's national legislature, and the name East Timor was provisionally adopted. The independent status of East Timor has yet to be formally established.

Independence: Proclaimed independence on August 17, 1945. On December 27, 1949, Indonesia became independent from the Netherlands.
Population : 224.8 million
Location/Size: Southeastern Asia/735,310 sq. mi., slightly less than three times the size of Texas
Major Cities: Jakarta (capital) , Surabaya, Bandung, Medan, Semarang, Palembang, Ujung Pandang
Languages: Bahasa Indonesia (official) , English, Dutch, local dialects including Javanese
Ethnic Groups: Javanese (45%) , Sundanese (14%) , Madurese (7.5%) , coastal Malays (7.5%) , other (26%)
Religions: Muslim (88%) , Protestant (5%) , Roman Catholic (3%) , Hindu (2%) , Buddhist 1%) , other (1%)
Defense: Army (220,000) , Navy (43,000) , Air Force (21,000) , Paramilitary forces: Police (177,000) , KAMRA (People's Security) (1.5 million trainees)

OIL AND GAS INDUSTRIES

Organizations: Perusahaan Pertambangan Minyak dan Gas Bumi Negara (Pertamina) - oil exploration, production, transportation, and marketing; Perum Gas Negara (PGN) -gas distributor and transmission company
Major Producing Oil Fields: Duri, Minas, Belida, Ardjuna, Arun, KG/KRA, Widuri, Nilam, Attaka
Oil Refineries (operating capacity-bbl/d, December 2009) : Cilacap, Central Java (380,000) ;

Pertamina-Balikpapan, Kalimantan (240,920) ; Musi, South Sumatra (109,155) ; EXOR-1, Balongan, Java (125,000) ; Dumai, Central Sumatra (114,000) ; Sungai Pakning, Central Sumatra (47,500) ; Pangakalan Brandan, North Sumatra (4,750) ; Cepu, Central Java (3,420)
Product Pipelines: Trans-Java (serving the Surabaya market)
Oil Tanker Terminals: Java: Cilegon, Cilacap, Surabaya, Ardjuna B (offshore) Sumatra: Pangkalan Brandan, Belawan, Dumai, Musi, Perlak, Palembang, Tanjung Uban (offshore) Kalimantan: Balikpapan Sulawesi: Ujung Pandang Irian Jaya: Sorong, Jaya Seram: Bula Natuna Sea: Ikan Pari
Major Gas Fields: Sumatra: Arun, Alur Siwah, Kuala Langsa, Musi, South Lho Sukon, Wampu East Kalimantan: Attaka, Badak, Bekapai, Handil, Mutiara, Nilam, Semberah, Tunu Natuna Sea: Natuna Java: Pagerungan, Terang/Sirasun Irian Jaya: Tangguh
Major Gas Pipelines: Sumatra: Pangkalan Brandan-Dumai
LNG Plants: Arun, Bontang

GEOGRAPHY

Location: Southeastern Asia, archipelago between the Indian Ocean and the Pacific Ocean
Geographic coordinates: 5 00 S, 120 00 E
Map references: Southeast Asia

Area:
total: 1,919,440 sq km
land: 1,826,440 sq km
water: 93,000 sq km

Area - comparative: slightly less than three times the size of Texas

Land boundaries:
total: 2,602 km
border countries: Malaysia 1,782 km, Papua New Guinea 820 km

Coastline: 54,716 km

Maritime claims: measured from claimed archipelagic baselines
exclusive economic zone: 200 nm
territorial sea: 12 nm

Climate: tropical; hot, humid; more moderate in highlands
Terrain: mostly coastal lowlands; larger islands have interior mountains

Elevation extremes:
lowest point: Indian Ocean 0 m
highest point: Puncak Jaya 5,030 m

Natural resources: petroleum, tin, natural gas, nickel, timber, bauxite, copper, fertile soils, coal, gold, silver

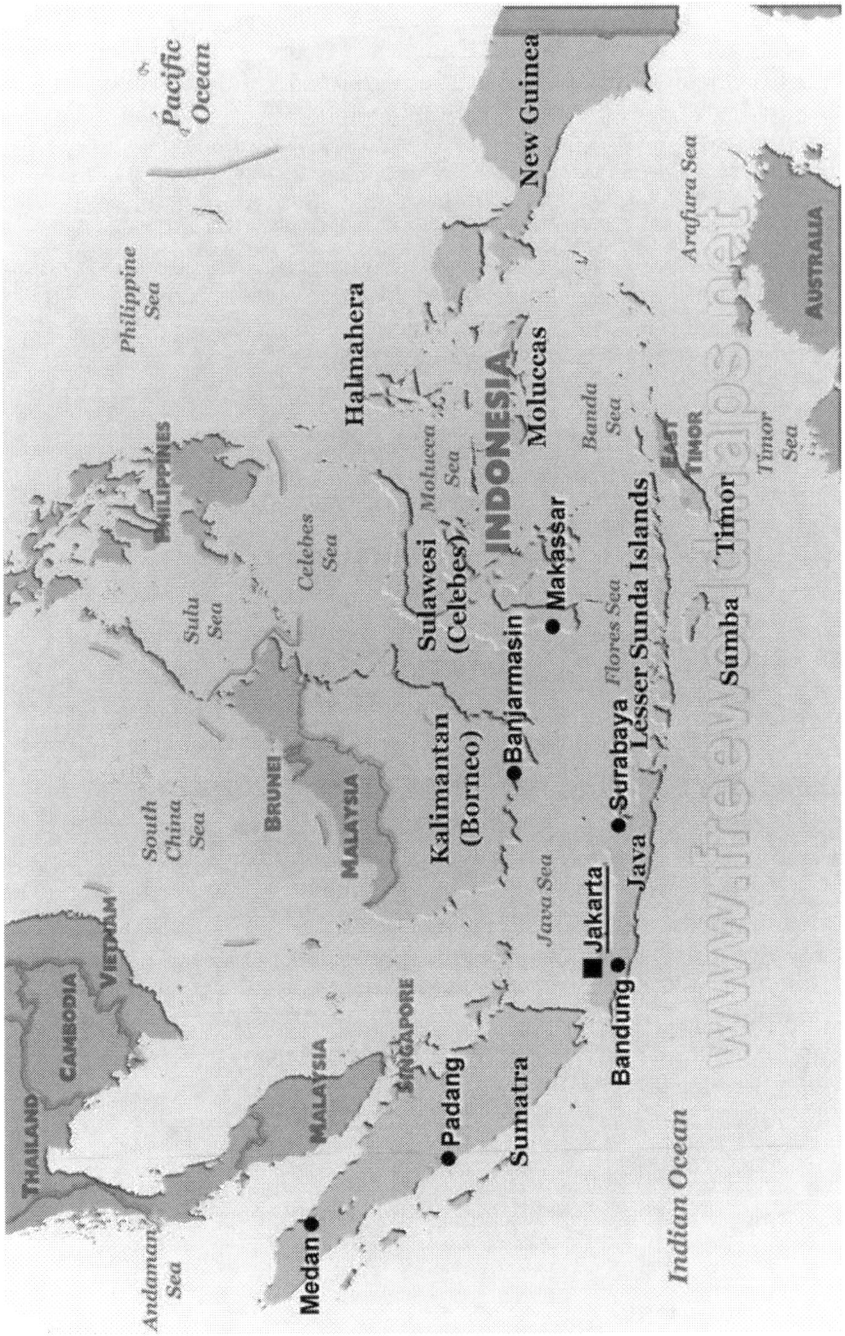

Land use:
arable land: 10%
permanent crops: 7%
permanent pastures: 7%
forests and woodland: 62%
other: 14%

Irrigated land: 45,970 sq km (1993 est.)
Natural hazards: occasional floods, severe droughts, tsunamis, earthquakes, volcanoes

Environment - current issues: deforestation; water pollution from industrial wastes, sewage; air pollution in urban areas; smoke and haze from forest fires

Environment - international agreements:
party to: Biodiversity, Climate Change, Desertification, Endangered Species, Hazardous Wastes, Law of the Sea, Nuclear Test Ban, Ozone Layer Protection, Ship Pollution, Tropical Timber 83, Tropical Timber 94, Wetlands
signed, but not ratified: Climate Change-Kyoto Protocol, Marine Life Conservation

Geography - note: archipelago of 17,000 islands (6,000 inhabited) ; straddles Equator; strategic location astride or along major sea lanes from Indian Ocean to Pacific Ocean

PEOPLE

Population: 224,784,210

Age structure:
0-14 years: 30.57% (male 34,932,102; female 33,783,603)
15-64 years: 64.96% (male 72,889,994; female 73,124,821)
65 years and over: 4.47% (male 4,413,268; female 5,640,422)

Population growth rate: 1.63%
Birth rate: 22.6 births/1,000 population
Death rate: 6.31 deaths/1,000 population
Net migration rate: 0 migrant(s) /1,000 population

Sex ratio:
at birth: 1.05 male(s) /female
under 15 years: 1.03 male(s) /female
15-64 years: 1 male(s) /female
65 years and over: 0.78 male(s) /female
total population: 1 male(s) /female

Infant mortality rate: 42.21 deaths/1,000 live births
Life expectancy at birth:
total population: 67.96 years
male: 65.61 years
female: 70.42 years
Total fertility rate: 2.61 children born/woman
Nationality:
noun: Indonesian(s)
adjective: Indonesian
Ethnic groups: Javanese 45%, Sundanese 14%, Madurese 7.5%, coastal Malays 7.5%, other 26%
Religions: Muslim 88%, Protestant 5%, Roman Catholic 3%, Hindu 2%, Buddhist 1%, other 1%

Languages: Bahasa Indonesia (official, modified form of Malay) , English, Dutch, local dialects, the most widely spoken of which is Javanese

Literacy:
definition: age 15 and over can read and write
total population: 83.8%

male: 89.6%
female: 78% (1995 est.)

GOVERNMENT

Country name:
conventional long form: Republic of Indonesia
conventional short form: Indonesia
local long form: Republik Indonesia
local short form: Indonesia
former: Netherlands East Indies; Dutch East Indies

Data code: ID
Government type: republic
Capital: Jakarta

Administrative divisions: 23 provinces (propinsi-propinsi, singular - propinsi) , 2 special regions(daerah-daerah istimewa, singular - daerah istimewa) , and 1 special capital city district*(daerah khusus ibukota) ; Aceh*, Bali, Bengkulu, Irian Jaya, Jakarta Raya**, Jambi, Jawa Barat, Jawa Tengah, Jawa Timur, Kalimantan Barat, Kalimantan Selatan, Kalimantan Tengah, Kalimantan Timur, Lampung, Maluku, Nusa Tenggara Barat, Nusa Tenggara Timur, Riau, Sulawesi Selatan, Sulawesi Tengah, Sulawesi Tenggara, Sulawesi Utara, Sumatera Barat, Sumatera Selatan, Sumatera Utara, Yogyakarta*; note - there may be a new province named Maluku Utara
note: following the 30 August 2009 provincial referendum for independence which was overwhelmingly approved by the people of Timor Timur and the October 2009 concurrence of Indonesia's national legislature, the name East Timor was adopted as a provisional name for the political entity formerly known as Propinsi Timor Timur until such time as the entity's independent status is formally established

Independence: 17 August 1945 (proclaimed independence; on 27 December 1949, Indonesia became legally independent from the Netherlands)
National holiday: Independence Day, 17 August (1945)
Constitution: August 1945, abrogated by Federal Constitution of 1949 and Provisional Constitution of 1950, restored 5 July 1959

Legal system: based on Roman-Dutch law, substantially modified by indigenous concepts and by new criminal procedures code; has not accepted compulsory ICJ jurisdiction

Suffrage: 17 years of age; universal and married persons regardless of age

Executive branch:

chief of state: President Susilo Bambang YUDHOYONO (since 20 October 2004) ; Vice President BOEDIONO (since 20 October 2009) ; note - the president is both the chief of state and head of government
head of government: President Susilo Bambang YUDHOYONO (since 20 October 2004) ; Vice President BOEDIONO (since 20 October 2009)
cabinet: Cabinet appointed by the president
elections: president and vice president elected for five-year terms (eligible for a second term) by direct vote of the citizenry; election last held on 8 July 2009 (next to be held in 2014)

election results: Susilo Bambang YUDHOYONO elected president; percent of vote - Susilo Bambang YUDHOYONO 60.8%, MEGAWATI Sukarnoputri 26.8%, Jusuf KALLA 12.4%

Legislative branch:
People's Consultative Assembly (Majelis Permusyawaratan Rakyat or MPR) is the upper house; it consists of members of the DPR and DPD and has role in inaugurating and impeaching the president and in amending the constitution but does not formulate national policy; House of Representatives or Dewan Perwakilan Rakyat (DPR) (560 seats, members elected to serve five-year terms) , formulates and passes legislation at the national level; House of Regional Representatives (Dewan Perwakilan Daerah or DPD) , constitutionally mandated role includes providing legislative input to DPR on issues affecting regions (132 members, four from each of Indonesia's 30 provinces, two special regions, and one special capital city district)
elections: last held on 9 April 2009 (next to be held in 2014)
election results: percent of vote by party - PD 20.9%, GOLKAR 14.5%, PDI-P 14.0%, PKS 7.9%, PAN 6.0%, PPP 5.3%, PKB 4.9%, GERINDRA 4.5%, HANURA 3.8%, others 18.2%; seats by party - PD 148, GOLKAR 107, PDI-P 94, PKS 57, PAN 46, PPP 37, PKB 28, GERINDRA 26, HANURA 17
*note:*29 other parties received less than 2.5% of the vote so did not obtain any seats; because of election rules, the number of seats won does not always follow the percentage of votes received by parties

Judicial branch: Supreme Court (Mahkamah Agung) , the judges are appointed by the president

Political parties and leaders: Crescent Moon and Star Party or PBB [Yusril Ihza MAHENDRA, chairman]; Development Unity Party or PPP (federation of former Islamic parties) [Hamzah HAZ, chairman]; Golkar [Akbar TANSUNG, general chairman]; Indonesia Democracy Party or PDI (federation of former Nationalist and Christian Parties) [Budi HARDJONO, chairman]; Indonesia Democracy Party-Struggle or PDI-P [MEGAWATI Sukarnoputri, chairperson]; National Awakening Party or PKB [Abdurrahman WAHID]; National Mandate Party or PAN [Amien RAIS, chairman]

International organization participation: APEC, AsDB, ASEAN, CCC, CP, ESCAP, FAO, G-15, G-19, G-77, IAEA, IBRD, ICAO, ICC, ICFTU, ICRM, IDA, IDB, IFAD, IFC, IFRCS, IHO, ILO, IMF, IMO, Inmarsat, Intelsat, Interpol, IOC, IOM (observer) , ISO, ITU, NAM, OIC, OPCW, OPEC, UN, UNCTAD, UNESCO, UNIDO, UNIKOM, UNMIBH, UNMOP, UNMOT, UNOMIG, UPU, WCL, WFTU, WHO, WIPO, WMO, WToO, WTrO

Diplomatic representation in the US:
chief of mission: Ambassador DORODJATUN Kuntjoro-Jakti
chancery: 2020 Massachusetts Avenue NW, Washington, DC 20036
telephone: (202) 775-5200
FAX: (202) 775-5365
consulate(s) general: Chicago, Houston, Los Angeles, New York, and San Francisco

Diplomatic representation from the US:
chief of mission: Ambassador Robert GELBARD
embassy: Medan Merdeka Selatan 5, Jakarta
mailing address: Unit 8129, Box 1, APO AP 96520
telephone: [62] (21) 344-2211
FAX: [62] (21) 386-2259
consulate(s) general: Surabaya

Flag description: two equal horizontal bands of red (top) and white; similar to the flag of Monaco, which is shorter; also similar to the flag of Poland, which is white (top) and red

ECONOMY

Indonesia, a vast polyglot nation, has grown strongly since 2010. During the global financial crisis, Indonesia outperformed its regional neighbors and joined China and India as the only G20 members posting growth. The government has promoted fiscally conservative policies, resulting in a debt-to-GDP ratio of less than 25% and historically low rates of inflation. Fitch and Moody's upgraded Indonesia's credit rating to investment grade in December 2011. Indonesia still struggles with poverty and unemployment, inadequate infrastructure, corruption, a complex regulatory environment, and unequal resource distribution among regions. The government also faces the challenges of quelling labor unrest and reducing fuel subsidies in the face of high oil prices.

GDP (purchasing power parity):
$1.285 trillion (2013 est.)
country comparison to the world: 16
$1.22 trillion (est.)
$1.149 trillion (est.)
note: data are in 2013 US dollars

GDP (official exchange rate):
$867.5 billion (2013 est.)

GDP - real growth rate:
5.3% (2013 est.)
country comparison to the world: 48
6.2% (est.)
6.5% (est.)

GDP - per capita (PPP):
$5,200 (2013 est.)
country comparison to the world: 158
$5,000 (est.)
$4,800 (est.)
note: data are in 2013 US dollars

Gross national saving:
31.5% of GDP (2013 est.)
country comparison to the world: 22
32.6% of GDP (est.)
33.1% of GDP (est.)

GDP - composition, by end use:
household consumption: 56%
government consumption: 9.4%
investment in fixed capital: 32.7%
investment in inventories: 2%
exports of goods and services: 23.5%
imports of goods and services: -25.8%
(2013 est.)

GDP - composition, by sector of origin:
agriculture: 14.3%
industry: 46.6%
services: 39.1% (2013 est.)

Agriculture - products:
rubber and similar products, palm oil, poultry, beef, forest products, shrimp, cocoa, coffee, medicinal herbs, essential oil, fish and its similar products, and spices

Industries:
petroleum and natural gas, textiles, automotive, electrical appliances, apparel, footwear, mining, cement, medical instuments and appliances, handicrafts, chemical fertilizers, plywood, rubber, processed food, jewelry, and tourism

Industrial production growth rate:
4.3% (2013 est.)
country comparison to the world: 66

Labor force:
120 million (2013 est.)
country comparison to the world: 5

Labor force - by occupation:
agriculture: 38.9%
industry: 22.2%
services: 47.9% (est.)

Unemployment rate:
6.6% (2013 est.)
country comparison to the world: 69
6.1% (est.)

Population below poverty line:
11.7% (est.)

Household income or consumption by percentage share:
lowest 10%: 3.3%
highest 10%: 29.9% (2009)

Distribution of family income - Gini index:
36.8 (2009)
country comparison to the world: 82
39.4 (2005)

Budget:
revenues: $137.5 billion
expenditures: $166 billion (2013 est.)

Taxes and other revenues:
15.8% of GDP (2013 est.)
country comparison to the world: 189

Budget surplus (+) or deficit (-):
-3.3% of GDP (2013 est.)
country comparison to the world: 133

Public debt:
24.2% of GDP (2013 est.)
country comparison to the world: 129
23% of GDP (est.)

Fiscal year:
calendar year

Inflation rate (consumer prices):
7.7% (2013 est.)
country comparison to the world: 193
4.3% (est.)

Central bank discount rate:
6.37% (31 December 2010)
country comparison to the world: 57
6.46% (31 December 2009)
note: this figure represents the 3-month SBI rate; the Bank of Indonesia has not employed the one-month SBI since September 2010

Commercial bank prime lending rate:
12.1% (31 December 2013 est.)
country comparison to the world: 71
11.8% (31 December 2012 est.)
note: these figures represent the average annualized rate on working capital loans

Stock of narrow money:
$82.99 billion (31 December 2013 est.)
country comparison to the world: 40
$87.04 billion (31 December 2012 est.)

Stock of broad money:
$325 billion (31 December 2013 est.)
country comparison to the world: 29
$342 billion (31 December 2012 est.)

Stock of domestic credit:
$336.2 billion (31 December 2013 est.)
country comparison to the world: 32
$350 billion (31 December 2012 est.)

Market value of publicly traded shares:

$396.8 billion (31 December 2012 est.)
country comparison to the world: 24
$390.1 billion (31 December 2011)
$360.4 billion (31 December 2010 est.)

Current account balance:
-$28.72 billion (2013 est.)
country comparison to the world: 184
-$24.07 billion (est.)

Exports:
$178.9 billion (2013 est.)
country comparison to the world: 29
$187.3 billion (est.)

Exports - commodities:
oil and gas, electrical appliances, plywood, textiles, rubber

Exports - partners:
Japan 15.9%, China 11.4%, Singapore 9%, South Korea 7.9%, US 7.8%, India 6.6%, Malaysia 5.9% (2012)

Imports:
$178.6 billion (2013 est.)
country comparison to the world: 28
$178.7 billion (est.)

Imports - commodities:
machinery and equipment, chemicals, fuels, foodstuffs

Imports - partners:
China 15.3%, Singapore 13.6%, Japan 11.9%, Malaysia 6.4%, South Korea 6.2%, US 6.1%, Thailand 6% (2012)

Reserves of foreign exchange and gold:
$83.45 billion (31 December 2013 est.)
country comparison to the world: 27
$112.8 billion (31 December 2012 est.)

Debt - external:
$223.8 billion (31 December 2013 est.)
country comparison to the world: 33
$224.1 billion (31 December 2012 est.)

Stock of direct foreign investment - at home:
$207.2 billion (31 December 2013 est.)
country comparison to the world: 27
$192.7 billion (31 December 2012 est.)

Stock of direct foreign investment - abroad:

$17.41 billion (31 December 2013 est.)
country comparison to the world: 48
$14.81 billion (31 December 2012 est.)

Exchange rates:
Indonesian rupiah (IDR) per US dollar -
10,341.6 (2013 est.)
9,386.63 (est.)
9,090.4 (est.)
10,389.9 (2009)
9,698.9 (2008)

ENERGY

Electricity - production:
173.8 billion kWh (est.)
country comparison to the world: 23

Electricity - consumption:
158 billion kWh (est.)
country comparison to the world: 24

Electricity - exports:
0 kWh (est.)
country comparison to the world: 150

Electricity - imports:
2.542 billion kWh (est.)
country comparison to the world: 51

Electricity - installed generating capacity:
39.9 million kW (est.)
country comparison to the world: 23

Electricity - from fossil fuels:
87% of total installed capacity (est.)
country comparison to the world: 85

Electricity - from nuclear fuels:
0% of total installed capacity (est.)
country comparison to the world: 108

Electricity - from hydroelectric plants:
9.9% of total installed capacity (est.)
country comparison to the world: 115

Electricity - from other renewable sources:
3.1% of total installed capacity (est.)
country comparison to the world: 57

Crude oil - production:
974,300 bbl/day (est.)
country comparison to the world: 23

Crude oil - exports:
338,100 bbl/day (est.)
country comparison to the world: 23

Crude oil - imports:
388,400 bbl/day (est.)
country comparison to the world: 23

Crude oil - proved reserves:
4.03 billion bbl (1 January 2013 es)
country comparison to the world: 27

Refined petroleum products - production:
935,300 bbl/day (est.)
country comparison to the world: 20

Refined petroleum products - consumption:
1.322 million bbl/day (est.)
country comparison to the world: 18

Refined petroleum products - exports:
142,400 bbl/day (est.)
country comparison to the world: 40

Refined petroleum products - imports:
473,400 bbl/day (est.)
country comparison to the world: 13

Natural gas - production:
76.25 billion cu m (est.)
country comparison to the world: 12

Natural gas - consumption:
39.56 billion cu m (est.)
country comparison to the world: 24

Natural gas - exports:
38.67 billion cu m (est.)
country comparison to the world: 12

Natural gas - imports:
0 cu m (est.)
country comparison to the world: 207

Natural gas - proved reserves:

3.069 trillion cu m (1 January 2013 es)
country comparison to the world: 13

Carbon dioxide emissions from consumption of energy:
426.8 million Mt (est.)
country comparison to the world: 16

COMMUNICATIONS

Telephones - main lines in use:	12.772 million
Telephones - mobile cellular:	46.91 million
Telephone system:	general assessment: domestic service fair, international service good domestic: interisland microwave system and HF radio police net; domestic satellite communications system international: country code - 62; satellite earth stations - 2 Intelsat (1 Indian Ocean and 1 Pacific Ocean)
Radio broadcast stations:	AM 678, FM 43, shortwave 82 (1998)
Television broadcast stations:	54 local TV stations (11 national TV networks; each with their own group of local transmitters)
Internet country code:	.id
Internet hosts:	170,834
Internet users:	16 million

TRANSPORTATION

Railways:
total: 6,458 km
narrow gauge: 5,961 km 1.067-m gauge (101 km electrified; 101 km double track) ; 497 km 0.750-m gauge

Highways:
total: 342,700 km
paved: 158,670 km
unpaved: 184,030 km (1997 est.)

Waterways: 21,579 km total; Sumatra 5,471 km, Java and Madura 820 km, Kalimantan 10,460 km, Sulawesi (Celebes) 241 km, Irian Jaya 4,587 km
Pipelines: crude oil 2,505 km; petroleum products 456 km; natural gas 1,703 km (1989)
Ports and harbors: Cilacap, Cirebon, Jakarta, Kupang, Palembang, Semarang, Surabaya, Ujungpandang
Merchant marine:
total: 586 ships (1,000 GRT or over) totaling 2,676,875 GRT/3,700,864 DWT
ships by type: bulk 38, cargo 346, chemical tanker 9, container 19, liquified gas 5, livestock carrier 1, passenger 7, passenger/cargo 13, petroleum tanker 114, refrigerated cargo 1, roll-on/roll-off 11, short-sea passenger 8, specialized tanker 9, vehicle carrier 5
Airports: 446

Airports - with paved runways:
total: 127
over 3,047 m: 4
2,438 to 3,047 m: 12
1,524 to 2,437 m: 39 *914 to 1,523 m:* 41 *under 914 m:* 31
Airports - with unpaved runways:
total: 319
1,524 to 2,437 m: 5 *914 to 1,523 m:* 33 *under 914 m:* 281
Heliports: 4

MILITARY

Military branches: Army, Navy, Air Force, National Police
Military manpower - military age: 18 years of age
Military manpower - availability:
males age 15-49: 62,948,286
Military manpower - fit for military service:
males age 15-49: 36,826,282
Military manpower - reaching military age annually:
males: 2,273,324
Military expenditures - dollar figure: $1 billion
Military expenditures - percent of GDP: 1.3%

TRANSNATIONAL ISSUES

Disputes - international: Sipadan and Ligitan Islands in dispute with Malaysia

Illicit drugs: illicit producer of cannabis largely for domestic use; possible growing role as transshipment point for Golden Triangle heroin

IMPORTANT INFORMATION FOR UNDERSTANDING INDONESIA

Official Name: Republic of Indonesia

PROFILE

Geography
Area: 2 million sq. km. (736,000 sq. mi.), about three times the size of Texas; maritime area: 7,900,000 sq. km.
Cities: Capital--Jakarta (est. 9.7 million). Other cities--Surabaya 2.8 million, Medan 2.1 million, Bandung 2.4 million.
Terrain: More than 17,500 islands; 6,000 are inhabited; 1,000 of which are permanently settled. Large islands consist of coastal plains with mountainous interiors.
Climate: Equatorial but cooler in the highlands.

People
Nationality: Noun and adjective--Indonesian(s).
Population (est.): 237,6 million.
Annual population growth rate (est.): 2.7%.
Ethnic groups (2000 census): Javanese 40.6%, Sundanese 15%, Madurese 3.3%, Minangkabau 2.7%, others 38.4%.
Religions (2000 census): Muslim 86.1%, Protestant 5.7%, Catholic 3%, Hindu 1.8%, others 3.4%.
Languages: Indonesian (official), local languages, the most prevalent of which is Javanese.

Education: Years compulsory--9. Enrollment--94.7% of eligible primary school-age children. Literacy--98.3% (2010).
Health: Infant mortality rate (2007)--34/1,000. Life expectancy at birth (est.)--70.76 years.
Work force: 117.4 million . Agriculture—39.3%, industry—20.9%, services—47.8%.

Government
Type: Independent republic.
Independence: August 17, 1945 proclaimed.
Constitution: 1945. Embodies five principles of the state philosophy, called Pancasila, namely monotheism, humanitarianism, national unity, representative democracy by consensus, and social justice.
Branches: Executive--president (head of government and chief of state) elected by direct popular vote. Legislative--The People's Consultative Assembly (MPR), which includes the 560-member House of Representatives (DPR) and the 132-member Council of Regional Representatives (DPD), both elected to 5-year terms. Judicial--Supreme Court is the final court of appeal. Constitutional Court has power of judicial review
Suffrage: 17 years of age universal, and married persons regardless of age.

Economy
GDP (2009): $539 billion; (2010): $707 billion; (est.): $834 billion.
Annual growth rate (2009): 4.5%; (2010): 6.1%; (est.): 6.5%
Inflation, end-period (2009): 2.8%; (2010): 7%; : 3.79%.
Per capita income (est., PPP): $4,668.1.
Natural resources (11.2% of GDP, 2010): Oil and gas, bauxite, silver, tin, copper, gold, coal.
Agriculture (15.3% of GDP, 2010): Products--timber, rubber, rice, palm oil, coffee. Land--17% cultivated.
Manufacturing (24.8% of GDP, 2010): Garments, footwear, electronic goods, furniture, paper products, automobiles.
Trade: Exports (2010)--$158 billion including oil, natural gas, crude palm oil, coal, appliances, textiles, and rubber. Major export partners--Japan, U.S., China, Singapore, Malaysia, and Republic of Korea. Imports (2010)--$136 billion including oil and fuel, food, chemicals, capital goods, consumer goods, iron and steel. Major import partners--Singapore, China, Japan, U.S., Malaysia, Thailand, South Korea.

PEOPLE

Indonesia's approximately 237.6 million people make it the world's fourth-most populous nation. The island of Java, roughly the size of New York State, is the most populous island in the world (136.6 million, 2010 est.) and one of the most densely populated areas in the world. Indonesia includes numerous related but distinct cultural and linguistic groups, many of which are ethnically Malay. Since independence, Bahasa Indonesia (the national language, a form of Malay) has spread throughout the archipelago and has become the language of most written communication, education, government, business, and media. Local languages are still important in many areas, however. English is the most widely spoken foreign language. Education is compulsory for children through grade 9. In primary school, more than 94% of eligible children are enrolled whereas 57% of eligible children are enrolled in secondary school.

Constitutional guarantees of religious freedom apply to the six religions recognized by the state, namely Islam (86.1%), Protestantism (5.7%), Catholicism (3%), Hinduism (1.8%), Buddhism (about 1%), and Confucianism (less than 1%). On the resort island of Bali, over 90% of the population practices Hinduism. In some remote areas, animism is still practiced.

HISTORY

By the time of the Renaissance, the islands of Java and Sumatra had already enjoyed a 1,000-year heritage of advanced civilization spanning two major empires. During the 7th-14th centuries, the Buddhist kingdom of Srivijaya flourished on Sumatra. At its peak, the Srivijaya Empire reached as far as West Java and the Malay Peninsula. Also by the 14th century, the Hindu Kingdom of Majapahit had risen in eastern Java. Gadjah Mada, the empire's chief minister from 1331 to 1364, succeeded in gaining allegiance from most of what is now modern Indonesia and much of the Malay archipelago as well. Legacies from Gadjah Mada's time include a codification of law and an epic poem. Islam arrived in Indonesia sometime during the 12th century and supplanted Hinduism by the end of the 16th century in Java and Sumatra. Bali, however, remains overwhelmingly Hindu. In the eastern archipelago, both Christian and Islamic proselytizing took place in the 16th and 17th centuries, and, currently, there are large communities of both religions on these islands.

Beginning in 1602, the Dutch slowly established themselves as rulers of Indonesia, exploiting the weakness of the small kingdoms that had replaced that of Majapahit. The only exception was East Timor, which remained under Portugal's control until 1975. During 300 years of rule, the Dutch developed the Netherlands East Indies into one of the world's richest colonial possessions, extracting natural resources through co-opted local elites but doing little to modernize Indonesia.

During the first decade of the 20th century, an Indonesian independence movement began and expanded rapidly, particularly between the two World Wars. Its leaders came from a small group of young professionals and students, some of whom had been educated in the Netherlands. Many, including Indonesia's first president, Soekarno (1945-67), were imprisoned for political activities.

During World War II, Japan invaded Indonesia (in early 1942), outclassing a combined American, British, Dutch, and Australian military command. Sizeable U.S. naval forces and smaller air and ground forces sent to defend Indonesia were defeated, with heavy losses in ships and large numbers of Americans killed or captured. The Japanese treated captured Allied troops and interned Western civilians with extreme cruelty. Because of local animosity toward Dutch colonial rule, there was no anti-Japanese guerilla movement as in the Philippines and Malaysia, and most Indonesians initially welcomed the Japanese as liberators. But increasingly harsh Japanese rule strengthened the prewar independence movement, and on August 17, 1945, 3 days after Japan's surrender to the Allies, a small group of Indonesians, led by Soekarno and Mohammad Hatta, proclaimed independence and established the Republic of Indonesia. They set up a provisional government and adopted a constitution to govern the republic until elections could be held and a new constitution written. Fighting soon broke out between Indonesian independence groups and Allied forces--mainly British, Indian, and Australian forces--sent to accept the Japanese surrender.

Dutch efforts later to reestablish complete control met resistance. Following the Philippines' independence in 1946, the U.S. was unwilling to see the Netherlands use post-war Marshall Plan support to indirectly fund the suppression of Indonesia's independence. Negotiations and on-and-off fighting in Indonesia continued until 1949. The stalemate, combined with reduced international support and a devastated economy in the Netherlands, led to the Dutch decision to withdraw from Indonesia. Colonial rule and its violent end left a legacy in Indonesia of mistrusting foreign motives, especially those of large powers. In 1950, Indonesia became the 60th member of the United Nations.

Shortly after hostilities with the Dutch ended in 1949, Indonesia adopted a new constitution, providing for a parliamentary system of government in which the executive was chosen by and accountable to parliament. Parliament was divided among many political parties before and after the country's first nationwide election in 1955, and stable governmental coalitions were difficult to achieve. The role of Islam in Indonesia was debated. Soekarno defended a secular state based on Pancasila, five principles of the state philosophy--monotheism, humanitarianism, national unity, representative democracy by consensus, and social justice--codified in the 1945 constitution, while some Muslim groups preferred either an Islamic state or a constitution that included a preambular provision requiring adherents of Islam to be subject to Islamic law. At the time of independence, the Dutch retained control over the western half of New Guinea (known as Irian Jaya in the Soekarno and Suharto eras and as Papua since 2000) and permitted steps toward self-government and independence.

Negotiations with the Dutch on the incorporation of Irian Jaya into Indonesia failed, and armed clashes broke out between Indonesian and Dutch troops in 1961. In August 1962, the two sides reached an agreement, and Indonesia assumed administrative responsibility for Irian Jaya on May 1, 1963. The Indonesian Government conducted an "Act of Free Choice" in Irian Jaya under UN supervision in 1969 in which 1,025 Papuan representatives of local councils agreed by consensus to remain a part of Indonesia. A subsequent UN General Assembly resolution confirmed the transfer of sovereignty to Indonesia. Opposition to Indonesian administration of Papua gave rise to small-scale guerrilla activity in the years following Jakarta's assumption of control. In the more open atmosphere since 1998, there have been more explicit expressions within Papua calling for independence from Indonesia.

Unsuccessful rebellions on Sumatra, Sulawesi, West Java, and other islands beginning in 1958, plus a failure by the constituent assembly to develop a new constitution, weakened the parliamentary system. Consequently, in 1959, when President Soekarno unilaterally revived the provisional 1945 constitution that provided for broad presidential powers, he met little resistance. From 1959 to 1965, President Soekarno imposed an authoritarian regime under the label of "Guided Democracy."

Soekarno favored a foreign policy of nonalignment, a stance supported by other prominent leaders of former colonies who rejected formal alliances with either the West or Soviet bloc. Under Soekarno's auspices, these leaders gathered in Bandung, West Java, in 1955 to lay the groundwork for what became known as the Non-Aligned Movement. In the late 1950s and early 1960s, President Soekarno moved closer to Asian communist states and toward the Indonesian Communist Party (PKI) in domestic affairs. The PKI represented the largest communist party outside the Soviet Union and China.

By 1965, the PKI controlled many of the mass civic and cultural organizations that Soekarno had established to mobilize support for his regime and, with Soekarno's acquiescence, embarked on a campaign to establish a "Fifth Column" by arming its supporters. Army leaders resisted this campaign. Under circumstances that have never been fully explained, on October 1, 1965, PKI sympathizers within the military, including elements from Soekarno's palace guard, occupied key locations in Jakarta and kidnapped and murdered six senior generals. Major General Suharto, the commander of the Army Strategic Reserve, rallied army troops opposed to the PKI to reestablish control over the city. Violence swept throughout Indonesia in the aftermath of the October 1 events, and unsettled conditions persisted through 1966. Right-wing groups killed tens of thousands of alleged communists in rural areas. Estimates of the number of deaths range between 160,000 and 500,000. The violence was especially brutal in Java and Bali. During this period, PKI members by the tens of thousands turned in their membership cards. The emotions and fears of instability created by this crisis persisted for many years as the communist party remains banned from Indonesia.

Throughout the 1965-66 period, President Soekarno vainly attempted to restore his political stature and shift the country back to its pre-October 1965 position. Although he remained President, in March 1966, Soekarno transferred key political and military powers to General Suharto, who by that time had become head of the armed forces. In March 1967, the Provisional People's Consultative Assembly (MPRS) named General Suharto acting President. Soekarno ceased to be a political force and lived under virtual house arrest until his death in 1970.

President Suharto proclaimed a "New Order" in Indonesian politics and dramatically shifted foreign and domestic policies away from the course set in Soekarno's final years. The New Order established economic rehabilitation and development as its primary goals and pursued its policies through an administrative structure dominated by the military but with advice from Western-educated economic experts. In 1968, the People's Consultative Assembly (MPR) formally selected Suharto to a full 5-year term as President, and he was reelected to successive 5-year terms in 1973, 1978, 1983, 1988, 1993, and 1998. In mid-1997, Indonesia suffered from the Asian financial and economic crisis, accompanied by the worst drought in 50 years and falling prices for commodity exports. As the exchange rate changed from a fixed to a managed float to fully floating, the rupiah (IDR or Rp) depreciated in value, inflation increased significantly, and capital flight accelerated. Demonstrators, initially led by students, called for Suharto's resignation. Amid widespread civil unrest, Suharto resigned on May 21, 1998, 3 months after the MPR had selected him for a seventh term. Suharto's hand-picked Vice President, B.J. Habibie, became Indonesia's third President. President Habibie reestablished International Monetary Fund (IMF) and donor community support for an economic stabilization program. He released several prominent political and labor prisoners, initiated investigations into the unrest, and lifted controls on the press, political parties, and labor unions.

In January 2009, Habibie and the Indonesian Government agreed to a process, with UN involvement, under which the people of East Timor would be allowed to choose between autonomy and independence through a direct ballot held on August 30, 2009. Some 98% of registered voters cast their ballots, and 78.5% of the voters chose independence over continued integration with Indonesia. Many people were killed by Indonesian military forces and military-backed militias in a wave of violence and destruction after the announcement of the pro-independence vote.

Indonesia's first elections in the post-Suharto period were held for the national, provincial, and sub-provincial parliaments on June 7, 2009. Forty-eight political parties participated in the elections. For the national parliament, Partai Demokrasi Indonesia Perjuangan (PDI-P, Indonesian Democratic Party of Struggle, led by Megawati Sukarnoputri) won 34% of the vote; Golkar ("Functional Groups" party) 22%; Partai Kebangkitan Bangsa (PKB, National Awakening Party, linked to the moderate Islamic organization Nadhlatul Ulama headed by former President Abdurrahman Wahid) 13%; and the conservative Islamic Partai Persatuan Pembangunan (PPP, United Development Party, led by Hamzah Haz) 11%. The MPR selected Abdurrahman Wahid as Indonesia's fourth President in November 2009 and replaced him with Megawati Sukarnoputri in July 2001.

The constitution, as amended in the post-Suharto era, now provides for the direct election by popular vote of the president and vice president. Under the 2004 amendment, only parties or coalitions of parties that gained at least 3% of the House of Representatives (DPR) seats or 5% of the vote in national legislative elections were eligible to nominate a presidential and vice presidential ticket.

The 2004 legislative elections took place on April 5 and were considered to be generally free and fair. Twenty-four parties took part in the elections. Big parties lost ground, while small parties gained larger shares of the vote. However, the two Suharto-era nationalist parties, PDI-P and Golkar, remained in the lead. PDI-P (opposition party during the Suharto era) lost its plurality in the House of Representatives, dropping from 33% to 18.5% of the total vote (and from 33% to 20% of the seats). The Golkar Party (Suharto's political party) declined slightly from 2009 levels, going from 22% to 21% of the national vote (from 26% to 23% of DPR seats). The third- and fourth-largest parties (by vote share) were two Islamic-oriented parties, the United Development Party (PPP) (8% of the votes, 10.5% of the seats) and National Awakening Party (PKB) (10.5% of the vote, 9.45% of the seats). Susilo Bambang Yudhoyono's nationalist Democratic Party (PD) won 7.45% of the national vote and 10% of the DPR seats, making it the fourth-largest party in the DPR. Seven of the 24 parties won no DPR seats; six won 1-2 seats, and the other six won between 2%-6% of the national vote (between 5-52 DPR seats).

The first direct presidential election was held on July 5, 2004, contested by five tickets. As no candidate won at least 50% of the vote, a runoff election was held on September 20, 2004, between the top two candidates, President Megawati Sukarnoputri and retired General Susilo Bambang Yudhoyono. In this final round, Yudhoyono won 60.6% of the vote. Approximately 76.6% of the eligible voters participated, a total of roughly 117 million people, making Indonesia's presidential election the largest single-day election in the world. The Carter Center, which sent a delegation of election observers, issued a statement congratulating "the people and leaders of Indonesia for the successful conduct of the presidential election and the peaceful atmosphere that has prevailed throughout the ongoing democratic transition."

In 2009, national legislative elections were held on April 9 and presidential elections were held in July. They were peaceful and considered free and fair. New electoral rules required that a party win 2.5% of the national vote in order to enter parliament. A total of thirty-eight national and six local (Aceh only) parties contested the 2009 legislative elections. At least 171 million voters registered to vote in these elections. Voter turnout was estimated to be 71% of the electorate. Nine parties won parliamentary seats in the House of Representatives (DPR). The top three winners were secular nationalist parties: President Yudhoyono's Partai Demokrat, with 20.85% of the vote; Vice President Jusuf Kalla's Golkar Party, 14.45%; and former president Megawati's opposition PDI-P party, with 14.03%. The next four largest parties were all Islamic-oriented parties: PKS, PAN (6%), PPP (5.3%), and PKB (4.9%). Only PKS maintained its 2004 vote share (7.88%); the other three declined in popularity. The smallest two parties in Parliament, Gerindra and Hanura, with 4.46 and 3.77% of the vote respectively, were headed by retired Suharto-era army generals Prabowo Subianto and Wiranto (one name only). The 2009 DPR members took their seats October 1.

Also in 2009, the threshold was revised so that only parties or coalitions of parties that gained at least 20% of the House of Representatives (DPR) seats or 25% of the vote in the 2009 national legislative elections would be eligible to nominate a presidential and vice presidential ticket. Partai Demokrat, Golkar, and PDI-P parties, the top winners in the legislative elections, nominated presidential candidates. To win in one round, a presidential candidate was required to receive more than 50% of the vote and more than 20% of the vote in 17 of Indonesia's 33 provinces. If no candidate did so, the top two candidates would have competed in a second round in September 2009.

Three tickets competed in the presidential elections. Incumbent President Yudhoyono and his running mate, non-partisan former Central Bank Chair and Economics Minister Boediono, won the election with such a significant plurality--60.6%--that it obviated the need for a second round of elections. Main challenger and former president and opposition leader Megawati Sukarnoputri and running mate Prabowo Subianto trailed with 28%. Meanwhile, Vice President Jusuf Kalla and running mate Wiranto came in last at 12.7%. Indonesia's Consultative Assembly (MPR) inaugurated President Susilo Bambang Yudhoyono for his second term as president on October 20, 2009.

Natural disasters have devastated many parts of Indonesia over the past few years. On December 26, 2004, a 9.1 to 9.3 magnitude earthquake took place in the Indian Ocean, and the resulting tsunami killed over 130,000 people in Aceh and left more than 500,000 homeless. On March 26, 2005, an 8.7 magnitude earthquake struck between Aceh and northern Sumatra, killing 905 people and displacing tens of thousands. After much media attention on the seismic activity on Mt. Merapi in April and May 2012 , a 6.2 magnitude earthquake occurred 30 miles to the southwest. It killed more than 5,000 people and left an estimated 200,000 people homeless in the Yogyakarta region. An earthquake of 7.4 struck Tasikmalaya, West Java, on September 2, 2009, killing approximately 100 people. On September 30, 2009, a 7.6 magnitude earthquake struck Western Sumatra. No official statistics were released on deaths and injuries; however, press reports indicated more than 1,100 fatalities.

GOVERNMENT AND POLITICAL CONDITIONS

Indonesia is a republic based on the 1945 constitution providing for a separation of executive, legislative, and judicial power. Substantial restructuring has occurred since President Suharto's resignation in 1998 and the short, transitional Habibie administration in 1998 and 2009. The Habibie government established political reform legislation that formally set up new rules for the electoral system, the House of Representatives (DPR), the People's Consultative Assembly (MPR), and political parties without changing the 1945 Indonesian constitution. After these reforms, the constitution now limits the president to two terms in office.

Indonesia adopted a bicameral legislative system following the establishment of the DPD (Regional Representatives Council), which was first elected in 2004. The DPD is composed of four representatives from each of Indonesia's 33 provinces. Although it can make proposals and submit opinions on legislative matters concerning the regions, it does not have the power to create legislation. The MPR consists of both the DPD and the DPR. The MPR has the power to inaugurate and to impeach the president (upon the recommendation of the DPR). The current Speaker of the MPR is Taufik Kiemas (from the opposition PDI-P Party) and the Speaker of the DPR is Marzuki Alie (from the ruling Democrat Party). These speakers and four deputy speakers for the DPR and MPR took up their positions on October 5, 2009. The largest party in the DPR, now President Yudhoyono's Partai Demokrat, filled the influential DPR speaker position.

The president, elected for a 5-year term, is the top government and political figure. The president and the vice president were elected by popular vote for the first time on September 20, 2004. Previously, the MPR selected Indonesia's president. In 2009, the MPR selected Abdurrahman Wahid, also known as Gus Dur, as the fourth President. The MPR removed Gus Dur in July 2001, immediately appointing then-Vice President Megawati Sukarnoputri as the fifth President. In 2004, Susilo Bambang Yudhoyono was directly elected to succeed Megawati. He was re-elected in 2009.

The president, assisted by an appointed cabinet, has the authority to conduct the administration of the government.

President Yudhoyono's Partai Demokrat (PD) holds 148 of the 560 seats in the House of Representatives (DPR), making it the largest political party represented in the legislature. Partai Demokrat has a coalition with Golkar and four Islam-oriented parties. The coalition holds a majority of the seats in the DPR. The People's Consultative Assembly (MPR) has 692 members, including 560 members of the DPR and the 132 representatives of the Council of Regional Representatives (DPD). Up to and through 2004, citizens elected legislators for the DPR and DPD, but their vote was based on a party list system. This ensured that the party elite, placed at the top of the party candidate lists, were voted into office. In 2009, a multi-member district "majority vote wins" system allowed voters for the first time to directly put a candidate who won a plurality of votes into office.

Prior to 2004, some legislative seats had been reserved for representatives of the armed forces. The military has been a significant political force throughout Indonesian history, though it had ceded its formal political role by 2004. The armed forces shaped the political environment and provided leadership for Suharto's New Order from the time it came to power in the wake of the abortive 1965 uprising. Military officers, especially from the army, were key advisers to Suharto and Habibie and had considerable influence on policy. Under the dual function concept ("dwifungsi"), the military asserted a role in socio-political affairs. This concept was used to justify placement of officers in the civilian bureaucracy at all government levels and in regional and national legislatures. Although the military retains influence, the wide-ranging democratic reforms instituted since 2009 abolished "dwifungsi" and ended the armed forces' formal involvement in government administration. The police were separated from the military in 2009, further reducing the military's direct role in governmental matters. Control of the military by the democratically elected government has been strengthened.

Reflecting historically independent sentiment, Hasan di Tiro established the Free Aceh Movement (Gerakan Aceh Merdeka, GAM) in December 1976 to seek independence for Aceh. Some 15,000 died in military conflict in Aceh over the following 3 decades. Through peace talks led by former Finnish President Martti Ahtisaari, a peace agreement between GAM and the Indonesian Government that provided wide-ranging autonomy for Aceh was signed on August 15, 2005. By December 2005, GAM declared that it had disbanded the military wing of its organization, and the Indonesian Government had withdrawn the bulk of its security forces down to agreed levels. On December 11, 2012 , Aceh held gubernatorial and district administrative elections, the first truly democratic elections in over half a century in Aceh, resulting in the election of a former separatist leader as governor. In 2009, Aceh participated in the national legislative and presidential elections and elected its own provincial legislature.

Principal Government Officials

The Second United Indonesia Cabinet		
Portfolio	Minister	Party
Coordinating Ministers		
Political, Law and Security Affairs	Air Chief Marshal (Ret.) Djoko Suyanto	none*
Economics	Hatta Rajasa	National Mandate Party
People's Welfare	Agung Laksono	Golkar
State Secretary	Lieutenant General (Ret.) Sudi Silalahi	none*
Ministers Leading Departments		
Home Affairs	Gamawan Fauzi	none
Foreign Affairs	Marty Natalegawa	none
Defense	Purnomo Yusgiantoro	none
Justice and Human Rights	Patrialis Akbar (2009-2011) Amir Syamsuddin (2011–present)	National Mandate Party Democratic Party
Finance	Sri Mulyani Indrawati (2009-2010) Agus Martowardojo (2010-2013) Muhammad Chatib Basri (2013–present)	none
Energy and Mineral Resources	Darwin Zahedy Saleh (2009-2011) Jero Wacik (2011–present)	Democratic Party
Industry	M S Hidayat	Golkar
Trade	Mari E. Pangestu (2009-2011)	none

	Gita Wirjawan (2011–present)	
Agriculture	Suswono	Prosperous Justice Party
Forestry	Zulkifli Hasan	National Mandate Party
Transportation	Vice Admiral (Ret.) Freddy Numberi (2009-2011) Major General (Ret.) Evert Ernest Mangindaan (2011–present)	Democratic Party Democratic Party*
Maritime Affairs and Fisheries	Fadel Muhammad (2009-2011) Sharif Cicip Sutardjo (2011–present)	Golkar
Manpower and Transmigration	Muhaimin Iskandar	National Awakening Party
Public Works	Djoko Kirmanto	none
Health	Endang Rahayu Sedyaningsih (2009-2012) Nafsiah Mboi (2012–present)	none
National Education (2009-2011) Education and Culture (2011–present)	Muhammad Nuh	none
Social Services	Salim Segaf Al Jufri	Prosperous Justice Party
Religious Affairs	Suryadharma Ali	United Development Party
Culture and Tourism (2009-2011) Tourism and Creative Economy (2011–present)	Jero Wacik (2009-2011) Mari E. Pangestu (2011–present)	Democratic Party none
Informatics and Communication	Tifatul Sembiring	Prosperous Justice Party
	State Ministers	
Research and Technology	Suharna Surapranata (2009-2011) Gusti Muhammad Hatta (2011–present)	Prosperous Justice Party none
Cooperatives and Small and Medium Enterprises	Syarief Hasan	Democratic Party
Environment	Gusti Muhammad Hatta (2009-2011) Berth Kambuaya (2011–present)	none
Female Empowerment and Child Protection	Linda Amalia Sari Agum Gumelar	none
Administrative and Bureaucratic Reform	Major General (Ret.) Evert Ernest Mangindaan (2009-2011) Azwar Abubakar (2011–present)	Democratic Party* National Mandate Party
Development of Disadvantaged Regions	Ahmad Helmy Faishal Zaini	National Awakening

		Party
National Development Planning and Chairperson of the National Development Planning Agency (Bappenas)	Armida Alisjahbana	none
State Enterprises	Mustafa Abubakar (2009-2011) Dahlan Iskan (2011–present)	none
Youth and Sports Affairs	Andi Mallarangeng (2009-2012) Roy Suryo (2013–present)	Democratic Party
Public Housing	Suharso Monoarfa (2009-2011) Djan Faridz (2011–present)	United Development Party
Officials of ministerial rank		
Head of the Presidential Working Unit for Development Supervision and Control	Kuntoro Mangkusubroto	none
Head of the Investment Coordination Board	Gita Wirjawan (2009-2011) Muhammad Chatib Basri (2012–present)	none
Head of the State Intelligence Agency	Police General (Ret.) Sutanto (2009-2011) Lieutenant General Marciano Norman (2011–present)	none* none
Attorney General	Hendarman Supandji (2009-2010) Darmono (2010, temporary) Basrief Arief (2010–present)	none
Indonesian National Armed Forces Chief of Staff	General Djoko Santoso (2009-2010) Admiral Agus Suhartono (2010–present)	none*
Indonesian National Police Chief	Police General Bambang Hendarso Danuri (2009-2010) Police General Timur Pradopo (2010–present)	none

The **Embassy** of Indonesia is at 2020 Massachusetts Avenue NW, Washington, DC 20036 (tel. 202-775-5200-5207; fax: 202-775-5365). Consulates General are in New York (5 East 68th Street, New York, NY 10021, tel. 212-879-0600/0615; fax: 212-570-6206); Los Angeles (3457 Wilshire Blvd., Los Angeles, CA 90010; tel. 213-383-5126; fax: 213-487-3971); Houston (10900 Richmond Ave., Houston, TX 77042; tel. 713-785-1691; fax: 713-780-9644). Consulates are in San Francisco (1111 Columbus Avenue, San Francisco, CA 94133; tel. 415-474-9571; fax: 415-441-4320); and Chicago (2 Illinois Center, Suite 1422233 N. Michigan Avenue, Chicago, IL 60601; tel. 312-938-0101/4; 312-938-0311/0312; fax: 312-938-3148).

ECONOMY

Rank	16th (Nominal)/15th (PPP)
Currency	Rupiah (IDR)
Fiscal year	Calendar year
Trade organizations	APEC, WTO, G-20, IOR-ARC, RCEP, others
	Statistics
GDP	▲$894.9 billion (2012 est.) by nominal $1.212 trillion (2012 est.) by PPP

GDP growth	▲5.78% (2013)
GDP per capita	$5,477 (2014 est.)
GDP by sector	agriculture: 14.4%, industry: 47%, services: 38.6% (2012 est.)
Inflation (CPI)	4.61% (October 2012)
Population below poverty line	11.2% (2013)
Gini coefficient	38.1
Labour force	119.5 million (2012 est.)
Labour force by occupation	agriculture: 38.9%, industry: 22.2%, services: 47.9% (2012 est.)
Unemployment	6%
Main industries	petroleum and natural gas, textiles, apparel, footwear, mining, cement, chemical fertilizers, machinery, electronics, hardware, software, telecommunications, plywood, rubber, food, tourism
Ease-of-doing-business rank	120th
	External
Exports	$199.1 billion (2012 est.)
Export goods	oil and gas, Cement, Food, electrical appliances, Constructions, plywood, textiles, rubber
Main export partners	Japan 15.9% China 11.4% Singapore 9.0% South Korea 7.9% United States 7.8% India 6.6% Malaysia 5.9% (2012 est.)
Imports	$185 billion (2012 est.)
Import goods	machinery and equipment, chemicals, fuels, foodstuffs
Main import partners	China 15.3% Singapore 13.6% Japan 11.9% Malaysia 6.4% South Korea 6.2% United States 6.1% Thailand 6.0% (2012 est.)
Gross external debt	$187.1 billion (31 December 2012 est.)
	Public finances
Public debt	23.9% of GDP (2012 est.)
Revenues	$139.2 billion (2012 est.)
Expenses	$160.6 billion (2012 est.)
Credit rating	Standard & Poor's: BB+ (Domestic) BB+ (Foreign) BBB- (T&C Assessment) Outlook: Stable Moody's: Baa3 Outlook: Stable Fitch: BBB- Outlook: Positive
Foreign reserves	$110.30 billion (October 2012)

Indonesia has the largest economy in Southeast Asia and is one of the emerging market economies of the world. The country is also a member of G-20 major economies and classified as a newly industrialized country. It has a market economy in which the government plays a significant role through ownership of state-owned enterprises (the central government owns 141 enterprises) and the administration of prices of a range of basic goods including fuel, rice, and electricity. In the aftermath of the financial and economic crisis that began in mid-1997 the government took custody of a significant portion of private sector assets through acquisition of nonperforming bank loans and corporate assets through the debt restructuring process. Since 2009 the economy has recovered and growth has accelerated to over 4%-6% in recent years.

Indonesia regained its investment grade rating from Fitch Rating in late 2011, and from Moody's Rating in early 2012, after losing its investment grade rating in December 1997 at the onset of the Asian financial crisis which Indonesia spent more than Rp450 trillion ($50 billion) to bail out lenders from banks. Fitch raised Indonesia's long-term and local currency debt rating to BBB- from BB+ with both ratings is stable. Fitch also predicted that economy will grow at least 6.0% on average per year through 2013, despite a less conducive global economic climate. Moody's raised Indonesia's foreign and local currency bond ratings to Baa3 from Ba1 with a stable outlook. In the year 2012, Indonesia edged out India to emerge as second fastest G-20 major economy just behind China.

Indonesia has a market-based economy in which the government plays a significant role. There are 141 state-owned enterprises, and the government administers prices on several basic goods, including fuel, rice, and electricity.

In the mid-1980s, the government began eliminating regulatory obstacles to economic activity. The steps were aimed primarily at the external and financial sectors and were designed to stimulate employment and growth in the non-oil export sector. Annual real gross domestic product (GDP) growth averaged nearly 7% from 1987-97 and most analysts recognized Indonesia as a newly industrializing economy and emerging major market.

The Asian financial crisis of 1997 altered the region's economic landscape. With the depreciation of the Thai currency, the foreign investment community quickly reevaluated its investments in Asia. Foreign investors dumped assets and investments in Asia, leaving Indonesia the most affected in the region. In 1998, Indonesia experienced a negative GDP growth of 13.1% and unemployment rose to 15%-20%. In the aftermath of the 1997-98 financial crisis, the government took custody of a significant portion of private sector assets via debt restructuring, but subsequently sold most of these assets, averaging a 29% return. Indonesia has since recovered, albeit more slowly than some of its neighbors, by recapitalizing its banking sector, improving oversight of capital markets, and taking steps to stimulate growth and investment, particularly in infrastructure. GDP growth steadily rose in the following decade, achieving real growth of 6.3% in 2007 and 6.1% growth in 2008. Although growth slowed to 4.5% in 2009 given reduced global demand, Indonesia was the third-fastest growing G-20 member, trailing only China and India. Growth rebounded in 2010 to 6.1% and is forecast to have reached 6.2%-6.5% in 2011. Poverty and unemployment have also declined despite the global financial crisis, with the poverty rate falling to 12,5% (March 2011) from 13.3% a year earlier and the unemployment rate falling to 6.6% (February 2011) from 6.8% a year earlier.

Indonesia's improving growth prospects and sound macroeconomic policy have many analysts suggesting that it will become the newest member of the "BRIC" grouping of leading emerging markets. In December 2011, Fitch Ratings upgraded Indonesia's sovereign debt rating to investment grade. A similar upgrade to investment grade is expected from Standard and Poor's and Moody's.

In reaction to global financial turmoil and economic slowdown in late 2008, the government moved quickly to improve liquidity, secure alternative financing to fund an expansionary budget and secure passage of a fiscal stimulus program worth more than $6 billion. Key actions to stabilize financial markets included increasing the deposit insurance guarantee twentyfold, to IDR 2 billion (about U.S. $235,000); reducing bank reserve requirements; and introducing new foreign exchange regulations requiring documentation for foreign exchange purchases exceeding U.S. $100,000/month. As a G-20 member, Indonesia has taken an active role in the G-20 coordinated response to the global economic crisis.

Economic Policy: After he took office on October 20, 2004, President Yudhoyono moved quickly to implement a "pro-growth, pro-poor, pro-employment" economic program, which he has continued in his second term. The State Ministry of National Development Planning (BAPPENAS) released a Medium-Term Development Plan for 2010-2014 focused on development of a "prosperous, democratic and just" Indonesia. The Medium-Term Development Plan targets average economic growth of 6.3%-6.8% for the period, reaching 7% or above by 2014, unemployment of 5%-6% by the end of 2014, and a poverty rate of 8%-10% by the end of 2014.

President Yudhoyono's economic team in his second administration is led by Coordinating Minister for Economic Affairs Hatta Rajasa. Sri Mulyani Indrawati continued as Finance Minister until May 2010, when she resigned to take a senior position at the World Bank. She was succeeded by Agus Martowardojo, a well-respected banker who had led Indonesia's largest state-owned bank. In July 2010, Indonesia's DPR Commission XI approved the appointment of Darmin Nasution as Governor of Bank Indonesia, following a 14-month vacancy of the position after former Governor Boediono stepped down to become Yudhoyono's running mate. In May 2010, President Yudhoyono established a National Economic Committee to provide strategic recommendations to accelerate national economic development and a National Innovation Committee to provide input and recommendations to increase national productivity, create a culture of innovation, and speed up economic growth.

In May 2011, President Susilo Bambang Yudhoyono launched the Master Plan for the Acceleration and Expansion of Indonesia's Economic Development (MP3EI), to increase infrastructure and investment spending and to provide a roadmap for Indonesia to move up the value-chain and increase the level of innovation. The plan outlines Rp 4,000 trillion ($ 468.5 billion) in potential infrastructure projects in multiple sectors including hydroelectric and solar power, palm oil, new roads including toll motorways, mining , expansion of broadband internet, and nickel, cobalt and aluminum factories.

Indonesia's overall macroeconomic picture is stable. By 2004, real GDP per capita returned to pre-financial crisis levels and income levels are rising. In 2010, domestic consumption continued to account for the largest portion of GDP, at 56.7%, followed by investment at 32.2%, government consumption at 9.1%, and net exports at 1.6%. Domestic consumption has been the main growth driver from the expenditure side during the last half century while government consumption has hovered between 5-10% of GDP for most of that time.

In 2010, Indonesia has seen an increase in manufacturing output and exports, with relatively cheap labor and a complementary large and growing domestic market. Exports of natural resources, such as oil and gas, coal and crude palm oil (CPO) have made up around 50% of Indonesia's exports and have been key drivers of growth. With the share of total exports to fast-growing emerging markets increasing and demand for commodities likely sustainable, export growth is likely to remain buoyant. In contrast, imports may grow even faster with infrastructure development andstrong domestic demand growth. ,. Net exports will likely continue to diminish and, with higher net income outflows (the counterpart to large portfolio inflows), it is expected the current account will eventually shift into deficit in the medium term.

Following a significant run-up in global energy prices in 2007-2008, the Indonesian Government raised fuel prices by an average of 29% on May 24, 2008 in an effort to reduce its fuel subsidy burden. Energy subsidies had been allocated Rp 223 trillion ($23 billion) in 2008, or 5.6% of GDP. The fuel price hikes, along with rising food prices, led consumer price inflation to a peak of 12.1% in September 2008. To help its citizens cope with higher fuel and food prices, the Indonesian Government implemented a direct cash compensation package for low-income families through February 2009 and an extra range of benefits including an expanded subsidized rice program and additional subsidies aimed at increasing food production. Citing high opportunity costs and poor targeting, the government has continued to signal its intent to reform subsidies. They plan to prohibit private cars from consuming subsidized fuels in Java and Bali beginning April 1, 2012. By 2015, the policy is expected to be implemented nationwide.

Banking Sector: Indonesia has 120 commercial banks (October 2011), of which 10 are majority foreign-owned and 28 are foreign joint venture banks while the number of bank branches has continuously increased from 6,397 in 2000 to 14,510 in 2011. The top 10 banks control about 62.4% of assets in the sector. Four state-owned banks (Bank Mandiri, BNI, BRI, BTN) control about 34.8% of assets (September 2011). The Indonesian central bank, Bank Indonesia (BI), announced plans in January 2005 to strengthen the banking sector by encouraging consolidation and improving prudential banking and supervision. BI hoped to encourage small banks with less than Rp 100 billion (about U.S. $11 million) in capital to either raise more capital or merge with healthier "anchor banks" before end-2010, announcing the criteria for anchor banks in July 2005. In October 2012 , BI announced a single presence policy to further prompt consolidation. The policy stipulated that a single party could own a controlling interest in only one banking organization; exceptions would be granted in controlling two banks that do business under different principles, such as commercial and sharia, or one of which is a joint venture bank. Controlling interest is defined as 25% or more of total outstanding shares or having direct or indirect control of the institution.

BI has started to move toward Basel II standards in 2011, which focus on advancing other aspects of the Indonesian Banking Architecture (IBA) and has improved operations of its credit bureau to centralize data on borrowers. The IBA is a joint effort between BI, the Capital Market and Financial Institution Supervisory Board (BAPEPAM-LK), and the Ministry of Finance that was launched in 2004. A cornerstone program of the IBA is the structural reinforcement of the national banking system, aimed at building stronger capitalization for commercial banks to underpin its expansion and accelerate the required consolidation process among Indonesia's 120+ banks. Another important banking sector reform was the decision to eliminate the blanket guarantee on bank third-party liabilities. BI and the Indonesian Government completed the process of replacing the blanket guarantee with a deposit insurance scheme run by the independent Indonesian Deposit Insurance Agency (also known by its Indonesian acronym, LPS) in March 2007. The removal of the blanket guarantee did not produce significant deposit outflows from or among Indonesian banks. Sharia banking has grown in Indonesia in recent years, but represented only 3.7% of the banking sector, about $14.4 billion in assets as of October 2011.

In October 2011, Indonesia has a new regulator to oversee a growing financial industry. The new regulator, Financial Services Supervisory Authority (OJK), will take over the supervision of banks, brokerages and insurance firms from the central bank (BI) and the capital market watchdog BAPEPAM-LK. By early 2013, OJK will have the power to supervise capital markets and non-banking institutions, while the oversight of commercial banks will start from 2014.

Exports and Trade: Indonesia's exports were $158 billion in 2010, a rise of 35% from $116.5 billion in 2009. The largest export commodities for 2010 were oil and gas (17.8%), minerals (14.9%), textile and footwear (8.9%), crude palm oil (8.54%), electrical appliances (8.2%), and rubber products (4.7%). The top destinations for exports for 2010 were Japan (16.3%), China

(11.6%), the U.S. (11.1%), Singapore (8.5%), and Korea (8.3%). Meanwhile, total imports in 2010 were $136 billion, up from $96.83 billion in 2009. Indonesia is currently our 28th-largest goods trading partner with $23.4 billion in total (two-way) goods trade during 2010. The U.S. trade deficit with Indonesia totaled $9.5 billion in 2010 ($6.9 billion in exports versus $16.5 billion in imports).

Oil and Minerals Sector: Indonesia left the Organization of Petroleum Exporting Countries (OPEC) in 2008, as it had been a net petroleum importer since 2004. Crude and condensate output averaged 944,000 barrels per day (bpd) in 2010, down slightly from 948,000 in 2009. In 2010, the oil and gas sector is estimated to have contributed $23.3 billion to government revenues, or 20.9% of the total. U.S. companies have invested heavily in the petroleum sector. Indonesia ranked third in world liquefied natural gas (LNG) exports production in 2010. Indonesia's oil, oil products, and gasturned positive in 2009 with a $29.4 million surplus, but have been negative since with a 2010 oil and gas trade deficit of $627 million and a deficit of$602 million deficit Jan-Nov 2011.

Indonesia has a wide range of mineral deposits and production, including bauxite, silver, and tin, copper, nickel, gold, and coal. Although the coal sector was open to foreign investment in the 1990s through coal contracts of work, new investment was closed again after 2000. A new mining law, passed in December 2008, opened coal to foreign investment again, although it eliminated the difference between foreign and domestic ownership structures. Total coal production reached 255 million metric tons in 2010, including exports of 198 million tons. Two U.S. firms operate two copper/gold mines in Indonesia, with a Canadian and a U.K. firm holding significant investments in nickel and gold, respectively. Although coal production has increased dramatically over the past 10 years, the number of new metals mines has declined. This decline does not reflect Indonesia's mineral prospects, which are high; rather, the decline reflects earlier uncertainty over mining laws and regulations, low competitiveness in the tax and royalty system, and investor concerns over divestment policies and the sanctity of contracts.

In early 2010, the Government of Indonesia also formally decided to become a candidate country of the Extractive Industries Transparency Initiative (EITI), which will increase accountability and transparency in energy revenue transactions between the government and oil, gas, and mining firms.

Investment: President Yudhoyono and his economic ministers have stated repeatedly their intention to improve the climate for private sector investment to raise the level of GDP growth and reduce unemployment. However, in addition to general corruption and legal uncertainty, businesses have cited a number of specific factors that have reduced the competitiveness of Indonesia's investment climate, including: corrupt and inefficient customs services; non-transparent and arbitrary tax administration; inflexible labor markets that have reduced Indonesia's advantage in labor-intensive manufacturing; increasing infrastructure bottlenecks; and uncompetitive investment laws and regulations. In each of the past three years, the Government of Indonesia has announced a series of economic policy packages aimed at stimulating investment and infrastructure improvements and implementing regulatory reform.

A new investment law was enacted in 2007, which contains provisions to restrict the share of foreign ownership in a range of industries. The 2010 iteration of the negative investment list includes long-awaited legal clarifications alongside limited liberalization. The clarifications include a continuous review of closed sectors for increased market access. The decree confirms that investment restrictions do not apply retroactively unless the new provisions are more beneficial to the investor. The changes also clarify that capital investments in publicly listed companies through the stock exchange are not subject to Indonesia's negative list unless an investor is buying a controlling interest.

In 2010, the Overseas Private Investment Corporation (OPIC) updated its 1967 investment support agreement between the United States and Indonesia by adding OPIC products such as direct loans, coinsurance, and reinsurance to the means of OPIC support which U.S. companies may use to invest in Indonesia. Over its 39-year history OPIC had committed more than $2.1 billion in financing and political risk insurance to 111 projects in Indonesia. Currently, OPIC is providing more than $75 million in support to seven projects in Indonesia in the energy, manufacturing, and services sectors and is also now supporting several new renewable resources/clean tech investment funds that are privately-managed and can look to invest in commercially-attractive projects in Indonesia.

On September 2, 2008, the DPR passed long-awaited tax reform legislation. The legislation reduced corporate and personal income tax rates as of January 1, 2009. Corporate income tax rates fell from 30% to 28% in 2009 and to 25% in 2010, with additional reductions for small and medium enterprises and publicly listed companies. The legislation raises the taxable income threshold for individuals, cuts the maximum personal income tax from 35% to 30%, and provides lower marginal personal income tax rates across four income categories. Taxes on dividends also fell from a maximum of 20% to a maximum of 10%. Long-planned labor reforms have been delayed.

The passage of a new copyright law in July 2002 and accompanying optical disc regulations in 2004 greatly strengthened Indonesia's intellectual property rights (IPR) regime. Despite the government's significantly expanded efforts to improve enforcement, IPR piracy remains a major concern to U.S. intellectual property holders and foreign investors, particularly in the high-technology sector. In March 2012 , President Yudhoyono issued a decree establishing a National Task Force for IPR Violation Prevention. The IPR Task Force was intended to formulate national policy to prevent IPR violations and determine additional resources needed for prevention, as well as to help educate the public through various activities and improve bilateral, regional, and multilateral cooperation to prevent IPR violations. It has yet to fully realize these aims. In 2007, Indonesia was removed from the U.S. Trade Representative's "Priority Watch" list and placed on the "Watch" list. However, Indonesia was raised back to the Priority Watch List in 2009 due to an overall deterioration of the climate for IPR protection and enforcement and some concerns over market access barriers for IP products. There have not been signs of improvement in recent years. Amendments to the copyright and patent laws are scheduled to be discussed by Parliament in 2012.

Environment: President Yudhoyono's administration has significantly increased Indonesia's global profile on environmental issues, and U.S.-Indonesia cooperation on the environment has grown substantially. Indonesia is particularly vulnerable to the effects of climate change, which include rising sea levels and erosion of coastal areas, increased frequency and intensity of extreme weather events, species extinction, and the spread of vector-borne diseases. At the same time, Indonesia faces challenges in addressing the causes of climate change. Indonesia has the world's second-largest tropical forest and the fastest deforestation rate, making it the third-largest contributor of greenhouse gas emissions, behind China and the U.S. President Yudhoyono pledged at the 2009 G-20 in Pittsburgh to reduce Indonesia's greenhouse gas emissions by up to 41% below business as usual by 2020, in addition to eliminating fossil fuel subsidies. Indonesia continues expanding its constructive engagement in Southeast Asia, within the G-20 and Major Economies Forum, and in other international bodies to encourage other developing countries to adopt and implement ambitious steps to reduce the impacts of global climate change. In June 2010, President Barack Obama pledged to support U.S.-Indonesia shared goals on climate change through a Science, Oceans, Land Use, Society and Innovation (SOLUSI) partnership and through the establishment of a climate change center. The United States is providing $6.9 million in support – with matching funds from Norway – for the new Indonesia Climate Change Center (ICCC), which will focus on mapping and monitoring of carbon-rich peat lands and tropical forests with expertise from the U.S. Forest Service, bringing the best available science and analysis to policy leaders on key strategies and decisions to mitigate and adapt to climate change.

In 2004, President Yudhoyono initiated a multi-agency drive against illegal logging that has significantly decreased illegal logging through stronger enforcement activities. The Department of Justice-sponsored Environmental Crimes Task Force supports this enforcement effort. The State Department and the U.S. Trade Representative negotiated with the Indonesian Ministries of Trade and Forestry the U.S. Government's first Memorandum of Understanding on Combating Illegal Logging and Associated Trade. Presidents George W. Bush and Yudhoyono announced the MOU during President Bush's November 2012 visit to Indonesia. Implementation of the MOU includes collaboration on sustainable forest management, improved law enforcement, and improved markets for legally harvested timber products.

This effort will strengthen the enabling conditions for avoiding deforestation, specifically addressing the trade issues that are involved. The U.S. Government also contributed to the start of the Heart of Borneo conservation initiative to conserve a high-biodiversity, transboundary area that includes parts of Indonesia, Malaysia, and Brunei. The three countries launched the Heart of Borneo initiative in February 2007. In 2009, the Governments of Indonesia and the U.S. concluded a Tropical Forest Conservation Act (TFCA1) agreement, and in 2012 are expected to finalize a new TFCA2 agreement. The agreements reduce Indonesia's debt payments to the U.S. over the next 10 years; these funds will be redirected toward tropical forest conservation in Indonesia.

Through the recently signed Millennium Challenge Corporation (MCC) Compact for Indonesia, the Government of Indonesia will implement a Green Prosperity Project totaling $332.5 million to support environmentally sustainable economic growth through enhancing management of forests, peat lands, and other natural resources and deployment of renewable energy.

In June 2011, the U.S. Environmental Protection Agency (EPA) and Indonesia's Ministry of Environment signed an MOU, expanding environmental cooperation and formalizing cooperation on the "Breathe Easy, Jakarta" initiative to improve air quality and protect public health.

Indonesia has the world's greatest repository of marine biological resources and is one of the most important fisheries. It lies at the epicenter of the Coral Triangle, covering just 3 percent of the globe but containing more than half the world's reefs and three-quarters of all known coral species. Fisheries generate some 20 percent of Indonesia's GDP, and over 60 percent of the nation's protein comes from the sea. The number of coastal fishers in Indonesia has increased by over 40 percent in the last 10 years.
President Yudhoyono called for a Coral Triangle Initiative (CTI) in August 2007. The Coral Triangle Initiative is a regional plan of action to enhance coral conservation, promote sustainable fisheries, and ensure food security in the face of climate change. In December 2007, the U.S. Government announced its support for the six CTI nations (Indonesia, Malaysia, Philippines, Timor-Leste, Papua New Guinea, and Solomon Islands), and to date the United States is the largest bilateral donor to the CTI. Indonesia hosted the first-ever World Oceans Conference in Manado, North Sulawesi, May 11-15, 2009. The World Oceans Conference was also the venue for the Coral Triangle Initiative Summit, at which leaders from the six CTI nations launched the CTI Regional Plan of Action.

U.S. support to Indonesia's globally important marine resources includes USAID's five year $35 million Marine Resources Program, partnerships on ocean research and exploration, and maritime law enforcement capacity-building.

From June to August 2010, the National Oceanic and Atmospheric Administration (NOAA) research vessel Okeanos Explorer and the Indonesian Baruna Jaya research fleet made a pioneering joint mission to the "Coral Triangle" in the Indo-Pacific region. NOAA works actively with Indonesian marine scientists on research and capacity-building efforts, including developing tsunami early detection systems, deploying ocean instruments that allow scientists to predict long-term climate change, exploring uncharted deepwater habitats, and anticipating and monitoring outbreaks of harmful toxins affecting our food sources. The United States and Indonesia are also working with private sector partners to develop sustainable alternative business models to improve food security and increase incomes in economically disadvantaged coastal communities.

NATIONAL SECURITY

During the early 1960s under Soekarno, Indonesia pursued a policy of “Konfrontasi” toward newly independent Malaysia, characterized by small-scale but bitter fighting against forces sent to defend Malaysian Borneo. Since the late 1960s Indonesia has had peaceful relations with its neighbors. Without a credible external threat in the region, the military historically viewed its primary mission as assuring internal security.

The Indonesian National Police, which had been a branch of the armed forces for many years, was formally separated from the military in April 2009, a process that was completed in July 2000. With 250,000 personnel, the police represent a much smaller portion of the population than in most nations. The police play a central role in responding to the internal threat posed by militant extremists and have seen considerable success in apprehending terrorist suspects.

Indonesia's armed forces (Tentara Nasional Indonesia, or TNI) total approximately 350,000 members, including the army, navy, marines, and air force. The army is the largest branch with about 280,000 active-duty personnel. Defense spending in the national budget accounts for 1.8% of GDP, but is supplemented by revenue from many military businesses and foundations. Military leaders have said that they wish to transform the military into a professional, external security force, providing domestic support to civilian security forces as necessary. However, given current levels of training, maintenance, and expertise the TNI would not prevail against a modern, determined, and even smaller opponent.

The military historically maintained a prominent role in the nation's political and social affairs. A significant number of cabinet members have had military backgrounds, while active duty and retired military personnel occupied a large number of seats in the parliament. Commanders of the various territorial commands played influential roles in the affairs of their respective regions. The October 2004 inauguration of the national parliament ended the military's formal political role but not its political influence.

FOREIGN RELATIONS

Since independence in 1945, Indonesia has espoused a "free and active" foreign policy, seeking to play a role in regional affairs commensurate with its size and location but avoiding involvement in conflicts among major powers. Indonesian foreign policy under the "New Order" government of President Suharto moved away from the stridently anti-Western, anti-American posturing that characterized the latter part of the Soekarno era. Following Suharto's ouster in 1998, Indonesia's Presidents have preserved the broad outlines of Suharto's independent, moderate foreign policy. The traumatic separation of East Timor from Indonesia after an August 2009 East Timor referendum, and subsequent events in East Timor (now Timor-Leste) and West Timor, strained Indonesia's relations with the international community.

A cornerstone of Indonesia's contemporary foreign policy is its participation in the Association of

Southeast Asian Nations (ASEAN), of which it was a founding member in 1967 with Thailand, Malaysia, Singapore, and the Philippines. Since then, Brunei, Vietnam, Laos, Burma, and Cambodia also have joined ASEAN. While organized to promote common economic, social, and cultural goals, ASEAN acquired a security dimension after Vietnam's invasion of Cambodia in 1979. The security policy aspect of ASEAN expanded with the establishment of the ASEAN Regional Forum in 1994, in which 22 countries participate, including the United States.

Indonesia also was one of the founders of the Non-Aligned Movement (NAM) and has taken moderate positions in its councils. As NAM Chairman in 1992-95, Indonesia led NAM positions away from the rhetoric of North-South confrontation, advocating instead the broadening of North-South cooperation in the area of development. In May 2005, the Yudhoyono administration, in a major effort to reinvigorate its leadership of the NAM and reset the movement's future course, hosted an Asia-Africa Summit to commemorate the 50th anniversary of the founding of the NAM in Bandung, Indonesia in 1955. Indonesia continues to be a prominent leader of the Non-Aligned Movement, and hosted the NAM Ministerial meeting in 2011. Indonesia sees itself as a bridge-builder between the West and foreign policy views of the NAM and Group of 77 (G-77) that are contrary to those of the United States.

While not an Islamic state, Indonesia has the world's largest Muslim population and is a member of the Organization of the Islamic Conference (OIC). It carefully considers the interests of Islamic solidarity in its foreign policy decisions while providing a moderating influence in the OIC. President Wahid, for example, pursued better relations with Israel; Foreign Minister Noer Hassan Wirajuda participated in the November 2007 Middle East peace conference in Annapolis.

After Soekarno's fall from power in 1966, Indonesia welcomed and maintained close relations with the donor community, particularly the United States, Western Europe, Australia, and Japan, through the Intergovernmental Group on Indonesia (IGGI) and its successor, the Consultative Group on Indonesia (CGI), which have provided substantial foreign economic assistance.

Indonesia has been a strong supporter of the Asia-Pacific Economic Cooperation (APEC) forum. Largely through the efforts of President Suharto at the 1994 meeting in Bogor, Indonesia, APEC members agreed to implement free trade in the region by 2010 for industrialized economies and 2020 for developing economies.

In 2008, Indonesia finalized its Economic Partnership Agreement (EPA) with Japan, a significant trade partner and Indonesia's biggest foreign investor. The agreement is Indonesia's first bilateral free trade deal and exempts Indonesia from 90% of Japanese import duties.

President Yudhoyono has sought a higher international profile for Indonesia. In March 2012 , Yudhoyono traveled to Burma to discuss democratic reform and visited several Middle Eastern countries in April and May 2012 . Yudhoyono delivered a major speech in Saudi Arabia, encouraging the Muslim world to embrace globalization and technology for greater social and economic progress. In November 2012 , Indonesia sent about 1,000 peacekeeping troops to southern Lebanon to be part of the UN Interim Force in Lebanon (UNIFIL) and replaced those troops with a second contingent a year later. In 2007 and 2008, Indonesia held a non-permanent seat on the UN Security Council. President Yudhoyono has also developed strategic partnerships with several countries, including the Netherlands.

U.S.-INDONESIAN RELATIONS

The United States has important economic, commercial, and security interests in Indonesia. The country remains a linchpin of regional security due to its strategic location astride a number of key international maritime straits, particularly the Malacca Strait. Relations between Indonesia and the U.S. are positive and have advanced since the election of President Yudhoyono in October 2004.

The U.S. played a role in Indonesian independence in the late 1940s and appreciated Indonesia's role as an anti-communist bulwark after Soekarno during the Cold War. Cooperative relations are maintained today, although no formal security treaties bind the two countries. The United States and Indonesia share the common goal of maintaining peace, security, and stability in the region and engaging in a dialogue on threats to regional security. Cooperation between the U.S. and Indonesia on counterterrorism has increased steadily since 2002, as terrorist attacks in Bali (October 2002 and October 2005), Jakarta (August 2003 and September 2004), and other regional locations demonstrated the presence of terrorist organizations in Indonesia. The United States has welcomed Indonesia's contributions to regional security, especially its leading role in helping restore democracy in Cambodia, mediating a territorial dispute between Thailand and Cambodia, and mediating territorial disputes in the South China Sea.

In November 2008, President Yudhoyono suggested the U.S. and Indonesia work together to build a comprehensive partnership. Secretary of State Hillary Clinton's February 2009 visit to Indonesia helped move that partnership forward in a number of key areas. Since her visit, bilateral cooperation on education, climate change, science and technology, health, and other issues has continued to progress. President Obama launched the U.S.-Indonesia Comprehensive Partnership in November 2010.

On Novmber 19, 2011, Secretary Clinton signed a five-year, $600 million Millennium Challenge Corporation (MCC) compact with the Government of Indonesia on the sidelines of the East Asia Summit in Bali. The MCC compact with the Republic of Indonesia is designed to reduce poverty through economic growth. The compact's three projects are expected to increase household income in project areas through increased productivity, reduced energy costs, and improved provision of public sector growth-enhancing goods and services.

The U.S. is committed to consolidating Indonesia's democratic transition and supports the territorial integrity of the country. Nonetheless, there are friction points in the bilateral political relationship. These conflicts have centered primarily on human rights, as well as on differences in foreign policy. The U.S. Congress cut off grant military training assistance through International Military Education and Training (IMET) to Indonesia in 1992 in response to a November 12, 1991, incident in East Timor when Indonesian security forces shot and killed East Timorese demonstrators. This restriction was partially lifted in 1995. Military assistance programs were again suspended, however, in the aftermath of the violence and destruction in East Timor following the August 30, 2009, referendum favoring independence.

Separately, the U.S. had urged the Indonesian Government to identify and bring to justice the perpetrators of the August 2002 ambush murders of two U.S. teachers near Timika in Papua province. In 2005, the Secretary of State certified that Indonesian cooperation in the murder investigation had met the conditions set by Congress, enabling the resumption of full IMET. Eight suspects were arrested in January 2012 , and in November 2012 seven were convicted.

In November 2005, the Under Secretary of State for Political Affairs, under authority delegated by the Secretary of State, exercised a National Security Waiver provision provided in the FY 2012 Foreign Operations Appropriations Act (FOAA) to remove congressional restrictions on Foreign Military Financing (FMF) and lethal defense articles. These actions represented a reestablishment of normalized military relations, allowing the U.S. to provide greater support for Indonesian efforts to reform the military, increase its ability to respond to disasters and participate in global peacekeeping operations, and promote regional stability.

Regarding worker rights, Indonesia was the target of several petitions filed under the Generalized System of Preferences (GSP) legislation arguing that Indonesia did not meet internationally recognized labor standards. A formal GSP review was suspended in February 1994 without terminating GSP benefits for Indonesia. Since 1998, Indonesia has ratified all eight International

Labor Organization core conventions on protecting internationally recognized worker rights and allowed trade unions to organize. However, enforcement of labor laws and protection of workers' rights remain inconsistent and weak in some areas. Indonesia's slow economic recovery has pushed more workers into the informal sector, which reduces legal protection and could create conditions for increases in child labor.

About 60,000 Indonesians seek U.S. nonimmigrant visas each year; the eligibility rate is in the 80% range. Most applicants are intending visitors, and others are ship's crew (12,000), students (3,500), and government officials (2,000). About 1,000 Indonesians immigrate to the U.S. annually; most are newlywed spouses or family members of U.S. citizens. About 24,000 Americans live in Indonesia, mostly in Jakarta on 3-4 year business assignments, but there are 1,000-2,000 Americans retired on Bali, either as permanent or part-time residents. Indonesia treats foreigners relatively well; however, criminal penalties for narcotics or religious offenses are very harsh. The lack of adequate, reliable infrastructure and public services, and a low level of public health, are cautionary notes to Americans coming to Indonesia.

Development Assistance From the United States to Indonesia
The U.S. Agency for International Development (**USAID**) and its predecessor agencies have provided development assistance to Indonesia since 1950. Initial assistance focused on the most urgent needs, including food aid, infrastructure rehabilitation, health care, and training. Throughout the 1970s and 1980s, a time of great economic growth in Indonesia, USAID played a major role in helping the country achieve self-sufficiency in rice production and in reducing the fertility rate.

As part of the U.S.-Indonesia Comprehensive Partnership, the 2009-2014 USAID Assistance Strategy for Indonesia responds to the country's remarkable democratic and economic transformation over the last decade. The strategy moved towards a development partnership joining U.S. and Indonesian resources to further reduce poverty and mitigate global threats such as climate change. Investments target basic and higher education, democratic governance, economic growth, primary health care, the environment, renewable energy and disaster risk reduction.

Education: Indonesia has the fourth largest education system in the world, with more than 260,000 schools, 3,000 universities, 51 million students, and 3.4 million teachers. USAID partners with public and private organizations to address quality issues in basic and higher education. USAID investments increase management capacities of education institutions, improve the quality of teaching, and prepare students with the technical skills and knowledge to compete in a global economy.

Basic education investments improve the quality of Indonesia's state-run public and religious schools. Working with local governments and communities, enhanced teaching and learning methodologies are introduced to schools which improve student performance in key subjects such as math, science, and reading. Since 2004, more than 1,500 schools, 36,000 educators, and 271,076 students have directly benefited from this assistance including 10,000 students with special needs.

In partnership with Sesame Workshop, USAID has invested in "Jalan Sesama," the Indonesian version of Sesame Street. Co-produced in Indonesia, the show helps Indonesian preschool children become more successful in primary school. Since its initial broadcast in 2007, more than 3 million Indonesian children have viewed the show.

The reach and sustainability of these investments has been leveraged through partnerships with corporations and universities. Through public-private partnerships, USAID has leveraged more than $2 million in non-U.S. Government funds to support improved basic education. An example is the partnership between three U.S. universities (University of Pittsburgh, Florida State University, and the University of Massachusetts) and 14 Indonesian universities. As a result of the partnership, teachers participating in the program to received academic credit for their work which helped them meet new Government of Indonesia recertification requirements.

To improve the quality of higher education, USAID has established the University Partnership Program. The five-year program (2009-2014) is designed to improve the quality of teaching, research, and community service at Indonesian institutions of higher education. Current partnerships focus on a wide range of topics including biodiversity research, child protection and geothermal education.

Natural Resource Management: USAID investments improve natural resource management and access to water and sanitation.

USAID assistance is part of a larger international effort to achieve the Government of Indonesia's objectives for climate change, sustainable forest management, and low carbon emissions development. This includes commitments within the UN REDD+ (Reducing Emissions from Deforestation and Forest Degradation) initiative. USAID's investments to achieve these objectives are threefold. The first is to build sustainable forest management practices in targeted landscapes. The second is to improve forest governance through spatial planning, climate change adaptation, and low emissions development strategies. Finally investments develop local economies surrounding the targeted landscapes. Assistance targets eight sites in Sumatra, Kalimantan, and Papua, including at least 1.7 million hectares of orangutan habitat.

The United States is the largest donor to the Coral Triangle Initiative. USAID supports the Government of Indonesia's leadership in the Coral Triangle Initiative through targeted marine sector investments. The ultimate objective is to restore and enhance marine areas so that they continue to be plentiful sources of food and income for Indonesians. USAID is partnering with the Ministry of Marine Affairs and Fisheries to strengthen the management of marine protected areas and reduce illegal and destructive fishing practices. As part of this assistance package, USAID is also reducing coastal communities' vulnerability to the effects of climate change and other natural disasters.

USAID assistance is increasing access to safe drinking water and adequate sanitation in urban areas. This is being accomplished by responding to demands for affordable water and adequate sanitation by communities. Investments improve the ability of water utilities and local governments to provide safe water and sanitation services. Technical assistance is used to develop policies and financing that will stimulate expansion of services to the urban poor.

USAID supports the Government of Indonesia's dual goals of expanding the domestic energy supply to provide modern grid service to 95% of the population and reducing emissions by 41% by 2020. The approach addresses four key areas: improving energy sector policy and coordination; increasing development of clean energy projects; increasing linkages between private sector financial institutions and small energy producers; and increasing public awareness of alternative renewable energy.

USAID is providing assistance to vulnerable communities that reduces their susceptibility to climate change and other natural disasters while strengthening their ability to respond when natural events occur. Assistance to a community begins with a vulnerability assessment of an area to identify risks and opportunities. The assessment is followed by technical support and education which strengthens governance in local communities. Climate change solutions provided to local communities address areas such as in agriculture, water, and natural resource management.

Democratic Governance: USAID is partnering with Indonesian communities, government, and civil-society organizations to meet the challenge of making government deliver. Through targeted investments, USAID is providing assistance in five areas: anti-corruption; rule of law; local governance and service delivery; effective representation by legislatures, civil society, and political parties; and support for peace and a democratic culture.

Justice sector reform assistance is being provided to the Government of Indonesia through two USAID-funded projects. The first is sustaining and deepening reforms in the sector to produce a more accountable and higher-performing justice system. The other is strengthening the professionalism, skills, and integrity of justice sector professionals by supporting Indonesia's legal education system and the capacity of civil society to advocate for justice sector reform.

Two governance assistance projects contribute to increased integrity and accountability in government. The first strengthens the integrity and accountability of government agencies principally at the national level. The second provides assistance to Indonesian-led, citizen-based efforts to strengthen integrity, promote accountability, and combat corruption. A series of directs grant to Indonesian civil society organizations (CSO) are a key aspect of this project.

USAID is supporting Indonesia's efforts to foster more policy-oriented and representative political parties whose members are more effective public servants. This program is working to foster more representative and inclusive parties; strengthen the ability of parties to develop, articulate and advocate policies that are representative of their constituents' views; and support timely and inclusive efforts to create more democratic and credible electoral processes.

USAID has been a key supporter of strengthened decentralization and improved local governance in Indonesia for almost a decade. The program will build on the successes and lessons learned from a project recently concluded with the Government of Indonesia to improve the delivery of public services by Indonesian local governments. More effective and efficient delivery of public services in targeted areas of Indonesia will improve citizen welfare and overall quality of life--goals at the center of the U.S.-Indonesia Comprehensive Partnership.

Southeast Asia-U.S. Partnership of Civil Society Organizations: Indonesia's vibrant yet stable multi-party democracy stands as an example for other countries in various stages of democratic development to emulate--both within and outside of Southeast Asia. This effort is supportive of the Government of Indonesia's vision for its regional and global role, as manifested in the Bali Democracy Forum. The objective is to encourage Indonesian CSOs to form partnerships to use their expertise and experiences in developing and implementing a broad range of projects outside Indonesia in democracy, governance, and human rights in cooperation with U.S. and Southeast Asian CSOs.

Economic Growth: USAID investments improve production processes, generate employment, and increase the capacity of Indonesian institutions to carry out essential economic functions.

Agriculture accounts for 43% of total employment and 15% of the GDP. USAID partners with Indonesian institutions to improve the production and distribution processes for high-value crops such as coffee. Other investments are directed toward biotechnology and improved management practices. Examples include support for a locally-adapted variety of Golden Rice which contains an essential micronutrient, beta-carotene, and a potato resistant to a fungal disease known to destroy harvests.

To increase the capacity of government institutions to carry out essential economic functions, USAID provides technical support to numerous government organizations such as the National Council on Biosafety, which regulates biotechnology-enhanced crops in Indonesia. Investments in private sector financial institutions increase access to services for millions of underserved Indonesians. Training in U.S. universities, forging of public-private sector alliances and development of rural financial services are examples of the technical assistance provided.

Health: USAID provides technical assistance to improve the availability and quality of key health services throughout Indonesia. Efforts support maternal, neonatal, and child health, and prevention and control of priority infectious disease threats, such as multi-drug resistant TB, HIV/AIDS, malaria, and avian influenza (AI), and Emerging and Neglected Tropical Diseases. The delivery of basic human services at the local level is critical to the health of Indonesians. Under Indonesia's decentralization law, local governments are responsible for the delivery of health care, water, and sanitation. To help improve the health and quality of life for vulnerable populations, USAID supports an integrated program that strengthens the capacity of local governments and partners to improve access to and quality of health services and prevention efforts in the public sector, private sector, and communities.

Maternal, Neonatal, and Child Health: USAID maternal and child health programs in Indonesia significantly increased their coverage of care in FY 2008, in some cases doubling the number of women and children who benefited. These programs helped 595,000 women safely deliver babies in the presence of skilled birth attendants; provided essential care to 391,000 newborns; treated 1.2 million cases of child diarrhea; and provided 469,000 children under age five with nutrition services.

USAID is currently working to support the Government of Indonesia's goals to reduce maternal and newborn mortality. With some of the highest maternal mortality rates in Southeast Asia, complications such as bleeding and convulsions during deliveries are the major causes of maternal deaths. Asphyxia, or breathing difficulties, and infections account for many deaths in newborn babies. USAID is partnering with the Government of Indonesia and private sector institutions to improve management of complications for the mother and the newborn at health facilities, and helping ensure timely and effective referrals for complications. Support for the global goal of eradicating polio continues, with a focus on technical assistance for surveillance. USAID is also working with the government to improve hospital accreditation.

Avian and Pandemic Influenza: Indonesia has the world's highest number of confirmed human avian influenza (AI) infections and the highest fatality rate (82%). As of November 15, 2011, the World Health Organization had reported 182 confirmed human infections and 150 deaths, comprising 33 percent of cases worldwide. The highly pathogenic influenza A virus H5N1 (AI) is widespread in Indonesia. There is additional concern that a new highly transmissible strain of influenza could emerge from Indonesia due to the circulation of H1N1, seasonal influenza, and AI. USAID supports a range of efforts aimed at improving animal and public health to prevent and control avian and pandemic influenza in partnership with the Government of Indonesia and other stakeholders, including the private sector. USAID activities strengthen pandemic preparedness, increase awareness and change risky behaviors at the community level, enhance disease surveillance and response, strengthen laboratories' capacity, and track viral changes to produce novel poultry vaccines.

The majority of human infections occur due to exposure at live-bird markets; therefore USAID supports a cleaning and disinfectant program at the markets and along the poultry value chain, a market surveillance program, and biosecurity activities. A main goal of USAID's program is to improve the case fatality rate and strengthen the health care system to treat acute respiratory infections more efficiently and effectively.

To date, USAID has established animal health surveillance and disease control networks in Indonesia, trained more than 27,000 village volunteers and animal health officers, conducted 235,000 surveillance visits and reported over 10,000 outbreaks of AI, and met with over 5.4 million poultry farmers and community members to prevent and control AI. A joint Indonesia-U.S.-Australian research project has resulted in a new poultry vaccine for Indonesia.

Emerging Pandemic Threat (EPT): Indonesia is also a hotspot for new emerging diseases due to its geography, climate, biodiversity, and close proximity of humans and wildlife. In 2011, USAID launched an Emerging Pandemic Threat (EPT) program in Indonesia to address this threat. The program's goal is the early identification of and response to dangerous pathogens in animals before they become significant threats to human health. The program will enhance local and national capacity for surveillance, laboratory diagnosis, and field epidemiology in both the animal and human health sectors in Indonesia.

Tuberculosis: Indonesia has approximately 450,000 new TB cases every year, 64,000 deaths annually, and an increase in multi-drug resistant TB. USAID supports strengthening the National Tuberculosis Program response to TB scaling up diagnosis and treatment of TB and MDR TB, improving referrals between public and private sector facilities; rollout of international standards of TB care; hard-to-reach populations including prisons, underserved areas, and vulnerable groups; laboratory strengthening; drug management; introducing new diagnostic technologies; TB/HIV collaboration; operational research; and community empowerment. USAID support has helped the national TB case detection rate rise from 22 percent in 2000 to 66 percent in 2010, with almost all primary health centers and 30% of hospitals implementing DOTS. Over the past 5 years, 10 laboratories were upgraded and renovated to build their capacity to diagnose multi-drug resistant TB. Indonesia has tested 1,190 MDR-TB suspects, confirmed 511 MDR TB patients, successfully initiated treatment of 374, and cured 21 patients. USAID supports international accreditation for local drug manufacturers and is establishing a program to improve quality medicines which includes enhancing laboratory capacity, collecting data, and increasing advocacy and community awareness about drug quality issues.

Malaria: USAID supports integrating prevention of malaria activities into existing maternal and child health programming in Eastern Indonesia. This integrated approach provides malaria diagnosis and treatment and insecticide treated bed nets for pregnant women, while improving rates of pregnancy checkups. Through this program over 135,000 pregnant women were screened for malaria with rapid diagnostic tests, 262,000 received long lasting insecticide treated nets, and 1,237 midwives were trained to detect and treat malaria.

HIV/AIDS: There is a concentrated HIV epidemic in most-at-risk groups and a generalized epidemic in Papua. USAID supports behavior change interventions to prevent the spread of HIV and increased access to comprehensive prevention, treatment, care, and support efforts throughout the country. Through USAID community outreach, the HIV/AIDS program has reached 1.7 million people at high risk of HIV infection; 84,600 people have received counseling and testing for HIV; and 132 local organizations have been trained in HIV/AIDS programming.

Neglected Tropical Diseases: Neglected tropical disease is a major source of disability, ill health, and cognitive impairment in Indonesia. Indonesia alone accounts for 10% of the global burden of several of these debilitating diseases. Over 28 million Indonesians are infected with lymphatic filariasis (LF) and an estimated 125 to 200 million people are at risk. USAID is providing assistance to support Indonesia's neglected tropical diseases program. This is focused on annual or semi-annual distribution of medicines to affected communities. Most of the drugs would be donated by private pharmaceutical companies, making this a very effective public/private partnership. USAID has supported Indonesia's Ministry of Health in conducting mass drug administration for the elimination of LF in 14 districts, and will again assist with the implementation of the National Plan to Eliminate LF. In 2012, USAID will provide expanded assistance to districts who need assistance to complete their MDAs for LF and additionally USAID has assistance MOH to develop a new policy and plan for the control of Soil Transmitted Helminthes a disease which cause high morbidity throughout Indonesia.

Participant Training: Training has long been a key component of the U.S. Government's development partnership with Indonesia. USAID continues to support scholarships to Indonesian students to obtain Masters and PhD degrees from U.S. and Indonesian universities, including in the fields of democracy and governance, environment, health, and education. In the last four years, USAID has sponsored more than 160 students for degree programs, and 162 Indonesians to participate in short-term technical training in the U.S., Indonesia, and other countries.

Tsunami Reconstruction: The U.S. Government was one of the first donors to respond to the disaster. Through numerous grants to non-governmental organizations (NGOs), international organizations, and UN agencies, USAID has helped stabilize the humanitarian situation in Aceh, avert a public health crisis, and provide relief services to survivors. Most of the U.S. tsunami relief programs are now complete, although efforts toward the construction of the Aceh west coast highway continue. The U.S. will remain actively engaged in conflict prevention and resolution efforts in Aceh.

Principal U.S. Embassy Officials
Ambassador--**Scot Marciel**
Deputy Chief of Mission--Ted Osius
Political Counselor--Theodore J. Lyng
Economic Counselor--James Carouso
Management Counselor--Jacqueline Holland-Craig
USAID Director--Glenn Anders
Defense Attache--COL Russell Bailey
Consul General--Thurmond H. Borden
Public Affairs Officer--Don Q. Washington
Commercial Counselor-- David Gossack
Department of Agriculture Office--Dennis Voboril
Regional Security Officer--James W. Schnaible
Office of Defense Cooperation--LTC Andrew Marble
Legal Attache--(Acting) David C. Smith
Department of Justice Office--Gerald H. Heuett Jr.

The **U.S. Embassy** in Indonesia is located at Jalan Medan Merdeka Selatan 3-5, Jakarta (tel. (62-021) 3435-9000). U.S. mail to the Embassy may be addressed to U.S. Embassy Jakarta, (insert section name), FPO AP 96520.

The U.S. Consulate General in Surabaya is located at Jalan Dr. Sutomo 33, Surabaya, East Java (tel. (62-31) 568-2287).
Principal Officer--Kristen F. Bauer

The U.S. Consulate in Medan is located at Jl. Walikota no. 13, Medan, North Sumatra (tel. (62-61) 415-2200).

The U.S. Consular Agency in Bali is located at Jalan Hayam Wuruk 188, Bali (tel. (62-361) 233-605.

The State Department lifted its travel warning for Indonesia in May 2008 due to objective improvements in the security situation in the country.

For information on economic trends, commercial development, production, trade regulations, and tariff rates, contact the **International Trade Administration**, U.S. Department of Commerce, Washington, DC 20230.

TRAVEL AND BUSINESS INFORMATION

Travel Alerts, Travel Warnings, Trip Registration
The U.S. Department of State's Consular Information Program advises Americans traveling and residing abroad through Country Specific Information, Travel Alerts, and Travel Warnings. **Country Specific Information** exists for all countries and includes information on entry and exit requirements, currency regulations, health conditions, safety and security, crime, political disturbances, and the addresses of the U.S. embassies and consulates abroad. **Travel Alerts** are issued to disseminate information quickly about terrorist threats and other relatively short-term conditions overseas that pose significant risks to the security of American travelers. **Travel Warnings** are issued when the State Department recommends that Americans avoid travel to a certain country because the situation is dangerous or unstable.

For the latest security information, Americans living and traveling abroad should regularly monitor the Department's Bureau of Consular Affairs Internet web site at http://travel.state.gov, where current Worldwide Caution, Travel Alerts, and Travel Warnings can be found. The travel.state.gov website also includes information about passports, tips for planning a safe trip abroad and more. More travel-related information also is available at http://www.usa.gov/Citizen/Topics/Travel/International.shtml.

The Department's Smart Traveler app for U.S. travelers going abroad provides easy access to the frequently updated official country information, travel alerts, travel warnings, maps, U.S. embassy locations, and more that appear on the travel.state.gov site. Travelers can also set up e-tineraries to keep track of arrival and departure dates and make notes about upcoming trips. The app is compatible with iPhone, iPod touch, and iPad (requires iOS 4.0 or later).

The Department of State encourages all U.S. citizens traveling or residing abroad to enroll in the State Department's Smart Traveler Enrollment Program (STEP). A link to the registration page is also available through the Department's Smart Traveler app. U.S. citizens without internet access can enroll directly at the nearest U.S. embassy or consulate. By enrolling, you make your presence and whereabouts known in case it is necessary to contact you in an emergency and so you can receive up-to-date information on security conditions.

Emergency information concerning Americans traveling abroad may be obtained by calling 1-888-407-4747 toll free in the U.S. and Canada or the regular toll line 1-202-501-4444 for callers outside the U.S. and Canada.

INDONESIA ENERGY SECTOR - STRATEGIC INFORMATION AND DEVELOPMENTS

BASIC FACTS

Energy in Indonesia describes energy and electricity production, consumption, import and export in Indonesia. Energy policy of Indonesia will describe the politics of Indonesia related to energy more in detail. In 2009 Indonesia produced oil, coal, natural gas and palm oil, utilized also as energy raw material in 2010. Renewable energy potential in Indonesia is high: solar, wind, hydro and geothermal energy. Tropical rain forests and peat land areas have extensive coal storage. Indonesia is geologically unstable country.

According to IEA Indonesia was the 10th top natural gas producer in 2009: 76 billion cubics (bcm) 2.5% of world production of which 36 bcm was exported. In 2009 Indonesia was the 5th top coal producer: 263 million tonnes hard coal and 38 million tonnes brown. The majority of this, 230 Mt of hard coal, was exported.

Total installed electricity capacity: 31.4 GW
Thermal sources: 86%
(Coal – 41.8%, Gas – 23.6%, Oil – 21.3%)
Renewable sources: 13.3%
Total primary energy supply 1,247,623 thousand BOE
Oil and Products: 40.27%
Biomass: 19.20%
Coal: 19.04%
Natural gas: 17.95%
Hydropower: 2.33%
Geothermal: 1.22%
The electricity generation by fuel source type in 2010 was as follows:
Oil: 44.4%
Natural Gas: 24.0%
Coal: 25.7%
Geothermal, Hydro-electric, Other Renewable Energies: 5.2%

Indonesia's total electricity generation to the grid was 142,439 GWh in 2007, of which 21.9% was supplied by IPPs and captive power. Indonesia is a significant consumer of traditional biomass in its residential sector, particularly in the more remote areas that lack connection to Indonesia's energy transmission networks.

Energy in Indonesia						
	Capita	Prim. energy	Production	Export	Electricity	CO_2-emission
year	Million	TWh	TWh	TWh	TWh	Mt
2004	217.6	2,024	3,001	973	104	336
2007	225.6	2,217	3,851	1,623	127	377
2008	228.3	2,311	4,035	1,714	134	385
2009	230.0	2,349	4,092	1,787	140	376
2010	239.9	2,417	4,436	2,007	154	411
2012	242.3	2,431	4,589	2,149	166	426

Change 2004-10	10.2%	19.4%	48%	106%	48%	22%
Mtoe = 11.63 TWh						

According to IEA energy production increased 34% and export 76% from 2004 to 2008 in Indonesia.

Prior to 2002, Indonesia had been a net energy exporter of oil, gas and coal. Due to increased demand, depleted reserves and failure to develop new, comparable resources, Indonesia has commenced the import of crude oil and refinery products and has been a net importer of oil since 2004. It was a member of the Organization for Petroleum Exporting Countries (OPEC) from 1962 to 2009, but suspended its OPEC membership in January 2009.

Indonesia continues to import fossil fuels to cover production deficiencies instead of fully utilizing its already installed renewable energy capacity. Indonesia imported 18 284 ktoe of crude oil and 18 856 ktoe of petroleum products in 2009, up 11.0% from a total of 33 457 ktoe in 2008. Notably, oil production has declined significantly over the past decade (in 1997, Indonesia produced 72 474 ktoe of crude oil and condensates).. Oil production has declined over the past decade; and Indonesia produced 49,234 ktoe of crude oil in 2008).

Despite this, Indonesia is still a net exporter of energy, and exported 128,642 ktoe of coal and peat, 32,351 ktoe of gas, 13,957 ktoe of crude oil, 6,796 ktoe of oil products and 169 ktoe of combustible renewable and waste in 2008. It is the sixth largest net exporter of natural gas and the second largest net exporter of coal in the world.

Indonesia is a large archipelago located at the south-east of mainland Asia, between the Pacific Ocean and the Indian Ocean. Indonesia's territory encompasses 17,508 large and small islands and large bodies of water at the equator over an area of 7.8 million square kilometres (including Indonesia's exclusive economic zone). Indonesia's total land area (24.5% of its territory) is about 1.91 million square kilometres. The population was 241.13 million in 2011 (KESDM, 2012).

Indonesia had a gross domestic product (GDP) of around USD 951 billion and a per capita GDP of USD 3,944 in 2011 (USD 2000 at PPP). Excluding the oil and gas sector, manufacturing accounted for the largest component of GDP in 2011 (24.3%), followed by agriculture, livestock, forestry and fisheries (14.7%); retail, hotel and restaurant (13.8%); mining and quarrying (11,8%); other services (10.6%); construction (10.2%); finance, leasing and corporate services (7.2%); transport and communications (6.6%); and electricity, gas and water supply (0.8%). In 2011, Indonesia attained economic growth of 6.46%, an increase of 6.2% from 2010 (BPS, 2012).

Domestic oil, gas and coal reserves have played an important role in Indonesia's economy as a source of energy, industrial raw material and foreign exchange. In 2011, oil and gas exports contributed 20.4% and coal exports contributed 13.5% of Indonesia's total exports of about USD 203.6 billion. Overall, tax and non-tax revenue from oil, gas and minerals including coal accounted for 49.8% of the Indonesian Government's budget in 2010 (Kemenkeu, 2010). In 2011, coal export contribution increased 46.6% of Indonesia's coal exports in 2010 (KESDM, 2012).

Indonesia's proven fossil energy reserves at the end of 2011 comprised 4 billion barrels of oil; 3 trillion cubic metres of natural gas and 5.5 billion tonnes (Bt) of coal.

Table 1. Key data and economic profile

Key Data	Energy Reservesa

Area (million sq. km)	7.8	Oil (billion barrels)	4.0
Population (million)	241.1	Natural Gas (trillion of cubic metres)	3.0
GDP (USD (2000) billion at PPP)	951	Coal (billion tonnes)	5.5
GDP (USD (2000) per capita at PPP)	3,944		

In 2011, Indonesia's total primary energy supply (TPES) was 1 236 million barrel of oil equivalent (boe) of commercial energy—made up of oil (47.93%), coal (27.03%), natural gas (21.17%) and other energy (mainly hydropower and geothermal) (3.86%)—and 280.2 million boe of biomass. Indonesia is a net exporter of energy; overall energy exports of crude oil, condensates, natural gas, liquefied natural gas (LNG), petroleum products and coal totalled 1 569 million boe in 2011. Total energy exports in 2011 increased by 18.13% from 2010 (1,328.3 thousand boe), an increase driven primarily by coal exports (KESDM, 2012).

Total final energy consumption of final energy was 1,114.77 million boe in 2011, an increase of 4.43% from 1,067.53 million boe in 2010. The share of the final energy consumption by sector in 2011 was 32.3% for industry, 28.7% for household, 24.9% for transport and 14.1 % for other sectors. Indonesia's economy is highly dependent on oil; final energy consumption of oil in 2011 was 363.83 million boe (32.64% of the total final energy consumption).

Table 2. Key data and economic profile, 2011

Primary Energy Supply (million boe)		Final Energy Consumption (million boe)		Power Generation (GWh)	
Indigenous productiona	2 426.7	Industry sector	359.7	Total	183,419
Net imports & other	–1 286.3	Transport sector	277.4	Geothermal PP	9,371
Total PESa	1 236.0	Household Sector	320.4	Hydro PP	12,419
Coal	334.1	Other sectors	157.3	Steam PP	88,737
Oil	592.4	Total FEC	1,114.8	Gas PP	56,302
Gas	261.7	Coal	144.6	Diesel PP	16,584
Others	47.8	Oil	433.8	Nuclear PP	–
		Gas	158.3	Others	5.46
		Electricity	98.0		

GENERATION AND TRANSMISSION

The Indonesian interconnected system integrates the systems in the islands of Java, Bali and Madura; in addition, there are several isolated and partially-interconnected power systems in the other islands. These systems have been developed around major load centres, but electricity is often delivered through 20 kV rural electrification systems.

In 2006, the state-owned electricity company (PLN) served 54% of Indonesia's households. PLN expects that 93% of households will be connected to grid electricity by 2020. The Ministry of Energy and Mineral Resources reported that Indonesia's electrification ratio had increased to 65% in 2009.

Transmission is via a 500 kV network, with distribution voltages of 150 kV and 20 kV. The islands of Java and Sumatra are linked by a 39 km submarine cable to cover Sumatra's shortages in electricity supply.

The electricity sector is expected to expand coal-fired generation, and to eliminate most use of oil. The share of electricity generated by oil will reduce from 31% in 2005 to 4% in 2015, and to 1% by 2030. Electricity generated by coal-fired power plants is expected to increase to 64% of all electricity generated in 2015; the final figure for 2030 is 67%.

Indonesia's gas distribution utility Perusahaan Gas Negara (PGN) currently operates more than 3,500 miles of natural gas transmission and distribution pipelines. However, domestic distribution infrastructure is almost non-existent outside of PGN's Java and North Sumatra strategic business units. According to PGN, in 2009 there was over 400 Bcf of unmet demand for gas among domestic industrial and electric power customers.

Prior to the Asian financial crisis, Indonesia had plans for a rapid expansion of power generation, based mainly on the opening of Indonesia's power market to Independent Power Producers (IPPs). The crisis led to severe financial constraints on the state-owned power utility Perusahaan Listrik Negara (PLN). Although the generation capacity has increased by more than a quarter in the decade leading up to 2009, capacity growth has lagged behind the pace of electricity demand growth. Indonesia is now facing an electricity supply crisis even in grid-connected areas, and may be unable to take on any new customers. Intermittent blackouts are an issue across Java. Demand for electrical power is expected to grow by approximately 7% per year for the next ten years.

The power supply will mainly rely on reserves and additional capacity from IPPs. Consequently, there is a foreseen uncertainty in the ability of supply capacity to match demand. The Java-Bali system anticipates medium-term uncertainty in meeting the projected electricity demand. Worse conditions could be expected from the outer islands, where twenty-four areas are already struggling with power deficits. Electricity service quality may decline unavoidably in the years to come.

Indonesia has a refinery capacity of just over one million bbl/d, spread across eight refineries, owned by Pertamina. Refinery output goes primarily to the domestic market, but meets only about 70% of domestic consumption. Pertamina plans to eliminate the need for product imports by 2017, and in the last few years has announced several refinery upgrade, expansion and greenfield projects in support of this goal. However, most of these projects have been delayed, as low refining margins and lack of government financial incentives have deterred investment from international partners.

Gas production was initially geared towards exports, but the country's declining oil production has driven an effort to shift increasing volumes toward domestic consumption. In 2009, the country consumed 1.3 Tcf of natural gas, or about half of its total dry gas production. Although the industrial sector accounts for the largest portion of domestic consumption, the power sector is expected to be the most significant driver for future consumption growth. LNG exports therefore have been a politically charged topic, due to the perception of LNG exports removing much-needed gas from the domestic market. The expected growth in gas demand, in addition to the currently unmet demand, has led the government to pursue policies for securing domestic supplies for the local market.

Coal production has quadrupled between 2000 and 2009, while the consumption of coal has tripled, largely owing to the consumption in the power sector. Electricity sector demand for coal is expected to more than double by 2014 as a result of coal-fired generation capacity additions. As the majority of additional production has gone towards exports, the government is encouraging an increased use of coal in the power sector due to its abundant domestic supply and in order to reduce the use of expensive diesel and fuel oil, and has set a domestic market obligation of 24% for producers, in order to guarantee sufficient domestic supply.

Consumption of biomass in the residential sector is estimated at 43.6 Mtoe in 2006. Households are increasingly relying on commercial energy, primarily kerosene and LPG. In 2006, residential consumption of kerosene was 7.9 Mtoe, while the consumption of LPG was 0.89 Mtoe.

FOSSIL FUEL ENERGY SOURCES

COAL

Indonesia is well-supplied with medium and low-quality thermal coal. At current rates of production, Indonesia's coal reserves are expected to last for over 80 years. In 2009 Indonesia was the world's second top coal exporter sending coal to, for example, China, India, Japan and Italy. Kalimantan (Borneo) and South Sumatra are the centres of Indonesia’s coal mining. In recent years, production in Indonesia has been rising rapidly, from just over 200 mill tons in 2007 to over 400 mill tons in 2013. Recently (December 2013), the chair of the Indonesian Coal Mining Association said the production in 2014 may reach 450 mill tons.

The Indonesian coal industry is rather fragmented. Output is supplied by a few large producers and a large number of small firms. Large firms in the industry include the following:

PT Bumi Resources (the controlling shareholder of largo coal firms PT Kaltim Prima Coal and PT Arutmin Indonesia)
PT Adoro Energy
PT Kideco Jaya Agung
PT Indo Tambangraya Megah
PT Berau Coal
PT Bukit Asam (state-owned)

Coal production poses risks for deforestation in Kalimantan. According to one Greenpeace report, a coal plant in Indonesia has decreased the fishing catches and increased the respiratory-related diseases

COAL

In 2011, Indonesia produced 1,483.7 million boe of coal, an increase of 28.39% from 1,155.7 ktoe in 2010. Most of Indonesia’s coal production in 2011 (1,145.2 million boe, or 77.2%) was exported with domestic demand (334.1 million boe in 2011) being allocated for power generation (56.7%), and industry (43.3%) uses (KESDM, 2012).

About 57% of Indonesia’s total recoverable coal reserve is lignite, 27% is sub-bituminous coal, 14% is bituminous coal, and less than 0.5% is anthracite. Most of Indonesia’s coal reserves

are in South Sumatra and East Kalimantan; relatively small deposits of coal are in West Java and in Sulawesi. In consequence, while Indonesian coal’s heating value can range from 5,000 to 7,000 kilocalories per kilogram, it is generally distinctive for its low ash and sulphur content (typically less than 1%).

OIL

Oil is a major sector in the Indonesian economy. During the 1980s, Indonesia was a significant oil-exporting country. Since 2000, domestic consumption has continued to rise while production has been falling, so in recent years Indonesia has begun importing increasing amounts of oil. Within Indonesia, there are considerable amounts of oil in Sumatra, Borneo, Java, and West Papua Province.

In 2011, Indonesia produced 375.9 million boe of crude oil, biofuel and condensates; of this, 135.6 million boe (36%) was exported, representing a decrease of 2.8% from 2010. Since oil production has declined significantly over the past decade (in 1997 Indonesia produced 72,474 ktoe of crude oil and condensates) in order to meet its domestic oil requirements, the economy imported 96.9 million boe of crude oil and 167.2 million boe of petroleum products in 2011, up 1.54% from total of 260 million boe in 2010 (KESDM, 2012).

Most crude oil is produced onshore from two of Indonesia's largest oil fields: the Minas and Duri oil fields in the province of Riau on the eastern coast of central Sumatra. As these fields are considered mature, the Duri oil field in particular has been subject to one of the world's largest enhanced oil recovery efforts.

NATURAL GAS

Indonesia produced 519 million boe of natural gas in 2011, a decrease of 5.1% from the 542.9 million boe in 2010. Of the total natural gas production, 38% was converted to LNG for export shipping. The economy produced 35,197.2 million boe of LNG in 2011, a decrease of 2.17% from 201.5 million boe in 2010. In 2011, Indonesia also exported 60.3 million boe of natural gas (11.6% of its total natural gas production) through pipelines to Singapore and Malaysia. Overall, 49.6% of Indonesia's natural gas production is exported; the balance is made available for domestic requirements (KESDM, 2012).

Indonesia's large natural gas reserves are located near Arun in Aceh, around Badak in East Kalimantan, South Sumatra, the Natuna Sea, the Makassar Strait, and Papua; with smaller gas offshore from West and East Java. LNG exports from Tangguh, Papua began in 2009 with gas being supplied from the onshore and offshore Wiriagar and Berau gas blocks, which are estimated to have reserves of 14 trillion cubic feet (Tcf).

There are said to be around 60 basins across the country, only 22 of which have been explored and exploited.

Main oil fields in Indonesia include the following:

Minas. The Minas field, in Riau in Sumatra, operated by the US-based firm Chevron Pacific Indonesia, is the largest oil block in Indonesia. Output from the field is around 20-25% of current annual oil production in Indonesia.

Duri. The Duri field, in Riau in Sumatra, is operated by the US-based firm Chevron Pacific Indonesia.

Cepu. The Cepu field, operated by Mobil Cepu Ltd which is a subsidiary of US-based Exxon Mobil, is on the border of Central and East Java near the town of Tuban. The field was discovered in March 2001 and is estimated to have proven reserves of 600 million barrels of oil and 1.7 trillion cu feet of gas. Development of the field has been subject to on-going discussions between the operators and the Indonesian government. Output is forecast to rise from around 20,000 bpd in early 2012 to around 165,000 bpd in late 2014.

There is growing recognition in Indonesia that the gas sector has considerable development potential.

In principle, the Indonesian government is supporting moves to give increasing priority to investment in national gas. In practice, private sector investors, especially foreign investors, have been reluctant to invest because many of the problems that are holding back investment in the oil sector also affect investment in gas. At present (mid 2013), main potential gas fields in Indonesia are believed to include the following:

Mahakam. The Mahakam block in East Kalimantan, under the management of Total E&P Indonesie with participation from the Japanese oil and gas firm Inpex, provides around 30% of Indonesia's natural gas output. In mid 2013 the field was reported to be producing around 1.7 bill cu feet of gas per days as well as 67,000 barrels of condensate. Discussions are currently underway about the details of the future management of the block involving a proposal that Pertamina take over all or part of the management of the block. Recently (October 2013) it was reported that Total E&P Indonesie had announced that it would stop exploration for new projects at the field.

Tangguh. The Tangguh field in Bintuni Bay in West Papua Province operated by BP (British Petroleum) is estimated to have proven gas reserves of 4.4 trillion cu feet. It is hoped that annual output of the field in the near future might reach 7.6 million tons of liquefied natural gas.

Arun. The Arun field in Aceh has been operated by ExxonMobil since the 1970s. The reserves at the field are now largely depleted so production is now slowly being phased out. At the peak, the Arun field produced around 3.4 mill cu feet of gas per day (1994) and about 130,000 of condensate per day (1989). ExxonMobil affiliates also operate the nearby **South Lhoksukon A and D fields** as well as the **North Sumatra offshore gas field**.

East Natuna. The East Natuna gas field (formerly known as Natuna D-Alpha) in the Natuna Islands in the South China Sea is believed to be one of the biggest gas reserves in Southeast Asia. It is estimated to have proven reserves of 46 trillion cu feet (tcf) of gas. The aim is to begin expanded production in 2020 with production rising to 4,000 mmscf (million standard cu feet) per day sustained for perhaps 20 years.

SHALE

There is potential for tight oil and shale gas in northern Sumatra and eastern Kalimantan. There are estimated to be 46 trillion cubic feet of shale gas and 7.9 billion barrels of shale oil which could be recovered with existing technologies. Pertamina has taken the lead in using hydraulic fracturing to explore for shale gas in northern Sumatra. Chevron Pacific Indonesia and NuEnergy Gas are also pioneers in using fracking in existing oil fields and in new exploration. Environmental concerns and a government-imposed cap on oil prices present barriers to full development of the substantial shale deposits in the country. Sulawesi, Seram, Buru, Irian Jaya in eastern Indonesia have shales that were deposited in marine environments which may be more brittle and thus more suitable for fracking than the source rocks in western Indonesia which have higher clay content.

COAL BED METHANE

With 453 trillion cubic feet Coal Bed Methane (CBM) reserve mainly in Kalimantan and Sumatra, Indonesia has potential to redraft its energy charts as United States with its Shale Gas. But with low enthusiastic to develop CBM project, the government only targeted 8.9 million metric standard cubic feet per day (mmscfd) for 2015.

ELECTRICITY INDUSTRY OVERVIEW

Total installed electricity capacity in 2011 was 28.6GW. Of this amount:

coal accounted for 44%;

oil accounted for 23%;

natural gas accounted for 21%;

hydropower accounted for 7%; and

geothermal accounted for 5%.

Despite an abundance of indigenous fossil fuel reserves and renewables capacity, Indonesia is not self-sufficient in energy supply.

It is estimated that just 65% of the population has access to electricity or modern forms of energy, with electrification rates lower in remote and rural areas. Despite significant growth in electricity infrastructure in the last decade, per capita capacity remains low.

An additional 54GW of electricity capacity will need to be developed by 2020 to meet continued growth in demand. Accordingly, the Government has sought to diversify the country's energy supplies with an increased renewable energy contribution while also attempting to foster competition in the electricity market.

Government

The Ministry of Energy and Mineral Resources (MEMR) is the overarching government body for the energy sector. MEMR oversees the electricity master plan, which lays out the future goals and targets for the sector

The Directorate General of Electricity and Energy Utilisation, which is under the Department of Energy and Mineral Resources (DEMR), is the chief regulator of the energy sector.

Generation

The state-owned electricity company, Perusahaan Listrik Negara (PLN), owns most of the generation assets, transmission and distribution infrastructure.

In 1994, PLN's status was changed from a general company to a corporation. It also became the only "authorised agency of electricity business" (PKUK)permitted to provide electricity to the public. However, in 2009 PLN's role was further reformed when it was restructured to become a state-owned enterprise (BUMN).

Two wholly owned subsidiaries of PLN, Pembangkit Jawa-Bali and Indonesian Power, are now the primary generators of electricity. Overall, PLN effectively controls 86% of generation assets. Independent Power Producers (IPPs) and various electricity cooperatives also generate electricity, however PLN is the sole buyer of electricity and retains the right of first refusal.

Distribution and transmission

The low rate of electrification in Indonesia is largely due to distribution and transmission infrastructure shortages. Experts have warned that unless Indonesia's electricity infrastructure is updated, then the country could face an energy crisis within a decade.

Five distribution entities now act semi-autonomously from PLN to distribute electricity to the east, west, central Java, Bali and Jakarta regions.

The Indonesian interconnected system encompasses Java, Bali and Madura. The islands of Java and Sumatra are connected by a 39km submarine cable.

Despite the increased privatisation of the electricity sector, PLN controls all transmission infrastructure.

Electricity laws

Electricity Law No. 15/1985 allowed limited private participation in the generation phase through IPPs who sold electricity to PLN. However, PLN was granted exclusive powers over the transmission, distribution and ultimate sale of electricity. The impact of the Asian financial crisis, however, significantly limited the IPPs operation. This law has been revoked and replaced by Electricity Law No. 30/2009.

Electricity Law No. 20/2002 sought to further privatise the industry by allowing market-driven tariffs and independent regulation whilst also amending Law No. 15/1985. However, Law No. 20/2002 was ruled as unconstitutional by the Constitutional Court. In its decision, the Court considered that electricity is an industry sector that the Indonesian Government must control because it is "important to the state and affects the lives of most people". As a result, Law No. 15/1985 was reinstated.

Electricity Law No. 30/2009 is now the key piece of legislation regulating the electricity sector. Under Law No. 30/2009, the Government (Central or Regional) has the authority to issue licenses that are needed by entities to conduct activities in the electricity sector.

Under Law No. 30/2009, PLN is treated as a BUMN and no longer has a monopoly power in the electricity sector. Law No. 30/2009 also provides for competition in the supply and distribution of electricity to end customers and establishes the basic framework for private entities to participate in the electric power business. Unlike Law No. 20/2002, which was intended to privatise the industry by allowing market-driven tariffs and independent regulation, Law No. 30/2009 still gives authority to the Government to control the energy sector. Energy Law No. 30/2007 established the National Energy Council.

ELECTRICITY IN INDONESIA- IMPORTANT DEVELOPMENTS

ELECTRICITY SITUATION

Indonesia had 39,885 MW of electricity generation capacity in 2011, which was owned by the state-owned electricity company (PLN), independent power producers (IPPs) and private power utilities (PPU). In 2011, 182,622 GWh of electricity had been generated, of which 22.2% was supplied by IPPs. In 2011, electricity production by energy mix was still dominated by coal (50.8%), followed by renewable energy (20.3%), natural gas (16.1%) and oil (12.8%) (KESDM, 2012).

To increase the national electrification ratio, the Government of Indonesia launched the 10,000 MW Fast Track Program Phase I (FTP1) in 2006 and Phase II (FTP2) in 2010. The impact of the program is the increase of national electrification ratio from 63.0% in 2006 to 72.95% in 2011. The FTP2 will add 10,047 MW capacity, of which 49% will be developed from geothermal, 30% from coal, 17% from hydropower, 3% from gas, and 1% from gassified coal. The government has been mandated the state-owned electricity company (*Perusahaan Listrik Negara*, PLN) to implement fast-track programs to accelerate development of generating facilities. In this case, PLN will focus on the development of medium scale renewable energy generation.

ELECTRICITY CURRENT CONDITION AND PLANNING

As of May, 2013, Indonesia had an installed electrical generating capacity of approximately 44,346 MW, of which 5% coming from gas/oil steam power plant, 13.0% from diesel power plant, 9.0% from hydro power plant, 28.9% from gas power plant, 3% from geothermal power plant, 41.0% from coal fired power plant, and 0.4% coming from renewable energy mini scale power plant. PLN owned about 73% of the total installed capacity, IPP about 23% and the rest belonging to PPU.

The Indonesia's electricity sector is facing numerous challenges in its development, including

(1) a mismatch between the availability of primary energy resources, which the supply of energy is located mostly in outside Java and Bali, and the demand for electricity, which came mostly from Java and Bali,

(2) In 2012, the percentage of oil for producing electricity (fuel mix) was still high (15.02%) despite the abundance of coal resources within the country and

(3) the limited availability of the government funds and other resources to finance the construction of new power plants as well as transmission and distribution networks. Regarding infrastructure development initiatives, the government has been placed a special emphasis on the development of power generation plants, particularly coal-fired power plants, in order to reduce dependence on oil for power generation. One of the government's ten model projects introduced at the Indonesia Infrastructure Conference and Exhibition (IICE) was a power generation project, namely the Central Java Coal-Fired Power Plant.

The Indonesian Goverment has been issued Presidential Regulation No. 78 Year 2010 regarding Infrastructure Guarantee for Public Private Partnership Projects through the Infrastructure Guarantee Agency, in which the Indonesia Government will give Goverment Guarantee to the 2 x 1,000 MW Central Java Coal-Fired Power Plant. The winner of the tender of Central Java Coal-Fired Power Plant had been announced by PLN since June 17, 2011.

The Central Java Coal-Fired Power Plant was divided into 2 units and expected to be operational in 2018.

In addition, the government assigned PLN to accelerate its construction of coal power plants with aggregate capacity of 10,000 MW and associated transmission line which called the 10,000 MW Fast Track Program Phase 1 (FTP1) in 2006. The total investment needed for constructing the power generation and transmission was estimated to Rp 95.89 Trillion, and for which the

government has pledged certain credit support. Currently, all of the power generation projects are entered to the construction stages, in which 11 power plants have been already in the commercial operation, and the rest of the projects are expected to be completed by the end of 2014. The aggregate value of the construction contracts is noted to approximately USD 10 billion (with assumption of 1 USD = Rp. 9,000 and 1 EUR = Rp. 12,153.86). As of June 2013, a total of 5,005 MW of power plants under the FTP1 were in the commercial operation (50% of the total capacity of projects).

Indonesia still needs high investment for developing power generation, transmission and distribution lines and facilities in order to meet the expected demand in the future. The government indicated that from year 2012 to 2021 the electricity sector will need the total investment approximately USD 107.13 billion, of which USD 77.38 billion for developing the additional capacity of power plant, USD 15.98 billion for transmission, and USD 13.77 billion for electricity distribution.

On December 19, 2012, the Ministry of Energy and Mineral Resources submitted the draft of the National Electricity General Plan (*Rencana Umum Ketenagalistrikan Nasional*, RUKN) 2012-2031 to the Chairman of the House of Representatives Commission VII of Republic of Indonesia for consultation before it stipulated by the government as mandated by the Electricity Law. The RUKN 2012-2031 is intended to coordinate electricity infrastructure development for the various regions of Indonesia. Based on this plan, Indonesia's demand for electricity is expected to grow at an average annual rate of 10.1% over the period 2012 to 2031. In order to meet high electricity demand growth, Indonesia will need the additional power plant capacity of approximately 237 GW by 2031. As a result, the electricity sector will need high investment in developing power generation, transmission and distribution lines and facilities.

Based on the PLN's 2012-2021 Electricity Supply Business Plan (*Rencana Usaha Penyediaan Tenaga Listrik*, RUPTL), the national electricity demand is expected to grow at an average annual rate of 8.65%. In order to meet this demand as well as to support the Masterplan for Acceleration and Expansion of Indonesia Economic Development (MP3EI) program, the government has been planned to increase the capacity of power plants about 57,250 MW by 2021 or an average about 5,725 MW per year. Noting that the demand for electricity is still expected to be concentrated in the Java-Bali electricity system and in Sumatera electricity system by 2021, the additional capacity of power plant portion will be developed more in Java-Bali electricity system (32,637 MW) and in Sumatera electricity system (14,513 MW).

Law of the Republic of Indonesia No. 30/2009 concerning Electricity had been passed on by the House of Representatives on Tuesday, 8 September 2009; signed into Law by the President on 23 September 2009, and had been promulgated in State Gazette number 133 of 2009. The Law replaced old electricity bill that was legislature on 1985 (Law No 13/1985). One of the major intentions of the new bill was to speed up the development of electricity networks in the country which currently only covers 68 % of the households. Government is targeting 85% of electrification ratio within three years time (in 2015).

The new bill incorporates some crucial matters in electricity business that have not been regulated in the old one, notably division of electricity business areas, application of regional tariffs differentiation, utilization of electricity networks for telecommunication, multimedia, and informatics, regulation of cross country electricity selling, and a recognition of regional autonomy.

The bill partially removes PLN monopoly and gives regional government a greater authority in providing direct electricity. Private sector, including foreign companies, can also now be participated in such provision. The new bill, however, does not allow for a free competition between state-owned electricity company and the private sector. As was mandated by the Constitution Court, state owned company (not necessarily be solely PLN in the future), is still given the first right for perusal in electricity provision. Private companies, cooperatives, regional-owned companies, and Non-Governmental Organizations are treated as the second option.

The government is still obliged to keep on subsidizing electricity for rural areas and low-income customers. With the application of regional tariff differentiation and division of business areas, the subsidy can now be more selective. Subsidy can also be given to private companies that provide electricity in rural and remote areas.

Supporting Government Regulations had been created as implementing regulations for this Law. Transitional period was taken up to two years. Some of Government Regulations are as follows:

1. Government Regulation on Electricity Provision Business

2. Government Regulation on Electricity Support Business

3. Government Regulation on Land Use for Electricity Business

4. Government Regulation on Electricity Business Areas

Major new features in the bill and its differences with the old one are as follows:

AUTHORITY OF ELECTRICITY PROVISION

Article 4 (1) in the new Law states that electricity provision now falls under authority of central and regional government, and are implemented by state-owned as well as regional-owned companies. Previous Law ascertained that such authority only belongs to central government and solely implemented by one state-owned company, which was PLN.

The new Law thus gives regional government new authority on electricity provision, with the power to set up regional tariffs, issuing permit, establishing regional-owned electricity company, etc, which are specifically regulated on the other parts of the Law.

The next section in article 4 establishes opportunities for private companies, cooperatives, and non-governmental organization to provide direct electricity to customers. Previously private parties can only sell its electricity generation to PLN through Independent Power Plant (IPP) scheme, not directly to customers.

However, state-owned companies are still given the first right of refusal in providing electricity. It means opportunities for private companies, cooperatives, NGOs, and regional-owned companies to directly sell electricity are only fully available in the areas where state-owned networks do not exist. In the areas where such networks exist, they can still directly sell electricity by renting PLN's network for electricity transmission and distribution. They also still can have the option for selling their electricity generation to PLN, as the usual IPP scheme.

ELECTRICITY BUSINESS

According to the Law, electricity business is divided into two broad categories:

1. Electricity power provision, which covers power generation, transmission, distribution, and selling.

2. Electricity Support & Services, which includes consultancy, construction, installation, testing and examination, installation operation, maintenance, research & development, education and training, and laboratory testing of electricity.

A company is allowed to integrate all segments in power provision, which are power generation, transmission, distribution, and direct electricity selling, in one company.

An integrative company in power provision business will be given exclusive right to provide direct electricity in one business area. Business area is to be appointed by government's decree, depends on the scope of the business. If the area surpasses provincial boundaries, it should be appointed by central government. If the area surpasses cities boundaries, it should be appointed by provincial government, and if it lies within a city, then it should be appointed by local government. The business areas do not necessarily follow administrative boundaries.

Permit for electricity provision is called Business Permit for Electricity Provision (IUPTK = *Izin Usaha Penyediaan Tenaga Listrik*) which will be given by local, provincial, or central government, depends on the scope of the business areas. Supervision and control will be done according to the level of government that gives the permit.

Permit holders have rights to pass river, lake, and sea, either above or under the surface. They also have rights to pass public road and railway, enter public and private places temporarily, use land and pass on it either on the surface or below, and cut plants that get in the transmission way. All must be done in line with existing regulations.

On the other hand, permit holders have obligations to provide a reliable and standardize electricity. They also must give best services to customers, fulfill electricity safety requirements, and prioritize domestic resources. Electricity safety requirements and national standards on electricity provision have been regulated by Director General of Electricity and Energy Usage in the Ministry of Energy and Mineral Resources. Foreign resources (labor and capital goods) can be used if similar resources could not be attained locally.

A company can also generate, transmit, and distribute power for internal use, using a different permit.

ELECTRICITY PRICING

Electricity pricing is regulated. Power price and rent cost of electricity networks are decided according to good competition principles. Permit holders must consult central or local government before applying the prices. They are not allowed to apply the prices without consent from central or local government.

Electricity tariff for customer is appointed by central or local government (according to its level of authority) with the approval of central or local parliament, respectively. In case local government could not decide on the tariff, it would be decided by central government with the approval from parliament. Tariff could vary between regions within one business area.

Opportunities

With current electrification ratio and electricity demand growing by 9.2 percent per year, Indonesia will need to add 16,183 megawatts every year in order to curb power shortages and meet increased demand. This will be followed by demand for transmission and distribution networks as well as electricity business supports that worth billions of dollars a year. The new bill opens up opportunities for private sector to be more directly involved in electricity business in Indonesia.

Indonesia possess tremendous energy potentials, both come from fossil as well as renewable energy resources. Indonesia produces 0,978 million BOEPD of oil and 1,34 million BOEPD of natural gas annually, and holds 19,3 billion tons of coal reserves. Indonesia also has hydro power potential equals to 75,67 GW, geothermal potential equals to 27 GW (40 % of world's potentials), biomass potential equals to 49,8 GW, and wind power equals to 9,3 GW.

Government has offered 10.000 MW power plant projects for investors, of which around 21 % is expected to come from renewable sources. Various taxes and non-taxes incentives has been given to investors, for instance through Ministry of Finance Regulation No. 154/2008 on October 2008 concerning custom exemption for import of capital goods for development of power plant industries.

Indonesian electricity condition is varied, creates more type of opportunities. Java-Madura-Bali-Sumatera islands have been well interconnected to high voltage transmission networks. Meanwhile, networks in outer islands comprise of smaller sub-system that have not been well interconnected (isolated).

Electricity demand and supply sometimes scattered around archipelago, makes it unattractive in terms of economies of scale. However, it also offers great opportunities for smaller scale of investment, for instance in mini-hydro and geothermal power plant with capacity below 10 MW.

ELECTRICITY MARKET, COMPETITION AND EFFICIENCY

ELECTRICITY MARKET

The initial steps in restructuring the Indonesian electricity industry took place in 1994, when PT Perusahaan Listrik Negara (PLN, www.pln.co.id) was converted from a state enterprise to a government-owned limited liability company.

Restructuring efforts continued in 1995 with the unbundling of PLN's Java, Bali and Madura generation, distribution and transmission assets. Generation assets were unbundled into two wholly owned subsidiaries of PLN: PJB (Pembangkit Jawa-Bali, www.ptpjb.com) and Indonesia Power (IP, www.indonesiapower.co.id). The distribution unit was separated into four distribution entities (East, West and Central Java, and Jakarta). Each distribution unit operates semi-autonomously, with an allocated budget to cover operational expenses in meeting the performance targets set out in its contract with PLN. The Java–Bali transmission business was transferred to the Java–Bali Electricity Transmission Unit and Load Dispatch Centre.

Apart from sharing the generation business with the IPPs and cooperatives, PLN is the sole buyer and seller of electricity in the power market, currently purchasing approximately 80% of the power produced by the IPPs. Outside Java, Bali and Madura, restructuring is taking place through the decentralisation of PLN's assets.

OIL AND GAS MARKET

The liberalisation of Indonesia's downstream oil and gas sector has been under discussion for several years. In October 2001, the Indonesian legislature passed the much-anticipated Oil and Gas Law 22/2001 which limited Pertamina's (www.pertamina.com) monopoly on upstream oil development (which requires it to be included in all Production Sharing Contracts (PSCs)) by the end of 2003. Also, Pertamina's regulatory and administrative functions were transferred to other entities, while its regulatory role was spun off to new state-owned legal bodiesy, BP Migas (www.bpmigas.com) as upstream regulator and BPH Migas as downstream regulator. The Ministry of Energy and Mineral Resources is responsible for entering into PSCs, while BP Migas serves as the upstream regulator that manages and implements these agreements. Legal changes adopted in 2005 have allowed for the extension of contracts beyond the previous 20-year limit, which helped to facilitate the deal with ExxonMobil for the development of the Cepu fields.

In addition to its upstream activities, Pertamina operates nearly all of Indonesia's refinery capacity, procures crude and products imports, and supplies products to the domestic market. While Pertamina's monopoly in the retail market ended in 2004, the company continued to be the sole distributor for subsidized fuels until early 2010. With the December 2010 decision to gradually remove fuel subsidies for private vehicles, Pertamina now competes more directly with product retailers.

The upstream oil sector is dominated by several international oil companies, in particular Chevron, Total, Conoco Phillips, Exxon and BP. Chevron is the largest single oil producer in Indonesia, accounting for more than 40% of the country's total crude production, followed by Pertamina, accounted for approximately 15% of 2009 crude and condensate production.

The upstream gas sector is dominated by international oil companies such as Total, Conoco Phillips, and ExxonMobil as well as Pertamina, which accounts for about 15% of natural gas production. However, natural gas transmission and distribution activities are carried out by a state-owned utility, Perusahaan Gas Negara (PGN).

COMPETITION

In September 2009, Electricity Law No. 30/2009 replaced Law No. 15/1985. A notable difference is the absence of the Holder of Electricity Business Authority (PKUK, Pemegang Kuasa Usaha Ketenagalistrikan). Under Law No. 15/1985, the government assigned the state-owned electricity company (PLN) as the sole PKUK. As the PKUK, PLN, on behalf of the government, was responsible for providing electricity across the whole of Indonesia, and for developing the electricity sector. Under the new electricity law, PLN will no longer have a monopoly on supplying and distributing to end customers. Instead, a licence to provide electricity for public use (IUPTL) may be granted to (among others) private business entities, subject to a 'right of first priority' provided to state-owned companies (i.e. PLN).

Under the provisions of the Oil and Gas Law No. 22 of 2001, Pertamina was transformed into a public liability company, with responsibilities including liberalisation of the upstream and downstream oil sectors. Senior Pertaminal officials have indicated plans to divest two of its subsidiaries by about 20% in an initial public offering, but details and timing have not been confirmed.

ENERGY EFFICIENCY

Consumption of refined products grew at a fairly steady 4.7% average annual growth of rate between 1995 and 2005, but declined by about 2% in both 2006 and 2007, largely as a result of the 126% increase in the price of subsidised fuel implemented in 2005. The 2005 price increase and the removal of fuel subsidies for large industrial consumers were part of President Yudhoyono's early attempts to gradually eliminate fuel subsidies. Consumption increased in both 2008 and 2009, exceeding 1.3 million bbl/d in 2009.

Indonesia's total primary energy consumption grew by nearly 30% between 1999 and 2008. Total final energy consumption (TFEC) was 948,112.455 thousand BOE in 2009. Oil continues to account for the most significant share of Indonesia's energy mix, at 51.9%, followed by natural gas at 18.4%, coal at 12.9%, electricity at 12.8% and LPG at 3.9%. The share of TFEC by sector in 2009 was 40.86% for industry, 36.8% for transport, 13.26% for households, 4.73% for commercial and 4.28% for other sectors. There is a need for improved energy efficiency in industry, and limiting petroleum consumption in the transport sector.

There is significant potential for energy savings. A move away from general subsidies on fuel, whilst continuing to assist those who would otherwise face energy poverty, could help promote more efficient use of energy, especially oil. Indonesia has a range of options for reducing the carbon content of its energy supply. These include increased development of its hydro-electric, geothermal and wind resources, and more efficient thermal electricity generating plants.

The efficiency of the power sector tends to fluctuate and has decreased slightly since 1990. It stood at 38% in 2009, compared with 40% in 1990. The reduction of the share of hydroelectricity and the development of coal power plants explain the deterioration of the ratio. In 2009, thermal power plants had an efficiency rate of 35%; that rate has remained relatively stable since 1990. The limited development of more efficient technologies, such as gas combined cycles and cogeneration, did not permit an increase in the ratio for thermal power generation. The rate for T&D losses in the Indonesian grid is above 10% of the distributed volumes, which is above the world average of 9%. Those losses tend to fluctuate significantly and have decreased since 2005.

Industrial energy consumption has soared by 6%/year since 1990, which is faster than the economic growth rate of 4.5%/year. Industrial energy intensity increased by 0.7%/year between 1990 and 2009. Since 2000, it has increased at the even faster pace of 1.2%/year.

The National Energy Conservation Master Plan (RIKEN) in 2005 identified the following sectoral energy saving potentials: 15-30% in industry; 25% in commercial buildings for electricity; and 10-30% in the households sector.

Industry

Energy Intensity reduction: 1%/yr until 2025, through fiscal incentives, educational programmes and audits.

Partnership Program on Energy Conservation – Government support in return for efficiency commitments from companies, 20% efficiency improvement targets.

Utilities

Energy Conservation Regulations – reductions in technical losses in T&D. Government initiatives to increase geothermal capacity to 5,000 MW (48%), with annual reductions in CO2e of 82 Mt. 12% new hydro capacity development. (2009-2014)

Transport

Mandatory 10% biodiesel content in road fuels

Clean and Lean Transportation Initiative, promoting efficiency in vehicle fleets.

Residential

National efficiency standards established for new buildings, and software tool provision for designers.Penduli program – rebates on CFL purchasing for households.

Public

Public-private partnerships with PLN, to stimulate the energy efficiency market.

Sustainable tourism initiatives, in partnership with the UNWTO.

Efficient street lighting refits.

BASIC REGULATIONS

On 16 November 2009, the Governmental Regulation (PP, Peraturan Pemerintah) No. 70/2009 on Energy Conservation, as called for by Law No. 30/2007 (the Energy Law) was issued. Regulatory measures included:

the formulation of a National Energy Conservation Master Plan (RIKEN, Rencana Induk Konservasi Energi Nasional), to be updated every five years or annually, as required;

the mandatory assignment of an energy manager, energy auditing, and an energy conservation program for users of final energy of 6000 toe (tonnes of oil equivalent) per year or more;

mandatory energy-efficiency standards and energy labelling;

government incentives, including tax exemptions and fiscal incentives for imports of energy-saving equipment and appliances, and special low interest rates for investments in energy conservation;

government disincentives, including written notices to comply, public announcements of non-compliance, monetary fines, and reductions in energy supply for non-compliance with the terms of the law.

In 2008, the Ministerial Regulation No. 32/2008 that was to make biofuel consumption mandatory, commencing in 2009, was passed.

ENERGY STRATEGY - IMPORTANT DEVELOPMENTS

GOVERNMENT AGENCIES

The central government, with approval from the parliament, will stipulate the tariff for electricity that is sold to end customers by the holder of a license to provide electricity for public use (IUPTL), whilst having regard as to national and regional interests, as well as to the end consumers' and power producer's interests. Meanwhile, for 'regional scope', the provincial government (Pemda), with approval from the regional parliament, will stipulate the tariff in accordance with tariff guidelines provided by the central government. The precise meaning of 'regional scope' in this context is not clear, and will hopefully be elaborated in the implementing regulations.

The government may stipulate different tariffs in each region within a business area. The holder of an IUPTL is prohibited from stipulating a tariff to end customers without approval from the central government or the Pemda. Failure to comply with this may result in an administrative sanction.

The central government and the Pemda, must also approve the sale price of electricity from the holder of an IUPTL (i.e. from the power producer to the holder of an integrated IUPTL) and the fee for use of a transmission line.

Energy consumers' level: The barriers to energy conservation include, the inadequacy of knowledge, lack of information, and high investment costs.

Other concerns are: lack of in-depth management skills; the concentration of decision making at the top level; and lack of expertise in proposal preparation.

Coordination (government) level: The government is constrained in implementing various EE&C programmes and activities due to limited budget. The absence of coordinating mechanisms among key government actors, decision making, and overlapping/unclear division of responsibilities among the institutions remain as barriers; and the situation has been further complicated by the decentralisation of some administration. There is a dearth of information on current and future trends in energy technology, cost and usage.

A significant portion of the country's geothermal resources are located in conservation forests. Indonesian law distinguishes between protected forests and conservation forests. Although some power plants and other projects are allowed in protective forests since February 2010, the only human intervention allowed in conservation forests is for education or research. The government has proposed that geothermal wells could be drilled in conservation forests, but geothermal power plants would have to be built outside of these areas. Further regulation is awaited.

The establishment of an independent regulatory agency could improve the confidence of private firms in the electricity sector in Indonesia.

While Indonesia needs to attract foreign investments to accelerate its effort in expanding renewable energy and has been attempted to create incentives for foreign investment in renewable energy (e.g. Geothermal Law 25/2007), it struggles to attract foreign investments due to some issues. The key issues include the difficulty of obtaining land-use permissions, subsidized tariffs and uncertain regulatory environment. PLN's status as the sole legal provider of electricity through Indonesia's power grid complicates the incentives for foreign developers and often limits the profitability of projects.

There are some perceived barriers to renewable energy development. The price regime has forestalled the development of cleaner energy infrastructure as there is a lack of financial incentives, combined with heavy subsidies for energy consumption. The country currently restricts foreign investment in power plants producing less than 10 MW, and therefore is unlikely to meet its targets for economic growth and renewable energy development without allowing participation by foreign companies, which often have the technical know-how and capital. It also needs to incentivise domestic finance for the renewable power sector s by providing a sovereign debt guarantee. There is also a lack of an after-sales service infrastructure and a strong transmission infrastructure for grid-connected renewable energy projects. While local governments have been given the authority to administer tenders for projects in an effort to implement policy initiatives efficiently, they often lack the capacity to manage tendering processes transparently. Furthermore, Indonesia lacks quality data; and there is a need to verify through third-party valuation.

The ***Ministry of Energy and Mineral Resources (MEMR***, www.esdm.go.id) provides the energy sector with a streamlined governance structure and has effectively become the responsible agency for formulating and implementing policies in all energy related sectors, including oil, gas, coal, and electric power.

Directorate General of New Renewable Energy and Energy Conservation was renamed from Directorate of New Renewable Energy and Energy Conservation in August 2010. It is a new unit of the Energy and Mineral Resources Ministry (ESDM) to formulate and implement policies and technical aspects of the new and renewable energy as well as energy conservation.

Directorate General of Electricity is under and responsible to the MEMR. It is tasked to formulate and implement policies and technical standardization in the field of electricity.

Due to the powerful position of the PLN, the electricity sector is also dependent on the administrative decisions by the ***Ministry of State-Owned Enterprises*** (***MSOE***, www.bumn-ri.com), the ***Ministry of Finance*** (***MOF***, www.depkeu.go.id) and the ***National Development Board*** (***BAPPENAS***, www.bappenas.go.id).

The MSOE oversees the shareholder interest of the state in the PLN, while the MOF is responsible for allocating government subsidies and loans to the electricity sector including the PLN. Furthermore, BAPPENAS is responsible for the development planning of the energy sector, and also possesses some jurisdiction over economic issues, natural resources and regional development.

Through the Directorate General of Electricity and Energy Utilisation (DGEEU, www.esdm.go.id/directorate-general-of-electricity-and-energy-utilization), the DEMR acts as the chief regulator of the sector, being responsible for the drafting of legislation, it's implementation, co-ordination, enforcement and compliance.

The DGEEU has the following responsibilities, set out in the establishing law:

To prepare and formulate policy for the DEMR in the fields of electricity and energy utilisation,

The implementation of said policy,

The preparation of standards, guidelines, criteria, and procedures for the fields of electricity and energy utilisation, and

To provide technical assistance and evaluation services in the fields for any who may require it.

ENERGY SECURITY

Energy security has been central to Indonesia's energy policy since its first General Policy on Energy (KUBE, Kebijakan Umum Bidang Energi) in the early 1980s, and thereafter in the Energy Policy of 2006. Basic policy principles include diversification from oil through the development of natural gas, coal and renewable energy resources; energy conservation in all economic sectors; and an intensified search for new energy resources to increase Indonesia's energy reserve base.

Indonesia recently took significant diversification measures. Rapid increases in geothermal power will begin in the next five years, including new capacity additions of 1,685 MW under the Accelerated Development of Power Generation program, Phase II. Indonesia has set a goal to have an installed geothermal capacity of 6,000 MW in 2020. The power development plan foreshadows the addition of at least 34,300 MW of non-oil power generation capacity from 2006 to 2015. By Ministerial Regulation No. 32/2008, the government has set out plans for a greater role for biodiesel and ethanol-blend fuel in transport.

The Upper Cisokan Pumped Storage Hydroelectric power plant (4 x 260 MW) project in West Java received a government loan from the World Bank/IBRD in late 2011. The project is expected to be completed in 2016. This plant will be the first of its kind in Indonesia. PLN has also secured financing for construction of the Jati Gede hydroelectric power plant (2 x 55 MW) in West Java, and the Merangin hydroelectric power plant (2 x 90 MW) in the province of Bengkulu, Sumatra.

PLN's 2009 Electricity Power Supply Business Plan (RUPTL) also expects the addition of 4740 MW to Indonesia's hydropower capacity during 2010–18; of this capacity, 3835 MW would be developed by PLN and 905 MW by IPPs. The hydropower capacity addition includes three pump-storage power plants in Java – specifically Upper Cisokan (1000 MW) in West Java, Matenggeng (885 MW) at the border of West and Central Java, and Grindulu (1000 MW) in East Java. These pump-storage plants are considered important for stabilising system frequency, to provide spinning reserves and system stability.

Under the 10 000 MW Accelerated Development of Electricity Generation – Phase II program for 2011–15, Indonesia is committed to building two hydropower plants – specifically Asahan III (174

MW) in north Sumatra, and Bakaru II (126 MW) in west Sulawesi. These hydropower plants would increase Indonesia's total large hydropower capacity to 3804 MW, or 5% of Indonesia's total hydropower potential of about 75 000 MW. It is worth noting that Indonesia's large hydropower potential is located in the eastern part of Indonesia, far from large demand centres.

To remove barriers to energy standards and labelling, Indonesia is currently participating in the UNDP/GEF-funded Barrier Removal to the Cost Effective Development and Implementation of Energy Efficiency Standards and Labelling Project (BRESL), which involves six developing Asian economies. BRESL has five major programs promoting energy standards and labelling: policy making; capacity building; manufacture support; regional cooperation; and pilot projects.

Since 2002, Indonesia has implemented a government-funded public–private partnership program in energy auditing for industry and commercial buildings. The program requires participating companies to implement measures identified in energy audits. Some 292 industries and commercial buildings have participated so far. Indonesia is an active participant in the ASEAN Energy Awards program (including the Best Practice for Energy Efficient Buildings and the Best Practice for Energy Management in Buildings and Industries).

The Indonesian Climate Change Trust Fund (ICCTF) is a financing mechanism for climate change mitigation and adaptation within Indonesia's policy framework. The ICCTF's two key objectives:

achieving Indonesia's goal of a low-carbon economy and a greater resilience to climate change through the facilitation and acceleration of investment in RE and EE, sustainable forest management and conservation; and reducing vulnerability in key sectors, such as coastal zones, agriculture and water resources;

enabling the government to increase the effectiveness and impact of its leadership and management in addressing climate change, by bridging the financial gap ;and increasing the effectiveness and impact of external finance for climate change work in Indonesia.

Indonesia is utilising solar PV systems to achieve its electrification target, which was 66% in 2009. Also in 2009, the government allocated IDR658.7 billion (about US$65 million) to provide new and renewable energy–based power generation for distributed power systems. The program provided electricity to around 94,000 households, particularly in 18 of the outermost islands, and in remote areas along the Indonesian border. In addition, the government allocated IDR841.3 billion for the extension of PLN's 20 kV rural electrification network and generation capacity.

In the 2010 budget, the government expects to allocate IDR561.5 billion for the electrification of 81000 households in very remote areas (based on new and renewable energy), and to allocate IDR591.5 billion to further extend PLN's rural electrification network.

Demand Side Management and Labelling: DSM focuses on clipping the peak load which is contributed primarily to by household lighting, street lighting, and household electric appliances. The idea is that the effective implementation of DSM and labelling costs less than building and operating new power plants to meet peaks in electricity demand.

Although the importance of DSM and labelling have been realised, current programmes have not yet been fully implemented due to financial constraints.

The Directorate General of Electricity and Energy Utilisation (DGEEU) is now planning/implementing the following three programmes:

the TERANG programme, aiming to reduce electricity demand by installing compact fluorescent lamps (CFLs) in households;

the PJU programme, aiming to reduce electricity demand by installing efficient lamps in street lighting;

the Socialisation programme, an awareness programme specifically for DSM.

ENERGY STRATEGY DEVELOPMENT

Until the time the National Energy Council (DEN) establishes a new National Energy Policy (KEN), the National Energy Policy of 2006 applies. The aim of this policy is to:

Achieve energy elasticity to GDP of less than one by year 2025

Realize an optimum primary energy consumption mix in 2025, with shares as follows:

oil - to become less than 20%

natural gas - to become more than 30%

coal - to become more than 33%

biofuels - to become more than 5%

renewable energy and other energy including nuclear - to become more than 10%

liquefied coal - to become more than 2%.

The details of the energy programs and targets of the National Energy Policy are elaborated in the Blue Print - National Energy Management 2005 to 2025. Indonesia's 2006 Energy Policy expects the combined share of renewable energy and nuclear in the overall energy mix in 2025 to have exceeded 17%. The policy has special emphasis on enhancing the share of biofuels. Renewable energy and other energy including nuclear (as in the list above) is expected to be made up of at least a 5% geothermal share and a combined share of biomass, hydropower, solar, wind and nuclear power to make up the remainder to 10% by 2025.

National Energy Conservation Master Plan (RIKEN) (2005) aims to achieve Indonesia's energy saving potential through energy efficiency and conservation (EE&C) measures, and so avoid wasteful energy use in the country. It also aims to reduce energy intensity by around 1%/year, on average, until 2025. Fiscal incentives (tax reductions and soft loans) together with other instruments such as training and educational programmes as well as energy audits are used to implement the plan.

National Biofuel Roadmap (Presidential Decree 10/2006) for 2006-2025, following the 2005 National Energy Blueprint, is to accelerate the use of biofuel as a replacement fossil-based fuel. It sets specific targets for production as well as utilization between 2006 and 2025, gradually increasing the utilization of biofuel from 2% to 5% of energy mix.

Government's '25/25 Vision' plans to have renewable energies fulfil 25% of the country's energy needs by 2025.

On 10 August 2007, Indonesia enacted the Energy Law (Law No. 30/2007). The Law elucidates principles for the utilisation of energy resources and final energy use, security of supply, energy conservation and protection of the environment with regard to energy use, pricing of energy, and international cooperation. The Energy Law sets out the content of the National Energy Policy (KEN, Kebijakan Energi Nasional) (2006); the roles and responsibilities of the central government and regional governments in planning, policy and regulation; development priorities for energy research and development, and the role of enterprises.

The Energy Law mandates the creation of a National Energy Council (DEN, Dewan Energi Nasional), the tasks of which are:

draft the National Energy Policy, to be endorsed and promulgated by the government, with due consent of parliament (the DPR),

draft the National Energy Master Plan (RUEN, Rencana Umum Energi Nasional), declare measures to resolve the conditions of energy crisis and energy emergency, provide guidance and management on the implementation of cross-sectoral policies on energy.

It is a national, independent, and permanent body and its members include 7 government officials and 8 stakeholders.

Under the Energy Law, the 2006 National Energy Policy will address the availability of energy, energy development, the utilization of domestic energy resources, and energy supply reserves. It states that Indonesia's goal is to achieve energy elasticity (the ratio of change of total primary energy supply over the rate of change of GDP) of less than 1 in 2025.

Until the new National Energy Policy is enacted, the Presidential Regulation No. 5/2006 on the National Energy Policy regulates the development of renewable energy. It mandates an increase in renewable energy production from 7% to 15% of generating capacity by 2025. It states that the contribution of new and renewable energy in the 2025 national primary energy mix is estimated at 17%, consisting of 5% biofuel, 5% geothermal power, biomass, nuclear, hydro and wind, and also liquefied coal at 2%. The government will take measures to add the capacity of micro hydro power plants to 2,846 MW by 2025, biomass of 180 MW by 2020, wind power of 0.97 GW by 2025, solar of 0.87 GW by 2024, and nuclear power of 4.2 GW by 2024. The total investment needed for this development of new and renewable energy sources up to the year 2025 is projected at $13,197 million.

The Green Energy Policy (2004): includes guidelines for the development of renewable energy, including regulatory instruments.

Biofuel Decree (Ministry of Energy and Mineral Resources Regulation No. 32/2008) settles a mandatory utilization framework in the transportation, industrial, commercial and power generation sectors for biodiesel, bioethanol and bio-oil from 2009 to 2025.

Biofuels subsidy: In February 2009, the Indonesian government announced that it would establish a biofuels subsidy to encourage investment in, and use of, biofuels made from palm oil and other feedstocks. The subsidy would only be paid if biofuel prices are higher than crude oil-based fuels.

The Geothermal Law (No. 27/2003): gives powers to regional governments to develop geothermal energy, in particular in respect of licensing, allowing investors to deal directly with regional governments and requiring projects to be competitively tendered. It also provides incentives for investment by establishing long-term licenses for land use (more than 30 years) and a regulated price for geothermal energy.

Tax incentives for geothermal exploration (Regulation 22/PMK011/2011).

Ministerial Regulation No.002/2006: on the commercialisation of middle scale renewable energy power plants.

Law No. 25/2007: concerning investment. Since 2008, the government has offered tax incentives for foreign investment including investors in renewable energy. Incentives include a 30% net income tax reduction for 6 years, free repatriation of investments and profits, and dispute settlement.

The Electricity Law (Law No. 30/2009): secures sustainable energy supplies, promotes conservation and use of renewable energy resources. The regulation was issued by the Ministry of Energy and Mineral Resources (ESDM) and referred to as "Purchasing Price by PT PLN (Persero) of Generated Electricity from Small and Medium Scale Renewable Energy Power Plant or Excess Power". The aim of this ministerial regulation is to enhance the electricity generated by small and medium scale of renewable energy power plant or excess power to be purchased by state owned company, regional owned company, cooperatives. The law does not make provisions for a feed-in tariff, but does provide for differing tariffs in different regions, more accurately reflecting the cost of supply. For rural development, the government set a bold electrification target of 90% by 2020.

Purchase of electricity from geothermal plants (Regulation 02/2011) of the Ministry of Energy and Mineral Resources regulates the power purchase tariff of electricity from geothermal sources. It assigns the PLN to purchase electricity from geothermal plants inside at a maximum price of 0.97/kWh.

Clean Technology Fund (CTF), a multilateral fund, aims to accelerate the country's initiatives to promote energy efficiency and renewable energy, and to help achieve the objective of increasing the electrification rate from 65% to 90% in 2020 as well as the long-term goal of reducing greenhouse gas emission by 26% by 2020. Under a new $400 million climate investment plan, endorsed by the CTF, Indonesia's geothermal power capacity is set to nearly double. The plan will use co-financing from the CTF to expand large-scale geothermal power plants and to facilitate energy efficiency and renewable energy by creating risk-sharing facilities and addressing financing barriers to small- and medium-scale investments. The CTF is slated to mobilise an additional $2.7 billion from a range of other sources.

Indonesia ratified the UN Framework Convention on Climate Change (UNFCCC) on 23 August 1994 and the Kyoto Protocol on 3 December 2004. As a non-Annex 1 party in the Kyoto Protocol, Indonesia has no obligation to reduce GHG emissions. However, the government is committed to participating in, and cooperating with, the global effort to combat climate change as Indonesia is the third largest emitter of greenhouse gases, mostly because of deforestation. It is also vulnerable to climate change as an island nation whose capital city, Jakarta, sits below sea level. This position was expressed by the Indonesian President in the G20 Finance Ministers and Central Bank Governors Summit held in September 2009 in the United States. The government has pledged to reduce GHG emissions from forestry and the energy sector by 26% through domestic effort by 2030, and by up to 41% through cooperation with other economies. To meet the goal, Indonesia will heavily invest in renewable energy and recommit to stopping deforestation.

RENEWABLE ENERGY

Indonesia has clearly established an overall legal framework for renewable energy by adopting Law No. 30 Year 2007 regarding Energy in order to realize its large potential of renewable energy. The Energy Law mandates the use of renewable energy sources be prioritized over non-renewable resources. It also defines the roles and responsibilities of the central government and local governments to prioritize renewable energy sources. In addition, a special legal framework for geothermal energy is supplied by the Geothermal Energy Law (Law No. 27 Year 2003).

In order to discharge the mandate of the Energy Law, the National Energy Council (DEN) was formed in 2008 with the tasks are drafting the National Energy Policy (KEN) and endorsement the National Energy Master Plan (RUEN). The assembly of the DEN members is chaired by the President of Indonesia. As an institution, the DEN is headed by the minister responsible for energy affairs. The DEN has 15 members: 7 ministers and high-ranking government officials responsible for the supply, transportation, distribution and use of energy and 8 stakeholder members from industry, academia, expert groups, environmental groups, and consumer groups. The selection and appointment of members of the DEN was finalized in late 2008. However, though the DEN is chaired by the President of Indonesia, policy formation and authorization occur slowly.

The Directorate General of New Renewable Energy and Energy Conservation (DGNREEC) was created within the Ministry of Energy and Mineral Resources (MEMR) in 2010 to focus on the implementation and regulation of renewable energy development. Other relevant government offices, including other directorate generals of MEMR, the Ministry of Environment, the state-owned electricity company, and state banks are also involved in policy implementation and regulation for renewable energy development. As renewable energy development inevitably entails many regulatory procedures at various governmental levels, the success of renewable energy implementation sometimes depends on the coordination between DGNREEC/MEMR and relevant offices not only within the central government but also local governments both at the provincial and municipal levels.

The contribution of renewable sources of energy to energy supply as a percentage of total primary energy supply in 2010 was 34.5%.Renewable generation sources supplied 16% of Indonesia's electricity in 2011. Indonesia has set a target of 26% of electricity generation from renewable sources by 2025.

Renewable energy plays a minor role in Indonesia's energy supply, however the Government is seeking an increase in renewable generation to 17% of the total energy consumption by 2025.

Current estimates place total renewable energy capacity at 2.9GW (on-grid) and 3.2GW (off-grid). Given a total renewable energy potential within the country of 163.3GW, this amounts to an undeveloped renewables potential of 96%.

The Government is aiming for 10.1GW of new capacity comprising of 3.9GW of geothermal and 1.2GW of hydropower by 2014.

Indonesia has introduced a Climate Change Trust Fund to ensure that climate change is integrated into the development planning. The Government is seeking to become a regional leader on climate change mitigation and in 2009 voluntarily committed to reduce greenhouse gas emissions by 26% by 2020.

The US$400 million Clean Technology Fund was created to promote renewable energy, improve electrification rates and enhance energy efficiency within Indonesia. The fund's predominant focus is on developing Indonesia's large-scaled geothermal projects.

The Directorate General of Renewable Energy and Energy Conservation, which sits under the MEMR, is the chief regulator in the renewable energy area.

Around 80% of Indonesia's geothermal resources lie in conservation forests, which are protected under Indonesian law. The Government has proposed to allow drilling in these conservation forests on the condition that the processing plant is located outside the conservation forest area. Still, developments within these areas require presidential approval and in May 2011 the Indonesian Government committed to a twoyear moratorium on forestry development as part of a US$1 billion forestry agreement with Norway. It will thus be difficult for Indonesia to balance its voluntary emission reduction targets which rely on reducing deforestation whilst also reaching its geothermal potential that is crucial for the country to avoid an energy crisis.

Commentators suggest that the Directorate General of Electricity and Energy Utilisation, which regulates the electricity industry, has no clearly defined role or free autonomy to balance the various interests in the energy sector. Additionally, the absence of an independent regulator, as well as infrastructure shortages, had led to cautiousness from foreign investors.

There is a large backlog of pipeline projects, particularly for geothermal projects. For instance, the World Bank has estimated that there are 580MW of geothermal projects in the pipeline and just 5MW of total capacity that have reached financial close.

The cost of electricity, like in Australia, is a politically sensitive issue. In 2011, the Government gave US$8 billion in subsidies to PLN to offset the renewable energy feed-in tariff scheme (see below). An ensuing proposal to raise the cost of electricity by 20% was met with mass strikes and demonstrations, ultimately defeating the proposal. It is estimated that only an 80% increase in electricity prices would cover production costs, which brings into question the financial viability of Indonesia's state-dominated electricity system.

RENEWABLES LAWS

Green Energy Policy 2004 is an MEMR Decision No. 2/2004 which aims to reduce the country's reliance on oil.

Geothermal Law No. 27/2003 gives power to the regional governments to develop geothermal energy and establishes a separate regime for geothermal licensing.

Governmental Regulation No. 70/2009 on Energy Conservation forms the National Energy Conservation Plan, promotes energy efficiency and offers incentives for the importation of energy-saving devices.

Presidential Regulation No. 5/2006 on National Energy Policy sets out the definition of renewable energy and the target for renewable energy production.

GOVERNMENT INCENTIVE PROGRAMS

Indonesia currently offers tax concessions for renewable energy projects, particularly geothermal energy projects. Incentives for geothermal projects include an investment credit of 20% of the qualifying capital investment, an extended tax loss carry forward period for up to 10 years, accelerated depreciation rates and a maximum dividend withholding tax of 10%.

The Ministry of Finance's Regulation No. 21/PMK. 011/2010 provides similar tax and also custom incentives for other renewable energy technologies.

Importantly, since 2002, the MEMR has also established a feed-in tariff scheme for renewable energy. However, international developers have not been as attracted by the feed-in tariff given the fixed end-user energy prices, which are regarded by many as unsustainably low.

BIOMASS

The installed capacity of biomass in 2010 was 5.7GW, while biomass potential is approximately 50GW.Indonesia is a leading exporter of palm oil. Remote villages in Indonesia rely heavily on traditional biomass sources for energy.

An estimated 55% of Indonesia's population, i.e. 128 million people primarily rely upon traditional biomass for cooking.

HYDROELECTRICITY

Indonesia has set a target of 2 GW installed capacity in hydroelectricity, including 0.43 GW micro hydro, by 2025. Indonesia has a significant hydropower potential of approximately 75GW, of which about half lies in isolated Papua. The current total installed capacity for hydropower is estimated by MEMR to be 4.3GW.

GEOTHERMAL ENERGY

Indonesia uses some geothermal energy. According to the Renewable Energy Policy Network's *Renewables 2013 Global Status Report*, Indonesia has the third largest installed generating capacity in the world. With 1.3 GW installed capacity, Indonesia trails only the United States (3.4 GW) and the Philippines (1.9 GW). However it leads Mexico (1.0 GW), Italy (0.9 GW), New Zealand (0.8 GW), Iceland (0.7 GW), and Japan (0.5 GW). Current official policy is to encourage the increasing use of geothermal energy for electricity production. Geothermal sites in Indonesia include the Wayang Windu Geothermal Power Station and the Kamojang plant, both in West Java.

The development of the sector has been proceeding rather more slowly than hoped. Expansion appears to be held up by a range of technical, economic, and policy issues which have attracted considerable comment in Indonesia. However, it has proved difficult to formulate policies to respond to the problems.

Geothermal energy is Indonesia's most feasible renewable energy source given that it accounts for roughly 40% of known geothermal capacity worldwide. Indonesia is currently the third largest geothermal generator.

Climate change activists like Al Gore have proclaimed that Indonesia can become the global geothermal superpower, while other analysts have called on the country to replicate its success in thermal coal production in the geothermal field.

Indonesia's total geothermal potential is reportedly 28.1GW, with current capacity of 1.2GW – a figure considered quite low given the country's potential.

A target of 5GW production by 2014 has been announced by the Government.

There is a separate legal regime for geothermal energy production in Indonesia compared to other forms of electricity. It is regulated by Geothermal Law No. 27/2003. A geothermal business requires a geothermal business licence, as well as the standard electricity supply business licence to generate electricity.

In April 2012, Indonesia signed an extensive geothermal cooperation agreement with New Zealand.

OCEAN ENERGY

With 54,716km of coastline, Indonesia has vast potential for ocean current, wave and tidal energy. However, at present the country has no installed ocean energy capacity and just one demonstration project has been developed.

The estimated potential of ocean energy in Indonesia is between 10MW and 35MW per kilometre of coastline.

WIND POWER

A very small amount of (off-grid) electricity is generated using wind power. For example, a small plant was established at Pandanmino, a small village on the south coast of Java in Bantul Regency, Yogyakarta Province, in 2011. However it was established as experimental plant and it is not clear whether funding for long-term maintenance will be available.

Wind speeds averaging between 3m and 6m/s have encouraged government investment in small and medium sized wind projects, however large-scale wind farms are not feasible with such wind speeds.

Much like in Mongolia and China, Indonesia's most productive wind sites are located far from population centres, thus requiring extensive transmission infrastructure.Indonesia is seeking a total installed wind capacity of 970MW by 2025.

SOLAR PHOTOVOLTAIC ELECTRICITY

The Indonesian solar PV sector is relatively underdeveloped but has significant potential. Its output is almost exclusively set aside for decentralised rural electrification. In 2011 it produced a relatively small amount of electricity over 2011 - only 22 GWh. In 2009, 77,433 50W solar PV home systems were distributed to Indonesian households.

MAJOR ENERGY COMPANIES IN INDONESIA

RENEWABLE ENERGY PRODUCERS

Ulumbu Geothermal Power Plant, which supplies electricity to neighbouring districts, has a capacity of 5MW.

There are a number of other geothermal plants that have recently become operational, including a 20MW plant in Lahedong.

Cisokan hydropower plant has a capacity of 1GW. The project is substantially funded by the World Bank.

As part of Indonesia's Power Transmission Development Phase II, PLN is undertaking extensive repairs to the transmission infrastructure on grids in Java-Bali, Sulawesi and Kalimantan. This is being funded by the World Bank.

The Itochu Corporation of Japan has been one of the largest sponsors of renewable energy projects in Indonesia

INDONESIAN FIRMS

Pertamina, the state-owned oil company
Perusahaan Listrik Negara, the state-owned electricity company.
Perusahaan Gas Negara, the state-owned gas company
PT Bumi Resources owned by the Bakrie Group
PT Medco Energi International, the largest publicly listed oil and gas company in Indonesia

FOREIGN FIRMS

US-based firm PT Chevron Pacific Indonesia is the largest producer of crude oil in Indonesia
Total E&P Indonesie which operates the East Mahakam field in Kaliamantan and other fields
ExxonMobil is one of the main foreign operators in Indonesia
Statoil, a Norweigian multinational firm, which has been operating in Indonesia since 2007, especially in Eastern Indonesia
BP which is a major LNG operator in the Tangguh gas field in West Papua.
ConocoPhilips which currently operates four production-sharing contracts including at Natuna and in Sumatra

NUCLEAR POWER IN INDONESIA

The program for **nuclear power in Indonesia** includes plans to build nuclear reactors in the country for peaceful purposes. The national legislative organ for nuclear energy, Badan Pengawas Tenaga Nuklir (BAPETEN), was founded in 1998. The national agency for atomic energy is BATAN. In 1954, research on atomic energy was begun in Indonesia. Apart from producing electricity, nuclear technology is also used for medical and agricultural purposes and for food safety. Plans for an atomic program were scaled back in 1997 due to the discovery of the

Natuna gas field, but have been revived since 2005. In December 2013 it was reported that Indonesia was planning to begin constructing a reactor in 2015. In February 2014 the Indonesian Government confirmed that it was planning to construct a 30MW reactor in the near future

According to Presidential Decree Five in 2006, Indonesia should have four nuclear-power plants built by 2025. Their total capacity will be at least 4,000 MW of electricity, about 1.96 percent of projected electricity demand in 2025 (200,000 to 350,000 MW). Indonesia has stated that the program will be developed in accordance with the International Atomic Energy Agency (IAEA), and Mohammed ElBaradei was invited to visit the country in December 2006. Protests against plans for nuclear power were held in June 2007 in central Java, and increased in mid-2007.

Indonesia has a greater depth of experience and infrastructure in nuclear technology than any other southeast Asian country except Australia.

A 30 MW nuclear power plant is planned to be built at Serpong, near Jakarta.

Indonesia's population of 242 million is served by power generation capacity of only 35 GWe, producing 182 billion kWh in 2011. Of this, 81 TWh (44%) came from brown coal, 42 TWh (23%) from oil, 37 TWh (20%) from gas, 12 TWh (7%) from hydro and 9 TWh (5%) from geothermal. It has per capita electricity consumption: 475 kWh/yr, but 36% of the population in 2013 have no access to electricity. The government has targeted a 26% reduction in CO2 emissions by 2020.

With an industrial production growth rate of 10.5%, electricity demand is estimated to reach 450 billion kWh in 2026. At present a low reserve margin with poor power plant availability results in frequent blackouts. The Java-Bali grid system accounts for more than three quarters of Indonesia's electricity demand – 132 TWh in 2012. PT PLN (Persero), the Indonesia Electricity Corporation, projects 55 GWe new capacity by 2021, 38 GWe of this coal-fired. It also plans major grid enhancements on Java, on Sumatra, and Kalimantan, with a HVDC link Sumatra to Java.

About 43% of Indonesia's electricity is generated by oil and gas, so as well as catering for growth in demand in its most populous region, the move to nuclear power will free up oil for export. However, in mid-2012 the National Energy Council (DEN) said that nuclear power was an unlikely last resort in the country.

Following earlier tentative proposals, in 1989 the government initiated a study focused on the Muria Peninsula in central Java and carried out by the National Atomic Energy Agency (BATAN – Badan Tenaga Nuklir Nasional) established in 1958. It led to a comprehensive feasibility study for a 7000 MWe plant, completed in 1996, with Ujung Lemahabang as the specific site, selected for its tectonic stability. Plans for the initial plant on the Muria Peninsula in central Java were then deferred indefinitely early in 1997. A National Nuclear Act was passed in 1997.

Then a 2001 power generation strategy showed that introduction of a nuclear plant on the 500 kV Java-Bali grid would be possible in 2016 for 2 GWe rising to 6-7 GWe in 2025, using proven 1000 MWe technology with investment cost $2000/kWe. The Java-Bali interconnected system accounts for more than three quarters of Indonesia's electricity demand.

Under the 2006 National Electricity Planning Scheme 2006-26 and Presidential Decree #43 in 2006 the project may be given to an Independent Power Producer to build and operate. Sites on the central north coast of Java were then under consideration, with access to the country's main grid infrastructure. Plans were to call tenders in 2008 for two 1000 MWe units, Muria 1&2, leading to decision in 2010 with construction starting soon after and commercial operation from 2016 and

2017, but these plans were put on hold. Fuel services would be purchased from abroad and fuel would preferably be leased. Used fuel would be stored centrally in the medium term.

The government has said that it has $8 billion earmarked for four nuclear plants of total 6 GWe to be in operation by 2025, and aimed to meet 2% of power demand from nuclear by 2017. It is anticipated that nuclear generation cost would be about 4 cents/kWh (US) compared with 7 c/kWh for oil and gas.

In mid-2010 three sites were being considered for main plants: Muria (central Java, actually 3 locations), Banten (west Java) and Bangka Island (off southern Sumatra to NE, 2 locations: West Bangka and South Bangka). All are on the north shores, away from the tectonic subduction zone. Over 2011-13 BATAN undertook a feasibility study for Bangka, and it signed an agreement with the Bangka-Belitung provincial government. Bangka is far from any active volcano, has low seismic hazard, no tsunami hazard (shallow sea), and low population. Site evaluation of Muntok, West Bangka and Permis, South Bangka, showed both to be suitable for some 10 GWe capacity meeting 40% of the demand in Sumatra, Java and Bali. Most of this capacity would be at the West Bangka site, with 600 MWe at Permis. After a change in provincial government, these Bangka sites seem in doubt.

BATAN's focus in 2013 shifted to West Kalimantan, using small reactor units suited to the relative lack of grid infrastructure there and where most electricity is imported from Malaysia. Six designs were being evaluated. In November 2013 the Research & Technology Ministry (RISTEK) affirmed its intention of building a small (eg 30 MWe) power reactor, at an unspecified place, possibly from 2015. In December 2013, on the 55th anniversary of founding BATAN, the minister said that a non-commercial power reactor (RDNK) and a gamma irradiation facility would be built by BATAN at Serpong, the site of its largest research reactor, and in February 2014 he was quoted as saying that a 30 MW nuclear power reactor would be built by BATAN at Serpong, near Jakarta.

In July 2007 Korea Electric Power Corp. and Korea Hydro & Nuclear Power Co. (KHNP) signed a memorandum of understanding with Indonesia's PT Medco Energi Internasional to progress a feasibility study on building two 1000 MWe OPR-1000 units from KHNP at a cost of US$ 3 billion. This was part of a wider energy collaboration.

In addition, BATAN has undertaken a pre-feasibility study for a small Korean SMART reactor for power and desalination on Madura island. However, this awaits the building of a reference plant in Korea. Also the province of Gorontalo on Sulawesi was reported to be considering a floating nuclear power plant from Russia, and late in 2007 a cooperation agreement with Japan was signed, envisaging its help in building and operating nuclear power plants.

The Japanese and Indonesian governments signed a cooperation agreement in November 2007 relating to assistance to be provided for the preparation, planning, and promotion of Indonesia's nuclear power development and assistance for public relations activities.

The IAEA has been reviewing the safety aspects of both Muria and Madura proposals, with Indonesia's Nuclear Energy Regulatory Agency – Badan Pengawas Tenaga Nuklir (BAPETEN). It was then looking at the Bangka sites. BAPETEN was established in 1998 and reports directly to the President.

In November 2009 the IAEA undertook an Integrated Nuclear Infrastructure Review mission to Indonesia. Against 19 parameters, 'no actions needed' on six, 'significant actions needed' on three, and the rest 'minor actions needed'. In respect to IAEA milestones, the country is at the first: 'ready to make a knowledgeable commitment'. During the 1980s many technical people were trained in anticipation of nuclear power development then, many of these are still available for the new project.

Russia is keen to export floating nuclear power plants (FNPP), on a fully-serviced basis, to Indonesia as a means of providing power to its smaller inhabited islands.

In December 2013 RISTEK reported the results of a poll conducted by an independent agency PT Iconesia Solution. Community acceptance in Indonesia is increasing, and 76.5% agreed with the development of nuclear science and technology, while 60.4% agreed with developing a nuclear power plant in the country.

Indonesia's safeguards agreement with the IAEA under the NPT entered force in 1980 and the Additional Protocol entered force in 1999. In 1997 it signed the Joint Convention on the Safety of Spent Fuel Management and Radioactive Waste Management.

Indonesia has a number of nuclear-related facilities in operation. BATAN operates three research reactors: in Serpong, Banten on the western outskirts of Jakarta (30 MW), Bandung, west Java (2 MW), and in Yogyakarta, central Java (100 kW).

The Serpong multipurpose reactor, which started up in 1987, is intended to support the introduction of nuclear power to the country. It is in the Litbangyasa Serpong Nuclear Zone, located in the Research Centre for Science and Technology (PUSPIPTEK), Serpong. The main facility here is the 30 MW Multipurpose Reactor GA. Siwabessy (RSG-GAS), but also there are the Centre for Reactor Technology and Nuclear Safety (PTRKN), Centre for Development of Nuclear Informatics (PPIN), Nuclear Device Engineering Centre (NEDC), Radioisotope Radiofarmaka Centre (PRR), Materials Technology Centre for Nuclear Fuel, Radioactive Waste Technology Centre (PTLR), Nuclear Industrial Materials Technology Centre (PTBIN), Centre for Standardization and Nuclear Quality Assurance (PSJMN), and the Centre for Nuclear Technology Partnership (PKTN).

A Government-owned company, PT Batan Teknologi, produces medical and industrial isotopes (including Mo-99) for domestic needs using the facilities in Serpong.

At Yogyakarta, as well as the 100 kW Kartini research reactor there is the Teknologi Accelerator and Process Materials Centre (PTAPB) and the College of Nuclear Technology (STTN). At Bandung the country's first research reactor was built in 1965, a small Triga mkII which was subsequently boosted to 2 MW, and the site also hosts the Nuclear Materials Technology and Radiometric Centre (PTNBR) where nuclear medicine in the country was established. Friday Market in Jakarta is a larger nuclear establishment, with Isotopes and Radiation Technology Applications Centre (PATIR), Technology Centre for Radiation Safety and Metrology (PTKMR), Nuclear Geology Development Centre (PPGN), Centre for Education and Training, and the Centre for Nuclear Science and Technology Dissemination (PDIN).

The country also has front-end capabilities in ore processing, conversion and fuel fabrication, all at a laboratory scale, though PT Batan Teknologi assembles fuel elements for the research reactors using imported US fuel. There have been no experiments in reprocessing, but BATAN operates a radwaste program including for spent fuel from the research reactors.

Uranium

There are some uranium resources in Kalimantan, and possibly West Papua. BATAN in September 2010 quoted 53,000 tonnes as high-cost resources: 29,000 t in West Kalimantan and

24,000 t in Bangka Belitung, including some associated with rare earths in monazite by-product from tin mining.

INDONESIA AND INTERNATIONAL NUCLEAR CO-OPERATION

Indonesia is a member of the IAEA, and a signatory to the NPT.

COMPREHENSIVE NUCLEAR-TEST-BAN TREATY

On December 6, 2011 the Indonesian Parliament ratified the Comprehensive Nuclear-Test-Ban Treaty (CTBT). Indonesia agreed not to conduct nuclear-weapons testing, but nuclear power plants are unaffected.

COOPERATION WITH OTHER COUNTRIES

In 2006 Indonesia signed treaties for nuclear cooperation with a number of countries, including South Korea, Russia, Australia and the United States. Australia has indicated its willingness to supply Indonesia with uranium for peaceful purposes. A well-publicized agreement with a Russian company to build a floating nuclear reactor in Gorontalo ran aground; Indonesia has since made it clear that it wants a higher-capacity nuclear power plant, and will construct a land-based plant.

IAEA APPRAISAL

The International Atomic Energy Agency (IAEA) considers Indonesia ready to develop nuclear energy in a statement issued in November 2009. Its appraisal considered four readiness aspects: human resources, stakeholders, industry and regulations. The Indonesian Nuclear Board (BATAN) has carried out research since the 1980s.

NATIONAL NUCLEAR ENERGY AGENCY

Perpustakaan BAPETEN
Phone : +62-21-63858269-70 Ext. 3100
Fax :+62-21-63858275
Email : perpustakaan@bapeten.go.id
Website : www.perpustakaan.bapeten.go.id

The **Badan Pengawas Tenaga Nuklir** (BAPETEN) is the Nuclear Energy Regulatory Agency of Indonesia. BAPETEN is a non-Department Government Institution (LPND) which is under and responsible to the President. BAPETEN has the tasks of implementing the surveillance of all activities of the use of nuclear energy in Indonesia through regulation, licensing and inspection in accordance with applicable laws and regulations. BAPETEN was founded on May 8, 1998 and began actively working on January 4, 1999

BATAN exported a small amount of molybdenum-99 fission material (Mo99) to Malaysia, the Philippines and Bangladesh (with a total value of Rp 50 billion ($5.5 million) in 2011. Due to a drop in world radioisotope production and increased demand from China and Japan, BATAN will invest Rp 100 billion ($11.1 million) to increase its production capacity to 900 Ci per week (from its present 40 Ci). Full capacity could be realized in 2013, but large-scale exports are possible by late 2012.

Nuclear Energy Act UU No. 10 of 1997 explains the important supervisory function in protecting public health and the health of the environment: the creation of regulations, licensing and inspections. This suspervisory function is the main activity carried out by BAPETEN.

Making Rules:

BAPETEN is responsible for the creation of rules and safety regulations. Representing the government, BAPETEN cooperate with the House of Representatives in making government regulations in relation to the safe use of radiation sources. BAPETEN also assists the President of the Republic of Indonesia in making a Presidential Decree.

As a regulatory agency, BAPETEN has issued more than 30 recommendations and safety instructions in the use of nuclear energy. All regulations and instructions can be found on the website of BAPETEN.

Licensing:

Uses of nuclear energy without prior permission of BAPETEN is prohibited. Until June 2005, BAPETEN has issued 3162 permits for industrial activities, 2958 permits for medical activities and 3383 working licenses working for Radiation Protection Officer.

BAPETEN continuous to try to raise awareness of radiation safety, this is important since many users of nuclear energy have not reported to BAPETEN of their ownership and uses of radioactive materials and their radiation emitting devices.

Inspection:

BAPETEN conducts supervisory inspections to ensure compliance of the users with the safety regulations and provisions relevant to the permit conditions.

SUPPORTING ACTIVITY MONITORING

Enforcement Regulations:

In the event of breach safety regulations, especially those that cause casualties and environmental damages, BAPETEN has the authority to enforce the rules in accordance with the laws and regulations.

Assessment of Monitoring System:

Monitoring system, namely regulation, licensing and inspections, continuously reviewed to enhance the effectiveness of supervisory functions.

Nuclear Preparedness:

If the case of nuclear accidents, emergency response team of BAPETEN will immediately take action to limit and minimize the impact of accidents and fatalities.

REACTOR LOCATIONS

For research purposes, the following experimental nuclear reactors have already been built in Indonesia:

1. Kartini nuclear research reactor in Yogyakarta, Central Java

2. MPR RSG-GA Siwabessy nuclear research reactor in Serpong, Banten
3. Triga Mark III nuclear research reactor in Bandung, West Java

According to an observer, Indonesia is capable of building a nuclear reactor in every province due to ample materials and geological support. As a tin-mining area, monazite exists throughout Bangka and Belitung island. Nearly 183 tons of monazite sediment was found in Mount Muntai, sufficient for nuclear power plants which will be built in West and South Bangka.

BATAN continues searching for uranium sources and suitable sites for nuclear power plants. Bangka Belitung is geologically stable and near the country's greatest electricity-consuming regions: Java and Sumatra. Local residents are more receptive to hosting a nuclear power plant, compared with other locations. Although local residents were opposed, BATAN was still considering the previously-studied locations of Mount Muria, Jepara, Central Java and Serang, Banten.

In July 2011, the Bangka Belitung governor requested the government to continue its plans for nuclear power plants in the Muntok and Permis areas between 2025 and 2030. The two plants will produce two gigawatts of electricity at a cost of Rp 70 trillion ($8.2 billion), producing 40 percent of electricity needs in Sumatra, Java and Bali.

Several locations have been proposed for nuclear reactors to generate electricity:

1. Cape Muria, Kudus (Central Java)
2. Gorontalo, in northernSulawesi
3. Bangka Belitung province (two plants with a total capacity of 18 GW)
4. Kalimantan

Concerning Fukushima's nuclear reactor blast, an Indonesian geodesics lecturer stated that while most of Japan was earthquake-prone, Indonesia has many quake-free areas in Kalimantan, Bangka-Belitung, northern Java (a populous area) and Irian

OIL AND GAS SECTOR: STRATEGIC INFORMATION

BASIC INFORMATION

Indonesia has the largest population in Southeast Asia and the fourth largest population in the world (behind China, India, and the United States). It is also the world's third-fastest growing economy. Although Indonesia has been a net importer of oil since 2004, it is the sixth largest net exporter of natural gas, and the second largest net exporter of coal. However, as a result of inadequate infrastructure and Indonesia's complex business environment, Indonesia has struggled to attract investment sufficient to meet its energy development goals.

Indonesia's total primary energy consumption grew by nearly 50 percent between 1999 and 2008. Oil continues to account for the most significant share of Indonesia's energy mix, at 44 percent in 2009. Coal consumption has tripled over the decade, accounting for 29 percent of total energy consumption in 2008, surpassing gas as the second most consumed fuel.

Indonesia is also a significant consumer of traditional biomass in its residential sector, particularly in the more remote areas that lack connection to Indonesia's energy transmission networks (power grids and pipelines, for example). The International Energy Agency estimates that combustible renewables and waste account for about a quarter of total primary energy supply.

OIL

Indonesia's oil production has declined in recent years, as older fields decline and new projects fail to replace them. Indonesia is currently a net importer of both crude oil and refined products. Indonesia's crude oil production has been declining since 1998, due to the maturation of the country's largest oil fields and failure to develop new, comparable resources. Indonesia was a member of the Organization of Petroleum Exporting Countries (OPEC) from 1962 to 2009. In 2004, the country became a net oil importer and in January 2009, suspended its OPEC membership.

SECTOR ORGANIZATION

Indonesia's upstream oil sector is dominated by several international oil companies - in particular Chevron, Total, Conoco Phillips, Exxon, and BP. Chevron is the largest single oil producer in Indonesia, accounting for more than 40 percent of the country's total crude production. PT Pertamina, Indonesia's state-owned integrated energy supply company, accounted for approximately 15 percent of 2009 crude and condensate production, making it the second largest producer in the country.

The Indonesia Ministry of Energy and Mineral Resources is responsible for entering into production sharing contracts (PSCs), while the state-owned legal entity BPMigas serves as the upstream regulator that manages and implements these agreements. Indonesia's 2001 Oil and Gas Law significantly restructured Indonesia's upstream oil and gas sector - most notably transferring the upstream regulatory role from PT Pertamina to BPMigas. Although PT Pertamina continues to be wholly state-owned, the 2001 law also led to its establishment as a limited liability corporation in 2003. Senior Pertamina officials have indicated plans to divest two of its subsidiaries by about 20 percent in an initial public offering, but details and timing have not been confirmed.

In addition to its upstream activities, PT Pertamina operates nearly all of Indonesia's refinery capacity, procures crude and products imports, and supplies products to the domestic market. While Pertamina's monopoly in the retail market ended in 2004, the company continued to be the sole distributor for subsidized fuels until early 2010.

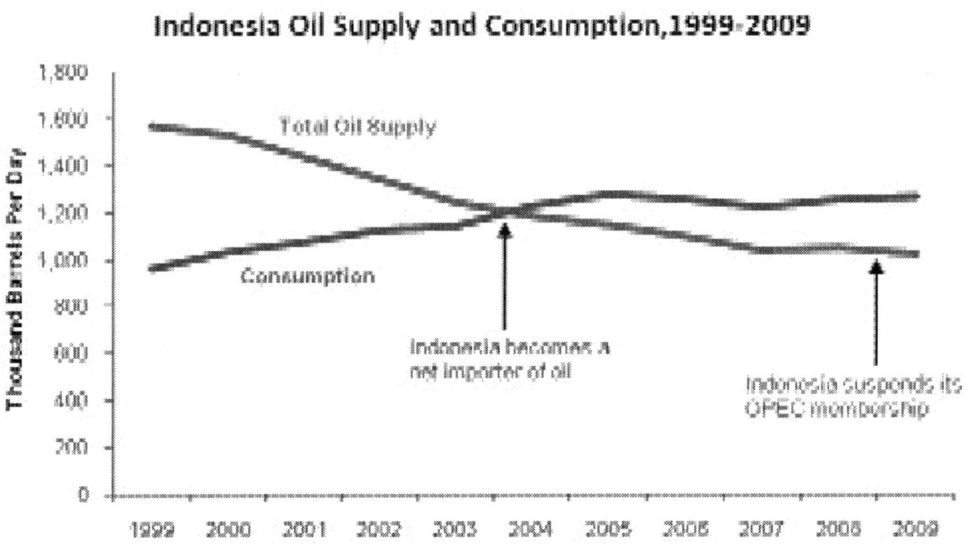

Source: EIA *International Energy Statistics*

Exploration and Production

According to Oil & Gas Journal, Indonesia had 3.9 billion barrels of proven oil reserves as of January 2011. In 2010, total oil supply averaged just over one million barrels per day (bbl/d). Of this total, about 943,000 bbl/d was crude oil and lease condensate production. Crude and condensate production has declined at an annual average rate of 4.1 percent between 2000 and 2010.

Indonesia's two largest producing oil fields are the Minas and Duri fields, located on the eastern coast of Sumatra. Chevron operates both fields with a 100 percent working interest under the Rokan Production Sharing Agreement. Producing since 1952 and 1955 respectively, production at both fields is in decline, even with the employment of enhanced oil recovery techniques at both fields to bolster production. Chevron uses steam injection enhanced oil recovery for 80 percent of the Duri field, one of the largest steamflood projects in the world.

The most significant recent discovery with the potential to counteract some of Indonesia's production decline is the Cepu Block of East and Central Java. Exxon Mobil is the operator of the Cepu PSC (45 percent interest), in a joint venture with PT Pertamina's E&P unit (45 percent working interest) and four local government companies (10 percent interest). Cepu is estimated to contain 600 million barrels of recoverable liquids, and to have a peak production of 165,000 bbl/d. Although discovered in 2001, the project has encountered several delays in the development process, and Exxon recently revised its goal for peak production from 2012 to 2014. Banyu Urip is currently the only producing field in the Cepu PSC, and as of January 2010, had reached a production level of about 18,000 bbl/d.

BPMigas and the Indonesian government have introduced policies aimed at increasing investment in the country's upstream sector - in particular via investment incentives and improving the flexibility of the PSC bidding process. However, the upstream investment environment is still considered to be risky, and 2009 and 2010 licensing rounds were disappointing. Recent events that caused particular concern were Parliament's attempts to mandate cost recovery caps for PSCs, as well as the government's cabotage rule - a shipping regulation requiring all marine vessels to carry the Indonesian flag. The government has since announced that cost recovery caps would be removed, and that implementation of the cabotage rule will except oil and gas vessels. However, Indonesia failed to meet its 2010 production goal of 965,000 bbl/d of crude and condensate production, and signed only 21 new oil and gas PSCs in 2010, relative to 34 in 2008.

REFINING

Indonesia has a refinery capacity of just over one million bbl/d, according to Oil & Gas Journal. Spread across eight refineries owned by PT Pertamina, the majority of the country's refinery capacity is located on Java and Sumatra. The three largest refineries are Cilacap (348,000 bbl/d) in Central Java, Balikpapan (260,000 bbl/d) in East Kalimantan, and Bolongan (125,000 bbl/d) in West Java. Refinery output goes primarily to the domestic market, but meets only about 70 percent of domestic consumption. Pertamina plans to eliminate the need for product imports by 2017, and in the last few years has announced several refinery upgrade, expansion and greenfield projects in support of this goal. However, most of these projects have been delayed, as low refining margins and lack of government financial incentives have deterred investment from international partners.

Pertamina signed an agreement with Kuwait Petroleum Corporation in August of 2010 to study an upgrade to the Bolongan refinery in West Java. The company is also in discussion with Saudi Aramco to build a new 200,000-300,000 bbl/d refinery in East Java, though Saudi Aramco indicated late in 2010 that Indonesia's government would need to provide more incentives in order to secure the company's participation. Pertamina's planned 300,000 bbl/d Banten Bay refinery in Bojanegoro is currently administered by a joint venture with Iran's Oil Refining Industries Development Company and Malaysia's Petrofield Refining - but the project was recently evaluated as being "economically infeasible" by Pertamina leadership.

The above-mentioned projects would all be supported by some level of guaranteed crude supply from Kuwait Petroleum Corporation, Saudi Aramco, and the National Iranian Oil Company respectively. These sources are vital for project viability as Indonesia's domestic crude production continues to lag. However, after years of delays and failed partnerships, there is little confidence in the mid-term completion of any of these projects, and none are expected to be online prior to 2015. Upgrades to existing refineries may continue to move forward - for example, Cilicap's 62,000 bbl/d residual fluid catalytic cracking unit still appears to be on track for 2013 or 2014 completion.

CONSUMPTION

Consumption of refined products grew at a fairly steady 4.7 percent average annual growth rate between 1995 and 2005, but declined by about two percent in both 2009 and 2010, largely as a result of the 126 percent increase in the price of subsidized fuel implemented in 2005. The 2005 price increases - and the removal of fuel subsidies for large industrial consumers - were part of President Yudhoyono's early attempts to gradually eliminate fuel subsidies. Consumption increased in both 2008 and 2009, exceeding 1.3 million bbl/d in 2009.

Fuel subsides will account for nearly ten billion dollars in 2010, or about ten percent of the government's tax revenues. In December of 2010 the Indonesian parliament approved a measure to remove fuel subsidies for all vehicles excluding motorcycles and public transportation vehicles. Cash transfers will be used to ease the impact on the economically disadvantaged. The policy was initially slated for April 2011 implementation in the greater Jakarta area, and nationwide by 2013. However, implementation was indefinitely delayed in March 2011 due to concerns over the effect on the rate of inflation.

NATURAL GAS

Natural gas production has increased by more than a quarter since 2005. Indonesia exported about half of its natural gas consumption in 2009 - but the share of domestic consumption is expected to increase. According to *Oil & Gas Journal*, Indonesia had 106 trillion cubic feet (Tcf) of proven natural gas reserves as of January 2011. Indonesia is the fourteenth largest holder of proven natural gas reserves in the world, and the third-largest in the Asia-Pacific Region. Although domestic consumption of natural gas has nearly doubled since 2004, Indonesia continues to be a major exporter of pipeline and liquefied natural gas (LNG).

Sector Organization

The regulatory structure that shapes Indonesia's upstream oil sector also forms the basis for the gas sector (see Oil: Sector Organization). BPMigas serves as the upstream regulator, and state-owned PT Pertamina - though still active in upstream exploration and production - no longer plays a regulatory role. Pertamina accounts for about 15 percent of natural gas production. International oil companies such as Total, ConocoPhillips, and ExxonMobil dominate the upstream gas sector, while natural gas transmission and distribution activities are carried out by the state-owned utility Perusahaan Gas Negara (PGN).

Production

In 2009, Indonesia produced 2.6 Tcf of dry natural gas. Production has grown at an average annual rate of about 1.5 percent over the previous two decades, and Indonesia's 2009 gas production was the eleventh-highest in the world. A little more than half of Indonesia's 2009 production came from offshore fields, although the government estimates that more than 70 percent of the country's conventional gas reserves may be located offshore. An increasingly large majority of Indonesia's natural gas production has come from non-associated fields in recent years, with associated gas accounting for about 18 percent of gross production in 2009.

Indonesia's geography presents a challenge to resource development, because the archipelago nation's most prolific blocks of conventional gas reserves are located far from its major demand markets. The most significant areas for current natural gas production are:

East Kalimantan's offshore fields, particularly the Mahakam PSCs operated by Total
South Sumatra, particularly the onshore Corridor PSC operated by Conoco Philips
Aceh and North Sumatra, including Exxon Mobil's declining offshore Arun field
South Natuna Sea, offshore Block B operated by Conoco Phillips

In 2009, East Kalimantan's offshore Mahakam PSCs accounted for about a third of gross production.

In addition to expansion of current projects, there are several major new gas projects in development for the next decade. Total and Petronas (Malaysia) recently joined ExxonMobil and

Pertamina to develop the Natuna D-Alpha field in the East Natuna Sea. Pertamina expects the project to start by 2021, and has estimated that East Natuna in total holds 46 Tcf of gas. However, the field contains about 70 percent carbon dioxide, significantly tightening production margins for its developers. Chevron is pursing the deepwater Gendalo-Gehem project with partners Eni (Italy) and Sinopec (China). At its peak, the project - which spans four PSC blocks - may produce 400 Bcf per year. The project is Indonesia's first deepwater gas project. According to BPMigas, the first stage of the project will come from the Bangka field, and may commence as early as 2014. Inpex (Japan) received permission to begin development in its Masela block, which is estimated to hold about 14 Tcf of natural gas. This offshore block in the Arafura Sea will serve export markets through the planned associated LNG terminal (see LNG section), and is expected to begin production in 2016.

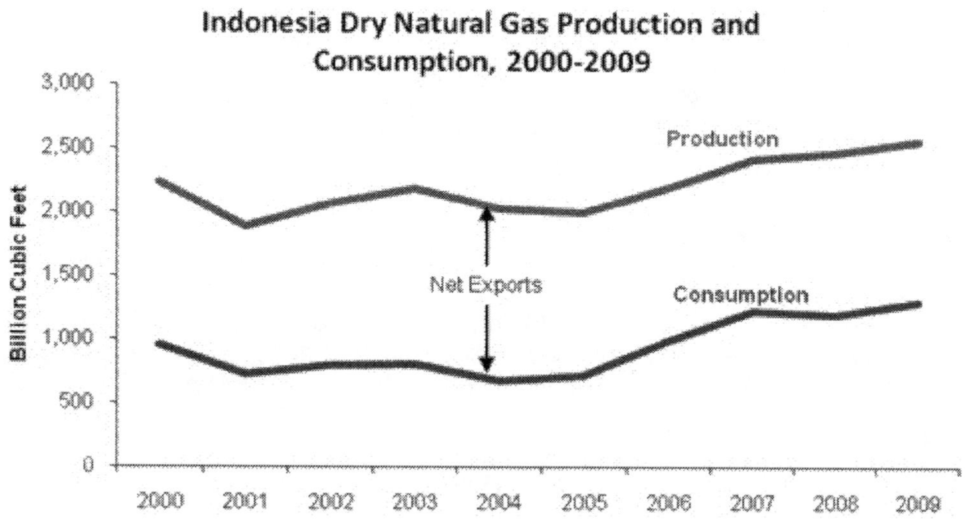

In addition to its considerable conventional gas resources, Indonesia also holds an estimated 453 Tcf of coalbed methane (CBM), a type of unconventional gas whereby methane - the primary component of natural gas - is extracted from coal beds. CBM reserves are located relatively close to Indonesia's population centers, primarily in South Sumatra and Kalimantan. The government began awarding PSC's in 2008, though all are still in exploration phase. However, first production is expected to begin in some of the PSCs held by local companies in late 2011 and 2012. Though local consortiums hold many of the existing CBM PSCs, foreign majors BP, Eni (Italy), and Dart Energy (Australia) also hold shares in Indonesia's CBM blocks.

Exports

Indonesia was the world's sixth largest net exporter of natural gas in 2009. Although the majority of Indonesia's gas exports are transported as LNG, Indonesia also exports about a quarter of its gas exports via pipeline to Singapore, with which it has two pipeline connections: one from its offshore fields in the West Natuna Sea, and the other from the Grissik gas processing plant in Sumatra. These pipelines have a combined capacity of approximately 400 Bcf/y and deliver gas to Singapore under long-term contracts, both set to expire around 2020. However, as part of its efforts to secure its own domestic supply, the Indonesian government has expressed an interest in negotiating a reduction in the volumes of these contracts.

The remaining three-quarters of Indonesia's gas exports - excluding a small amount of pipeline exports to Malaysia - are exported as LNG. Japan is the major destination for Indonesia's LNG exports, accounting for about 65 percent of total LNG exports, but South Korea and Taiwan are also significant importers (see Liquefied Natural Gas for more information).

Consumption and Distribution

Indonesian gas production initially was geared towards exports, but the country's declining oil production has driven an effort to shift increasing volumes toward domestic consumption. In 2009, Indonesia consumed 1.3 Tcf of natural gas, or about half of its total dry gas production. Although the industrial sector accounts for the largest portion of domestic consumption, the power sector is expected to be the most significant driver of future consumption growth.

Indonesia's gas distribution utility Perusahaan Gas Negara (PGN) currently operates more than 3,500 miles of natural gas transmission and distribution pipelines. However, domestic distribution infrastructure is almost non-existent outside of PGN's Java and North Sumatra strategic business units. According to PGN, in 2009 there was over 400 Bcf of unmet demand for gas among domestic industrial and electric power consumers.

PGN began operations of the South Sumatra-West Java pipeline in 2008, providing an important link between the gas producing region of South Sumatra and the densely populated market of West Java. The Grissik-Duri pipeline is another important domestic transmission pipeline, as it provides gas to Chevron's Duri (see Oil: Exploration and Production) field for its steamflooding and power generation activities. A pipeline problem in September of 2010 resulted in a production outage, which has since been cited as a contributing factor to Indonesia's failure to meet its 2010 oil production goals.

LIQUEFIED NATURAL GAS

Indonesia was the third-largest exporter of liquefied natural gas in 2009, following only Qatar and Malaysia. There are three operational liquefaction terminals in Indonesia, with a combined production capacity of about 1.6 Tcf (32 million metric tons (MMT)) per year. In 2009, Indonesia exported about 950 Bcf of LNG.

The Bontang LNG terminal in East Kalimantan is the largest in Indonesia at about 1.1 Tcf/y (22.6 MMT/y), and one of the largest in the world. Bontang delivered its first cargo in 1977, and was followed shortly thereafter (1978) by the Arun liquefaction terminal in Northern Sumatra. Due to a lack of sufficient additional gas reserves in the Arun field, LNG exports from the Arun plant have been declining in recent years, and are expected to stop altogether by 2014. The newest addition, BP-operated Tangguh in Western Papua, came online in July of 2009 and exported almost 330 Bcf in 2010. BP had originally intended to add a third train to Tangguh's first two trains, but has not yet proceeded with plans for this project.

The next anticipated addition to Indonesia's liquefaction capacity is the Donggi-Senoro LNG plant in Central Sulawesi. The project developers (Mitsubishi, Kogas, Pertamina, and Medco) signed a final investment decision in early 2011, and the 370 Bcf/y (2.5 MMT/y) plant is expected to be commercial in 2014. Inpex, a Japanese company, received government approval at the end of 2010 for the Masela liquefaction terminal in the Arafura Sea, but has delayed the expected startup date of the floating terminal until 2018.

LNG exports have been a politically charged topic in Indonesia, due to the perception of LNG exports removing much-needed gas from the domestic market. The expected growth in gas demand, in addition to the currently unmet demand, has led the government to pursue policies for securing domestic supplies for the local market. The Donggi-Senoro plans received government approval only after 30 percent of the output was designated for domestic consumption. Similarly, a third of the output from Inpex's planned Masela floating LNG liquefaction terminal will be designated for the domestic market, according to regulator BPMigas. Indonesia also plans to reorient the output from the Bontang LNG plant - which is relatively centrally located - into the domestic market over the next decade. Though the plant will remain operational, it will send LNG within Indonesia - and no longer serve export markets by 2020.

Indonesia's Liquefaction Capacity

	Location	Capacity (Bcf/y)	Commercial Operations
Operational			
Arun	Aceh	102	1978
Bontang	East Kalimantan	1101	1977
Tangguh	Papua	370	2009
Planned			
Donggi-Senoro	Sulawasi	102	2014
Masela	Arafura Sea, Maluku Province	122	2016+

Sources: FACTS Global Energy, BP, Inpex

In order to have more flexibility to secure supplies of both domestic and foreign LNG, plans for several LNG receiving terminals are underway in Indonesia. The first regasification terminal, a 143 Bcf/y (3.0 MMT/y) joint-venture between Pertamina and PGN, will supply the Jakarta market starting in the first quarter of 2012. Pertamina also plans to invite bids for an additional receiving terminal of similar size to serve East Java. In addition, Pertamina and PLN (Indonesia's state electricity firm) have announced plans to develop eight LNG receiving "mini terminals" by 2015, with a total capacity of 67 Bcf/y (1.4 MMT/y). These terminals will be scattered throughout the eastern region of the island nation, and are intended to guarantee supply of natural gas for electricity plants.

COAL

Indonesia's coal exports have more than quadrupled in the last decade, making it the second-largest coal exporter in the world, and a critical player in Asian coal markets. According to EIA estimates, Indonesia has 4.8 billion short tons of recoverable coal, of which the vast majority is located in Sumatra and East and South Kalimantan. Government and industry estimates suggest that the resource base may be considerably higher than this amount. Indonesian coal production - which is primarily bituminous or sub-bituminous in rank - has approximately quadrupled between 2000 and 2009, reaching 333 MMst in 2009. In the same year, Indonesia consumed 77 MMst of coal, which was less than a quarter of its production, but more than three times the consumption level in 2000. Power plants accounted for nearly two-thirds of 2009 total coal sales. Electricity sector demand for coal is expected to more than double by 2014 as a result of coal-fired generation capacity additions.

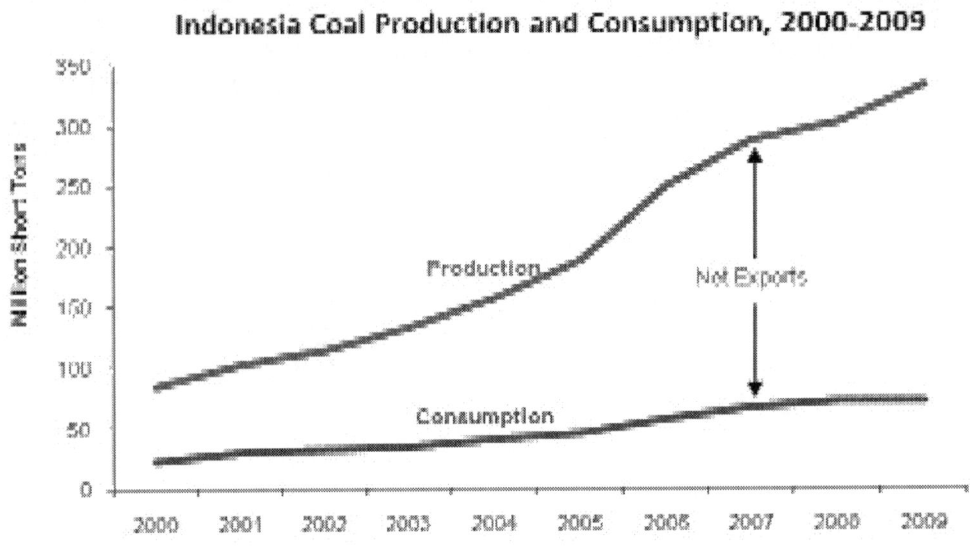

Source: EIA *International Energy Statistics*

Although coal consumption has grown significantly in the last decade, the majority of additional production has gone towards exports. In order to guarantee sufficient domestic supply, the Indonesian government has set a domestic market obligation of 24 percent for producers. Indonesia is now the second-largest exporter of coal (after Australia) on the international market, and is the largest exporter of thermal coal used for power plants. Indonesia's coal exports primarily serve Asian markets, with about 70 percent of 2009 total exports being sent to China, Japan, Taiwan, and other Asian markets. In 2010, Indonesia was the leading source of Chinese coal imports.

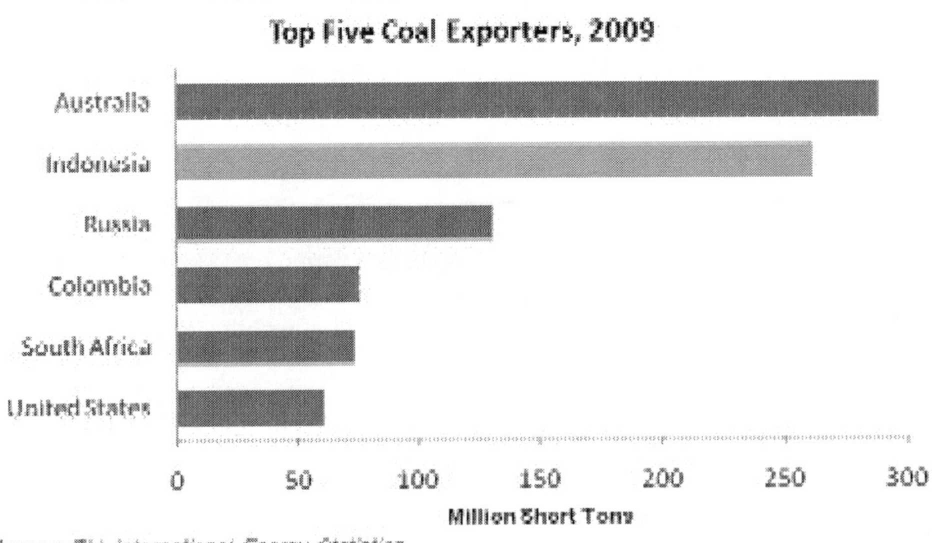

Source: EIA *International Energy Statistics*

ELECTRICITY

Although Indonesia generates 86 percent of its electricity from conventional thermal sources (coal, gas, and oil), it was the third-largest generator of geothermal power in 2009.

Indonesia had 27.8 gigawatts of installed generating capacity in 2008. Indonesia generated 122 billion kilowatthours (Bkwh) of electricity during 2008, of which 86 percent came from conventional thermal sources (oil, natural gas, and coal), eight percent from hydroelectric, and six percent came from geothermal and other renewable sources. Of the conventional thermal total, coal accounts for the largest share at 47 percent, followed by gas at 33 percent, and oil at 19 percent.

State-owned electric utility PT PLN (Perusahaan Listrik Negara) is the most significant company in the electric power sector. PLN owns and operates 86 percent of the country's generating capacity through its subsidiaries, and maintains an effective monopoly over distribution activities. Although the most recent 2009 Electricity Law ends PLN's distribution monopoly, sufficient implementing regulations to support this law have yet to come into effect.

Indonesia's generating capacity has increased by more than a quarter in the decade leading up to 2009. However, as of 2009, only 65 percent percent of Indonesia's population has access to electricity. In addition, because capacity growth has lagged behind the pace of electricity demand growth, grid-connected areas have also suffered from power shortages. Investment in Indonesia's power sector had lagged for several reasons, including inadequate supporting infrastructure, difficulty obtaining land-use permissions, subsidized tariffs, and an uncertain regulatory environment.

In order to address the capacity shortage, in 2009, the government embarked upon the first stage of its "fast track" plan, designed to add 20 additional gigawatts to the grid by 2014. The first stage, which includes 10 gigawatts of primarily coal-based generation, was initially set for completion in 2010 - though subsequent delays have led to a revised completion date of 2013. The second phase, which includes an additional 10 gigawatts to be completed by 2014, includes more cleaner sources of generation such as natural gas and renewables. Unlike in many other countries, Indonesia's government is encouraging an increased use of coal in the power sector, due to its relatively abundant domestic supply, and in order to reduce the use of expensive diesel and fuel oil.

Top Generators of Geothermal Electricity, 2008

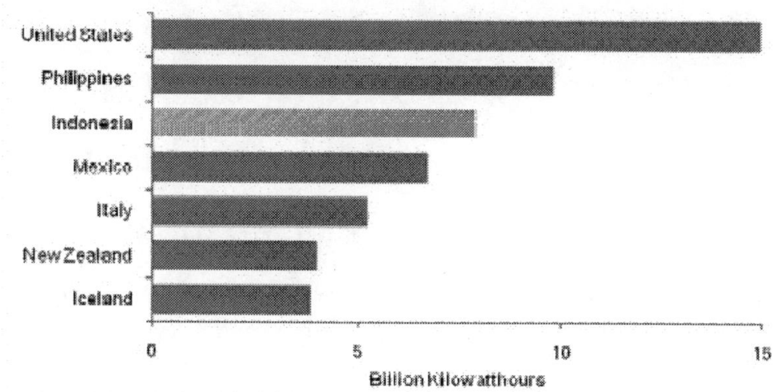

Source: EIA *International Energy Statistics*

Indonesia's power sector is notable for its significant level of geothermal power, and was the third-largest geothermal generator in the world in 2008. However, Indonesia's 2008 capacity of approximately one gigawatt falls substantially below its available resource, which the government has estimated to be sufficient to generate 28 gigawatts. Government plans to increase the use of renewable energy to 15 percent of the electricity portfolio by 2025 are centered on development of geothermal resources. The second phase of the fast track plan includes additional geothermal capacity of nearly four gigawatts by 2014, most of which will be operated by independent power producers.

The Government through the Ministry of Energy and Mineral Resources to change the paradigm of energy management supply side management to demand side management. Some of the things done by the government, among others, by doing so achieving energy conservation energy efficient needs. It was delivered by the Secretary of the Directorate General of Electricity Ir Arief Indarto MM in Indonesia opening Electro Summit (IES) in 2014 and the National Seminar organized by the Energy Revolution Student Association School of Electrical Engineering (STT), Monday (5/5). According to Arief, yet energy efficient needs. "The energy needs are met by fossil energy with any cost and even subsidized, renewable energy as an alternative only," said Arief. In other words paradigm for the management of our energy is more likely to supply side management.

In his speech Arief said that the theme of the seminar is the national IES 'National Energy Sovereignty Brought to Maximize Potential Energy Natural Gas and Geothermal Energy' was also a concern of government. "The theme of the national seminar by the brothers of these students were brought in line with the spirit of the National Energy Policy (KEN) which was approved in January 2014 the House of Representatives a month ago," he said.

According to Arief, KEN mandated to meet the availability of energy to the national needs through the efforts of one of them is reducing fossil waterwheel exports gradually mainly gas and coal. Priority energy development is done through the development of energy by considering the balance of energy economics, energy supply security and environmental conservation are based on the principles include maximizing the use of renewable energy by taking into account the economic level and optimize the utilization of natural gas and new energy.

In the next ten years is projected needs of the power plant about 60 GW, of which about 15% consists of a power plant / Combined Cycle Power Plant / PLTMG and approximately 10% is composed of geothermal power plants. Power plant / Combined Cycle Power Plant / PLTMG plant will be operated as a follower load or peak load and geothermal power plants as base load plants bearer. With the development of gas-fired generation is the requirement of gas for power will continue to increase, up to 2022 is expected to reach 5.3 TSCF. (PSJ)

ENERGY SECTOR: STRUCTURE

MINISTRY OF ENERGY AND MINERAL RESOURCES

The Structure Organization of The Ministry of Energy and Mineral Resources

The Ministry of Energy and Mineral Resources (MEMR) has responsibilities to manage the part of government affairs in the fields of energy and mineral resources. MEMR has been divided into the following directorate generals and agencies, namely:

1 Directorate General of Oil and Gas

Directorate General of Oil and Gas has the duty to formulate and implement policies and technical standardizations in the fields of oil and gas. In order to perform her tasks, the Directorate General of Oil and Gas has the functions in preparing and implementing the policies in the fields of oil and gas, preparing the standards, norms, guidelines, criteria, and procedures in the fields of oil and gas, providing technical guidance and evaluation, and conducting the administration affairs of the Directorate General of Oil and Gas.

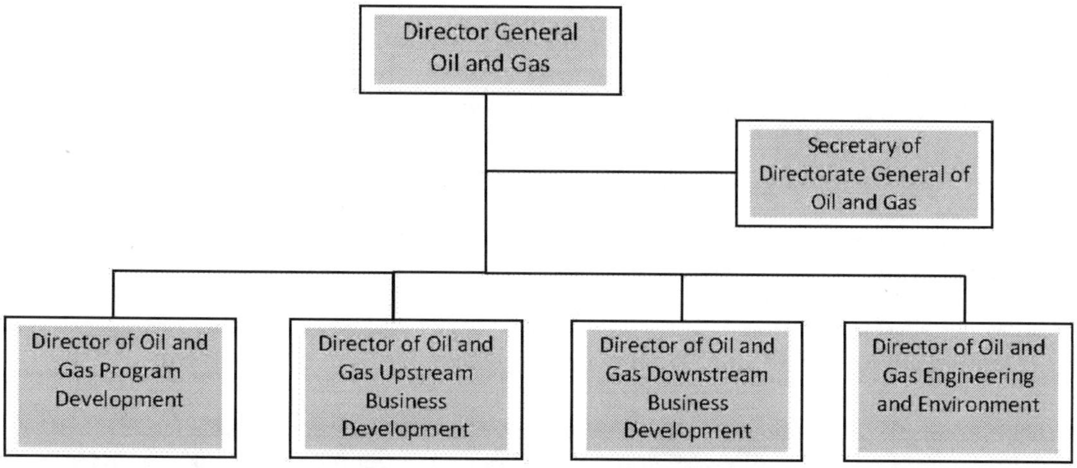

The Structure Organization of Directorate General of Oil and Gas

2 Directorate General of Electricity

Directorate General of Electricity has the duty to formulate and implement policies and technical standardizations in the fields of electricity. In order to perform her tasks, the Directorate General of Electricity has the functions in preparing and implementing the policies in the fields of electricity, preparing the standards, norms, guidelines, criteria, and procedures in the fields of electricity, providing technical guidance and evaluation, and the administration affairs of the Directorate General of Electricity. There are three directorates and one secretariat under the Directorate General of Electricity which have different duty and function as shown in the organization chart below:

3 Directorate General of Mineral and Coal

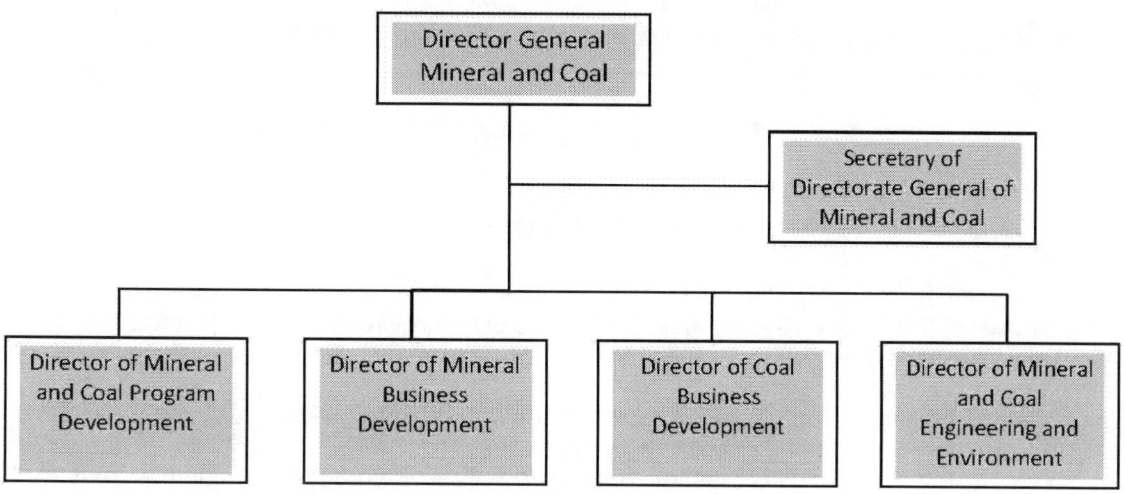

The Structure Organization of Directorate General of Mineral and Coal

Directorate General of Mineral and Coal has the duty to formulate and implement policies and technical standardizations in the fields of mineral and coal. In order to perform her tasks, the Directorate General of Mineral and Coal has the functions in preparing and implementing the policies in the fields of mineral and coal, preparing the standards, norms, guidelines, criteria, and procedures in the fields of mineral and coal, providing technical guidance and evaluation, and conducting the administration affairs of the Directorate General of Mineral and Coal.

4 Directorate General of New Renewable Energy and Energy Conservation (DGNREEC)

Directorate General of New Renewable Energy and Energy Conservation has the duty to formulate and implement policies and technical standardizations in the fields of new, renewable energy and energy conservation. In order to perform her tasks, the Directorate General of New Renewable Energy and Energy Conservation has the functions in preparing and implementing the policies in the fields of new, renewable energy and energy conservation, preparing the standards, norms, guidelines, criteria, and procedures in the fields of new, renewable energy and energy conservation, providing technical guidance and evaluation, and conducting the

administration affairs of the Directorate General of New Renewable Energy and Energy Conservation. There are four directorates and one secretariat under the Directorate General of New Renewable Energy and Energy Conservation which have different duty and function as shown in the organization chart below:

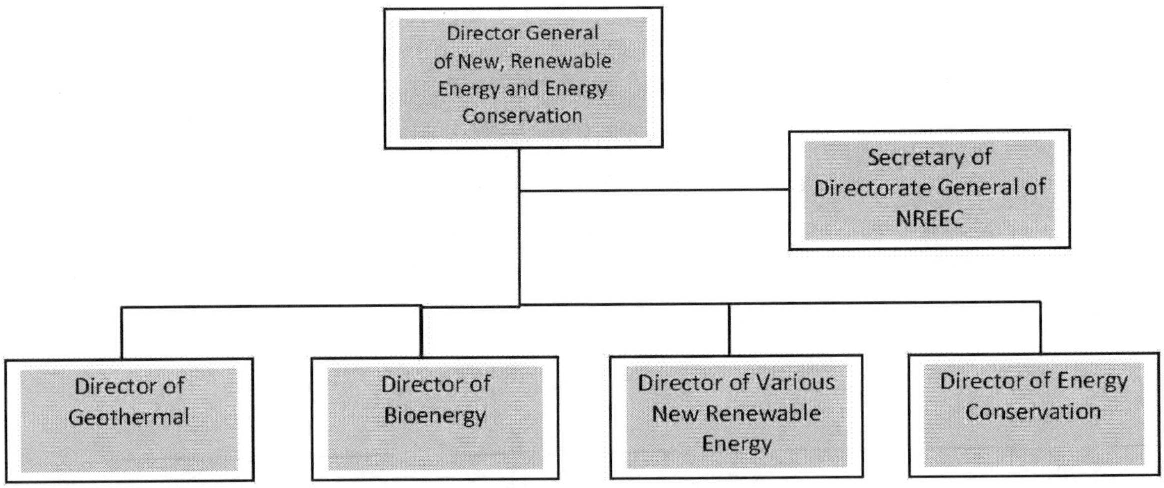

Figure 5. The Structure Organization of Directorate General of New, Renewable Energy and Energy Conservation

5 Geology Agency

Geology Agency has the duty to conduct research and service in the fields of geology. In order to perform her tasks, the Geology Agency has the functions in formulating policies in the fields of geology, formulating research and service plans and programs, fostering and implementing research and service, servicing in geological surveys, as well as researching and servicing in the fields of geology resources, volcanology and geological disaster mitigation and environmental geology, providing recommendation and presentation of information on the results of surveys, researches and services, evaluating the implementation of research and service in the fields of geology, and conducting the administration affairs of the Geology Agency.

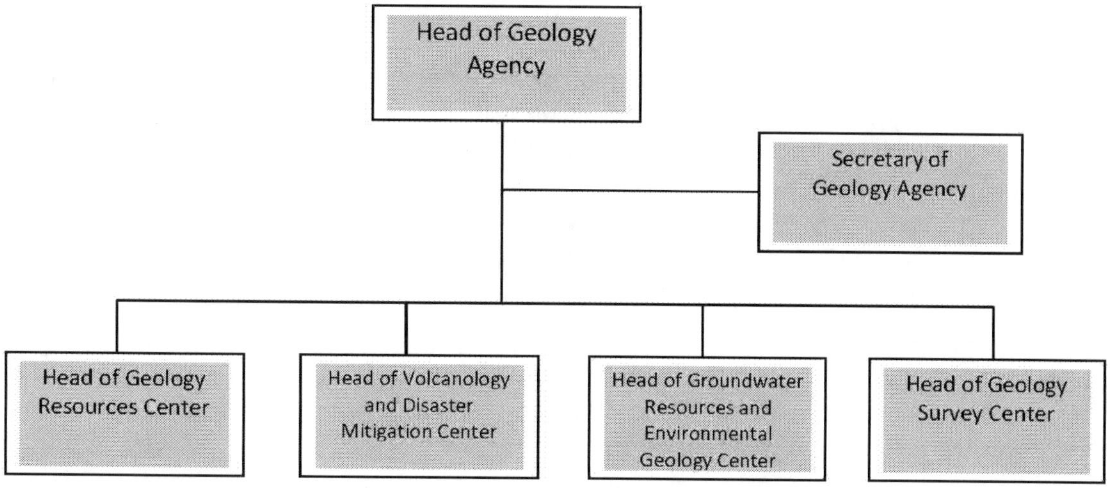

Figure 6. The Structure Organization of Geology Agency

6 Research and Development Agency of Mineral and Energy Resources

Research and Development Agency of Mineral and Energy Resources has the duty to conduct research and development in the fields of energy and mineral resources. In order to perform her tasks, the Research and Development Agency has the functions in preparing technical policies, plans and programs of research and development in the fields of energy and mineral resources, implementing research and development in the fields of energy and mineral resources, monitoring, evaluating and reporting the implementation of research and development in the fields of energy and mineral resources, and conducting the administration affairs of the Research and Development Agency.

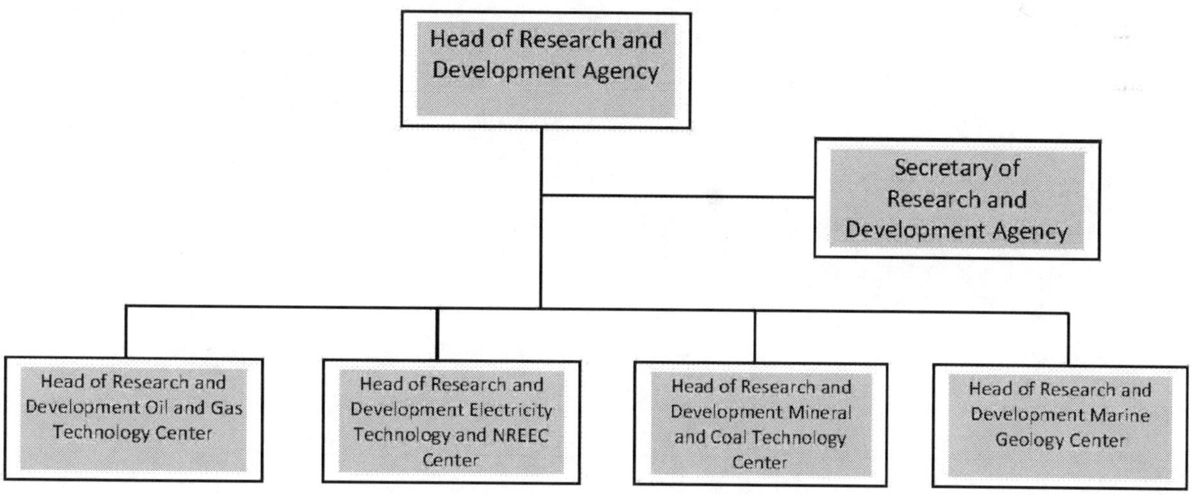

The Structure Organization of Research and Development Agency

7. Geology Agency

Geology Agency has the duty to conduct research and service in the fields of geology. In order to perform her tasks, the Geology Agency has the functions in formulating policies in the fields of geology, formulating research and service plans and programs, fostering and implementing research and service, servicing in geological surveys, as well as researching and servicing in the fields of geology resources, volcanology and geological disaster mitigation and environmental geology, providing recommendation and presentation of information on the results of surveys, researches and services, evaluating the implementation of research and service in the fields of geology, and conducting the administration affairs of the Geology Agency.

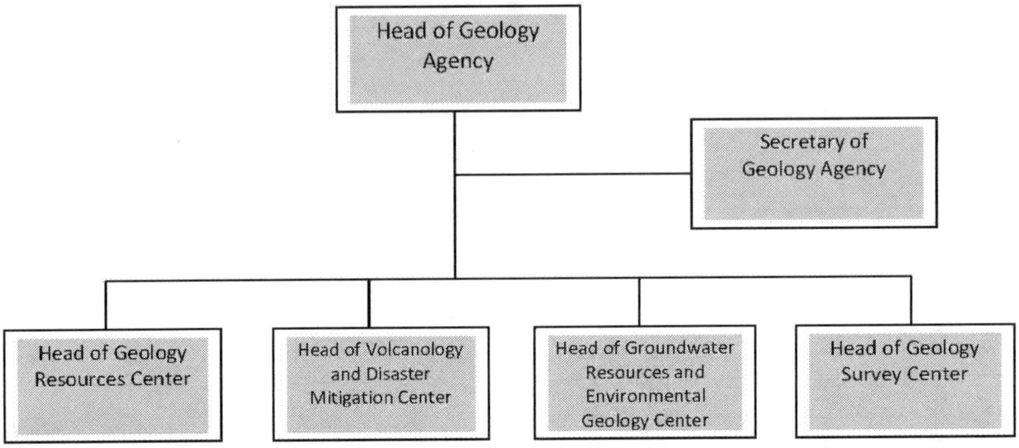

The Structure Organization of Geology Agency

8. RESEARCH AND DEVELOPMENT AGENCY OF MINERAL AND ENERGY RESOURCES

Research and Development Agency of Mineral and Energy Resources has the duty to conduct research and development in the fields of energy and mineral resources. In order to perform her tasks, the Research and Development Agency has the functions in preparing technical policies, plans and programs of research and development in the fields of energy and mineral resources, implementing research and development in the fields of energy and mineral resources, monitoring, evaluating and reporting the implementation of research and development in the fields of energy and mineral resources, and conducting the administration affairs of the Research and Development Agency.

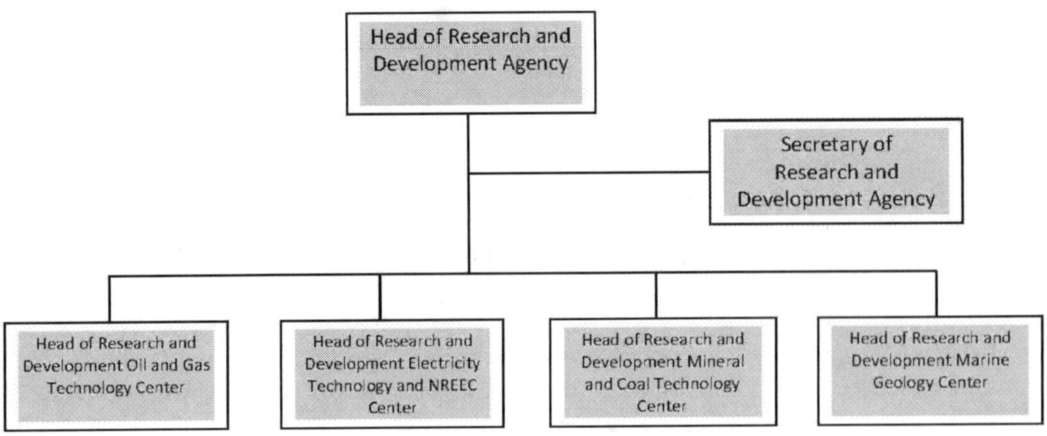

The Structure Organization of Research and Development Agency

9. EDUCATION AND TRAINING AGENCY OF MINERAL AND ENERGY RESOURCES

Education and Training Agency of Mineral and Energy Resources has the duty to conduct education and training in the fields of energy and mineral resources. In order to perform her tasks, the Education and Training Agency has the functions in preparing technical policies, plans and programs of education and training in the fields of energy and mineral resources, implementing education and training in the fields of energy and mineral resources, monitoring, evaluating and reporting the implementation of education and training in the fields of energy and mineral resources, and conducting the administration affairs of the Education and Training Agency.

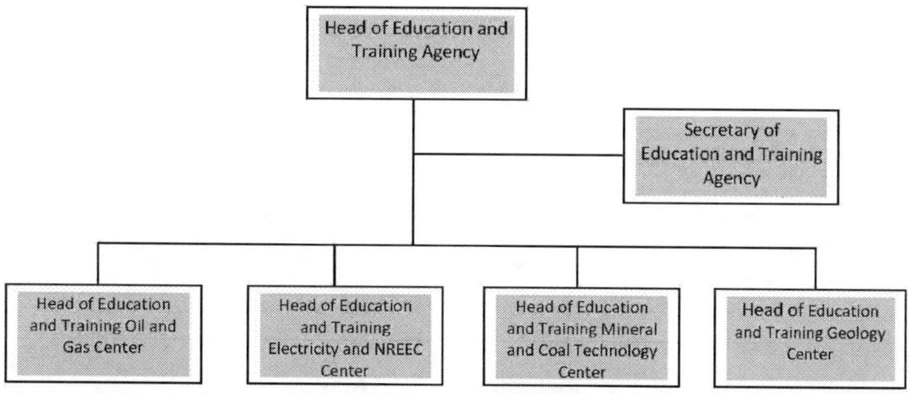

The Structure Organization of Education and Training Agency

STAKEHOLDERS

The successful implementation of policies and programs of the government in the energy sector is highly dependent on the support and cooperation of the stakeholders (associations, people, and public and private companies). Therefore, a good and clear communication, and transparency in policies and regulations making between government and stakeholders should be strengthened and expanded. The stakeholders who have had a good relationship and cooperation with the government in the energy sector are as shown in the table below.

STAKEHOLDERS ON NEW RENEWABLE ENERGY

Stakeholders

I	STATE OWNED COMPANY	
1	State Owned Electricity Company PT Perusahaan Listrik Negara	PLN
2	State Owned Oil and Natural Gas Mining Company Perusahaan Pertambangan Minyak dan Gas Bumi	Pertamina
3	State Owned Gas Company PT Perusahaan Gas Negara	PGN
II.	NRE (New and Renewable Energy) CORE BUSINEESS ASSOCIATION	
4	Indonesian Biofuel Producers Association Asosiasi Produsen Biofuels Indonesia	Aprobi
5	Indonesian Bioethanol Producers Association	APBI

	Asosiasi Produsen Bioethanol Indonesia	
6	Indonesian Palm Oil Association Gabungan Pengusaha Kelapa Sawit Indonesia	Gapki
7	Indonesian Geothermal Association Asosiasi Panas Bumi Indonesia	API
8	Indonesian Minihidro Entrepreneur Association Asosiasi Pengusaha Minihidro Indonesia	APMI
9	Hydro Power Plant Entrepreneur Association Asosiasi Pengusaha Pembangkit Listrik Tenaga Air	-
10	Indonesian Ethanol and Spiritus Industry Association Asosiasi Industri Ethanol dan Spiritus Indonesia	Asendo
11	Indonesian Hydro Association Asosiasi Hidro Indonesia	AHI
12	Indonesia Solar Module Manufacturers Association Asosiasi Pabrikan Modul Surya Indonesia	Apamsi
13	Indonesian Ocean Energy Association Asosiasi Energi Laut Indonesia	ASELI
14	Indonesian Network Microhydro Jejaring Mikrohidro Indonesia	JMI
III.	NRE BUSINESS ASSOCIATION SUPPORT	
15	New Renewable Energy Support Business Association Asosiasi Perusahaan Usaha Penunjang EBT Indonesia	Appebti
16	Association of Indonesia Automotive Industries Indonesia Gabungan Industri Kendaraan Indonesia	Gaikindo
IV.	NRE PROFESSIONAL ASSOCIATION	
17	Indonesian Association of Bioenergy Scientist and Technologist Ikatan Ahli Bioenergi Indonesia	IKABI
18	Indonesian Biofuel Expert Association Ikatan Ahli Bahan Bakar Indonesia	IABI
19	Indonesian Biodiesel Forum Forum Biodiesel Indonesia	FBI
20	Indonesian Solar Energy Engineering Association Ikatan Ahli Teknik Energi Surya Indonesia	IATESI
21	Indonesian National Committee on Large Dams Komite Nasional Indonesia Bendungan Besar	KNI-BB
22	Nuclear Power Experts Association Komisi Ahli Tenaga Nuklir	KATN
V	NRE COMMUNITY	
23	Indonesian Renewable Energy Society Masyarakat Energi Terbarukan Indonesia	METI
24	Indonesian Wind Energy Society Masyarakat Energi Angin Indonesia	MEAI
25	Indonesian Palm Oil Board Dewan Minyak Sawit Indonesia	DMSI
26	Indonesian Biogas Community Masyarakat Biogas Indonesia	MBI
27	Community for Energy and Environment	MPEL

	Masyarakat Peduli Energi dan Lingkungan	
VI	ENERGY CONSERVATION BUSINESS ASSOCIATION SUPPORT OR CARBON SERVICE	
28	Indonesian Association of Energy Conservation Asosiasi Perusahaan Konservasi Energi Indonesia	Apkenindo
29	Indonesian Lamp Manufacturers Association Asosiasi Produsen Lampu Indonesia	Aperlindo
30	Indonesian Electronics Association Gabungan Elektronika Indonesia	GABEL
31	Indonesian Electrical Industry Luminaries Association Asosiasi Industri Luminer Kelistrikan Indonesia	AILKI
VII.	PROFESSIONAL ASSOCIATION OF ENERGY CONSERVATION OR CARBON SERVICE	
32	Energy Conservation Expert Himpunan Ahli Konservasi Energi	HAKE
33	Indonesian Building Utility & Building Physics Engineers Association Ikatan Ahli Fisika Bangunan Indonesia	IAFBI
34	Indonesian Hotel Engineer Association Asosiasi Ahli Tekhnik Hotel Indonesia	ASATHI
35	Building Engineer Association	BEA
VIII	ENERGY CONSERVATION COMMUNITY OR CARBON SERVICE	
36	Energy Saving Society Forum Forum Komunikasi Masyarakat Hemat Energi	FKMHE
37	Green Building Council Indonesia	GBCI
38	Indonesian Business Council For Sustainable Development	IBCFSD

INDONESIA ENERGY STRATEGY - IMPORTANT DEVELOPMENTS

NATIONAL ENERGY POLICY - IMPORTANT DEVELOPMENTS

The basis for renewable energy development in Indonesia is the Presidential Regulation (Peraturan Presiden) no. 5/2006 on National Energy Policy (Kebijakan Energi Nasional, KEN). It sets a national target for the optimal energy mix in 2025 to be: (i) less than 20% from oil; (ii) more than 30% from gas; (iii) more than 33% from coal; (iv) more than 5% from biofuel; (v) more than 5% from geothermal; (vi) more than 5% from other renewable especially biomass, nuclear, microhydro, solar and wind; and (vi) more than 2% from liquefied coal.

In the same year, a Presidential Instruction (Inpres) no. 1/2006 was issued for biofuel development. The roadmap for biofuel has been prepared which aims to develop pro-growth, pro-poor, and pro-job biofuel projects in Indonesia. The government seeks to use investment in biofuel projects to provide a sustainable source of energy supply while at the same time helping to alleviate poverty and supporting national development.

The Energy Blueprint issued by the MEMR to guide the development of energy in Indonesia is based on Perpres 5/2006. The Blueprint includes a number of programs including a subsidy phase-out in the energy sector, introduction of new incentives, and introduction of a carbon tax. The Blueprint also includes roadmaps for energy development including solar, geothermal, and nuclear.

In 2004, MEMR launched the National Energy Policy (Kebijakan Energi Nasional) to provide strategic direction to energy development till 2020. That policy specifically included community empowerment to self-sufficiently develop energy supply especially in rural areas.

Utilize local energy resources in order to reduce the costs.

Develop the capacity of rural communities to manage rural energy services

Implement new approaches which promote the growth of rural electrification

Provide subsidies and other incentives for strategic action

Establish a new system for subsidy – “Social Electricity Development Fund”

ENERGY SUBSIDY REMOVAL PROGRAM

To reduce the burden of energy subsidies in the state budget, the government plans to increase the price of energy, particularly for oil and electricity. This is an important item in the economic restructuring program, which aims to mitigate the economic crisis, as detailed in PROPENAS (Program Pembangunan Nasional, National Development Program) 2001 Law.

Elimination of the oil subsidy will open the opportunity for developing other types of energy. Unsubsidized oil prices and electricity tariffs will help encourage industry and residential and commercial consumers to practice energy conservation. Renewable energy options, which have so far had difficulty penetrating the market because it cannot compete with subsidized oil prices, will become least-cost options in many locations.

OIL AND GAS LAW

The government has moved steadily to promote greater capacity and efficiency in the downstream oil and gas sectors. Oil and Gas Law No. 22/2001 stipulates that downstream oil and gas operations, including processing, transportation, storage, and marketing of oil fuel products shall proceed in a transparent manner, and that transparent pricing mechanisms based on market prices shall be established. It also ensures investors and other actors equal regulatory and legal treatment.

For the upstream sectors, oil and gas exploration and exploitation will be managed by an Executive Agency with Parliament approval. These activities will also be conducted in a transparent manner and private companies and state-owned enterprises (such as Pertamina) will have the same role in upstream activities.

REGIONAL AUTONOMY LAW

In order to strengthen democracy and pay more attention to community aspirations, regional autonomy laws have been enacted. Under Laws No. 22/1999 and No. 25/1999, regional governments will have a greater role in natural resource management in their respective regions, particularly in the mining and energy sectors. Each region will be able to adjust energy sector management to conform to its own socio-economic condition and the availability of energy resource in the particular region. The energy sector in Indonesia is dominated by four key policies and objectives:

Diversification: A key objective of the GOI is to reduce dependence on oil by expanding the use of coal, gas and renewable energy sources.

Rational Energy Pricing: The GOI recognizes that it can no longer sustain uniform pricing for electricity and petroleum products across the country, and has begun to eliminate subsidies.

Energy Sector Reform: The combination of decentralization of government decision making to give greater involvement to regional authorities and the need to attract capital for investment in the energy sector makes energy sector reform that introduces greater transparency to planning and decision-making a critical priority for the government.

Rural electrification: The GOI wants to bring electricity to 90 percent of the population by 2020.

The "Indonesia Energy Outlook 2010 Study" conducted in 2002 recognized the many problems the country faces resulted from the policy of subsidizing oil fuels adopted and maintained by the government over the past few decades. Prices were set lower than market prices to enable all strata of the community to easily procure. The policy has led to an approximately 75% reliance on oil fuels in the current national energy mix.

The oil price subsidy policy put other energy options at a financial disadvantage, hampering programs for energy conservation and energy diversification. Rural electrification programs developed based on diesel generators rather than lower cost renewable energy options. As domestic oil production declined and international oil prices increased, the economic and financial cost of rural electricity service jumped dramatically. The subsidy program also encouraged economic inefficiencies. Smuggling of oil fuels to other countries proliferated during the financial crisis in 1997 when the exchange rate of rupiah to US dollars was weak and the subsidized price of domestic oil fuels was below the international market price.

To reduce the burden of energy subsidies in the state budget, the government has committed to increase the price of energy, particularly for oil and electricity. Elimination of the subsidies will open the opportunity for developing other types of energy. Unsubsidized oil prices and electricity tariffs help encourage industry and residential and commercial consumers to practice energy conservation. Renewable energy options, which have so far had difficulty penetrating the market because they cannot compete with subsidized oil prices, will become least-cost options in many locations.

ELECTRICITY LAW

The financial viability of national utility Perusahaan Listrik Negara (PLN) was severely damaged by the economic crisis in the late 1990s and efforts to restructure the electricity sector were initiated in 1998. The restructuring aimed to separate the commercial, social and regulatory functions of PLN. Electricity Law No. 20 was passed in 2002 to facilitate the liberalization and privatization of the electricity sector by introducing a multi-sellersmulti-buyers system in which private investors could sell power to PLN which would control the transmission and distribution system. However, on December 15, 2004, the Constitutional Court decided that:

> The Electricity Law No. 20 of 2002 was not in line with Article 33 of the Indonesian Constitution of 1945 and it was annulled.
>
> All contracts issued based on Law No. 20/2002 are still valid until the end of the contracting period.
>
> The Electricity Law No. 15/1985 has been reenacted to replace the Electricity Law No. 20/2002.
>
> The GOI has to draft a new electricity law in line with Article 33 of the Indonesian Constitution of 1945.

The largest implication of this decision is to return the electricity sector to a monopoly with centralized control. With the liberalization of the energy sector, the GOI gave the Ministry of Energy and Mineral Resources (MEMR) more responsibilities to oversee market activities especially in the oil and electricity markets. In addition, there is greater emphasis on transparency and decentralization in response to delegation of greater authority to regional governments in managing energy resources in their respective regions.

Table 1. Relevant Decrees, Regulations, Policies and Laws Affecting the Indonesian Energy Sector

Law	Purpose
Presidential Regulation No.5 on National Energy Policy (2006)	• Set energy diversification targets for 2025 that include 5% biofuels, 5% geothermal and from other renewables. • For energy conservation sets target to reduce energy intensity by 1% per year.
Blue Print of National Energy Implementation Program 2005-2025 (2005)	• Delineate measures for Indonesia to enhance energy supply security. • Provides development “roadmaps” for various sectors • Describes programs to phase our subsidies and improve of energy efficiency.
	• Supports law No. 15/1985 on electricity, which was reenacted late 2005 following Constitutional Court ruling that annulled Law No. 20/2202 on electricity.

Government Rule No. 3	• Private sector must partner with PLN to develop electricity
year 2005 on supply of	projects.
Electricity (2003)	• However, companies generating power for their own use or those using renewable energy can set-up plants independently without having to partner with PLN.
Oil and Gas Law No. 22 (2001)	• Stipulates that downstream oil and gas operations, including processing, transportation, storage, and marketing or oil fuel products shall proceed in a transparent manner. • Pricing mechanisms are to be based on market prices • Provides investors equal regulatory and legal treatment.
	• Seeks to attract private investments in exploration. • Allows an additional 20 percent cost recovery for
MEMR Regulation 8 (April 2005)	developers of marginal oil and gas fields. • Offers more favorable split to contractors for oil and gas development in the 2005 round of bidding.
	• Reduce burden of energy subsidies in the state budget,
PROPENAS 2001 Law	
(2001)	• Increase the prices of energy, particularly for oil and electricity.
Law No. 27 on Geothermal Power (2003)	• Provide certainty of law to the geothermal industry. • Defines new rules under which exploration and development can occur using a competitive bidding process to grant three-year licenses for exploration with possible two year extensions to complete feasibility studies.
Presidential Decree No.1 on Supply and Use of Biofuels (January 2006)	• Set targets for biofuels utilization.
Ministerial Decree No.	• Requires PLN to purchase electricity generated from
1122/K/30/MEM on	renewable energy sources by non-PLN producers for
Small-Scale Power	projects of up to 1 MW capacity.
Purchase Agreement	• Institutions eligible to participate are cooperatives, private
(June 2002)	and government companies. • Purchase tariffs will be calculated at 80% for medium voltage and 60% for low voltage of PLN's announced "Electricity Base Price" which is supposed to be its marginal production cost at the location where the plant is to be built.
Ministerial Regulation No. 2 on Medium Scale Power Generation from Renewable Energy Sources (January 2006)	• Extends the same price guidelines as MD No. 1122/K/30/MEM for projects from 1 MW to 10 MW • Sets a minimum contract period of ten years.
Ministerial Decree No. 141/Men-LH (2003)	• Make motor vehicles compliant with Euro 2 by January 2005 for new types of motor vehicles and by January 2007 for current motor types.
Ministry of Environment Mandatory Disclosure of Automotive Emissions Program (MDOAE) (2003)	• Complements Ministerial Decree No. 141/Men-LH • Help implement Clean Air Act of Indonesia • To support development and sale of environmentally friendly vehicles and prevent import of "dirty" vehicles. • Vehicle manufacturers must have their vehicles tested to measure emissions of CO, HC+NOx, and particulates using procedures developed by the program.

Regional Autonomy Law Laws No. 22 and No. 25 (1999)	• Give regional governments greater role in natural resource management in their respective regions • Each region can adjust energy sector management to conform to its own socio-economic condition and the availability of energy resource in the particular region.

NATIONAL PROGRAMS AND TARGETS FOR RENEWABLE ENERGY

Renewable energy projects have received significant attention from policy makers in Indonesia. The electricity regulations for small- and medium- scale power generation from renewable energy resources described in the table above specifically seek to increase the use of renewable energy. Other electricity regulations for selling power to the grid include:

Ministry Regulation No. 001/2006 (Revision of Regulation No. 09/2005) on Procedures for Electric Power Purchase and/or Transmission Line Leasing for Public Electricity Supply

Ministry Regulation No. 010/2005 on Procedures for Electric Power Business Licenses for Inter-Provinces, Regions, or for National Grid Connections

Ministry Regulation No. 1122K/30/MEM/2002 on Scattered-Small Power Generation Using Renewable Energy

Ministry Regulation No.1213K/31/MEM/2005 on National Electricity General Plan (RUKN)

Ministry Regulation No. 2059K/31/MEM/2005 on Establishment of PLN's Power

Supply Plan for 2006-2015. Despite efforts to clarify procedures and terms of sale, negotiation of power purchase agreements with PLN that achieve acceptable rates of return on investment for renewable energy projects remains challenging even in locations where the renewable option is significantly cheaper than the conventional diesel generators used by PLN.

RENEWABLE ENERGY POLICIES

GREEN ENERGY POLICY

To encourage development and utilization of renewable energy and to improve efficiency on energy utilization, a "Policy on Renewable Energy and Energy Conservation" – also called the "Green Energy Policy" was promulgated by the Ministry of Energy and Natural Resources on 23 December 2003. This policy provides the reference for renewable energy development and energy conservation in Indonesia to support sustainable development.

Under the Green Energy Policy, renewable energy in Indonesia has been classified into three types: (a) already developed commercially (biomass, geothermal and hydro energy); (b) already developed but still limited (solar, wind); and (c) still at the research stage (ocean energy).

The Green Energy Policy defines the following renewable energy strategies:

Determining energy price in accordance with the economic aspect;

Encouraging the development of renewable energy and energy conservation infrastructure which is supported by the society;

Prioritizing the use of renewable energy;

Implementing energy saving principles in energy management;

Applying energy saving life style;

Increasing the role of stakeholders in the utilization of renewable energy and energy conservation;

Improving cooperation at the national, regional, and international levels especially in the context of information access, financing, and transfer of technology;

Encouraging the use of domestic services and goods in the field of renewable energy and energy conservation;

Improving the quality of human resources in the field of renewable energy and energy conservation;

Improving the renewable energy and energy conservation supporting ventures in the country;

Improving the society's access to energy;

Formulating the priority scale of renewable energy development based on available potential and technology, financial, and social feasibility.

The Green Energy Policy defines actions steps consisting of formulation of more specific policies and programs. These include policies for: (a) investment and funding; (b) incentives; (c) energy pricing; (d) human resources; (e) information dissemination; (f) standardization and certification; (g) research and development; and (i) institutional development.

DISTRIBUTED SMALL SCALE POWER GENERATION (PSK TERSEBAR)

In June 2002, MEMR issued a new regulation for small renewable energy projects interested in selling power to PLN known as PSK Tersebar (Ministerial Decree on Small-Scale Power Purchase Agreement (MD No.1122/K/30/MEM/2002). The regulation requires PLN to purchase electricity generated from renewable energy sources by non-PLN producers for projects of up to 1 MW capacity. Institutions eligible to participate are cooperatives, private and government companies. Purchase tariffs will be calculated at 80% for medium voltage and 60% for low voltage of PLN's announced "Electricity Base Price' which is supposed to be its marginal production cost at the location where the plant is to be built.

The first project was completed in December 2003 with PLMTH Kalimaron, a community based organization in Seloliman, East Java, which owns a 25 kW micro-hydro power plant. PLN agreed to buy its excess generation of 12 kW. This provided extra earning of IDR 3,000,000 (US$300) per month. The following shows the other plants operating under PSK Tersebar as of September 2005.

Table 2. Plants operating under PSK Tersebar in 2005

	Installation	Price	Comments
1	Micro Hydro Curung Agung, 12 kW, Subang, West Java	Rp 432/kWh	60% marginal production cost for low voltage Yearly extendable contract
2	Micro Hydro Cinte Mekar, 100 kW, Subang, West Java	Rp 432/kWh	60% marginal production cost for low voltage Yearly extendable contract
3	Micro Hydro Waikelosawah, 14 kW, West Sumba, NTT	Rp 218/kWh	Not in line with PSK Tersebar In process of adjusting price
4	Micro Hydro, 25 kW, Mojokerto, East Java	Rp 341/kWh	80% marginal production cost for medium voltage Marginal production cost reduced from first year

Despite the regulation, it remains challenging to negotiate agreements. The following table shows additional projects under negotiation (DGEEU, 2005). Although the number of projects participating in the program has increased, the extendable one year contracts make it difficult to obtain financing from financial institutions. Program growth is also constrained by the lack of clarity in how to calculate the tariff and the need to negotiate each deal with the local PLN office where the project will be located.

Table 3: Plants in Negotiation for PSK Tersebar Contracts in 2005

	Installation	Price	Comments
1	Micro Hydro Dompyong, 25 kW, Trenggalek, East Java	Rp 302/kWh	60% Marginal production cost for low voltage Yearly extendable contract
2	Micro Hydro Santong, West Lombok, NTB		Low voltage Rehabilitation of existing plant
3	Micro Hydro Anggrek Mekar Sari, 900 kW, Pasaman, West Sumatra		Medium voltage Under construction
4	Micro Hydro Kalumpang, 1 MW, Luwuk, Central Sulawesi	Rp 470/kWh	Medium voltage Commissioning test Price not in line with PSK Tersebar
5	Micro Hydro Hanga Hanga, 2x1 MW, Luwuk, Central Sulawesi	Rp 470/kWh	Medium voltage Price not in line with PSK Terseber 20 year contract
6	Micro Hydro Wangan Ali, 120 kW, Wonosobo, Central Java		Medium voltage Under construction
7	Micro Hydro Ulu Danau, 224 kW. South Sumatera		Medium voltage Under construction
8	Micro Hydro Cipayung, 250 kW, Bogor, West Java		Medium voltage Rehabilitation
9	Micro Hydro Cileunca, 900 kW, Pangalengan, West Java		Medium voltage Land Acquisition
10	Micro Hydro Curug Gede, 120 kW, Luwuk, Central Sulawesi	Rp 470/kWh	Medium voltage Commissioning test Price not in line with PSK Tersebar 20 year contract
11	Wind TurbineNusa Penida, 80 kW, Bali		Medium voltage Under construction
12	Wind Farm, 960 MW, Soe, NTT		Medium voltage Documentation in process

In January 2006, MEMR issued Ministerial Regulation No. 002/2006 on Medium Scale Power Generation from Renewable Energy Sources that applies the same price guidelines for projects from 1 MW to 10 MW and sets a minimum contract period of ten years.

BIOFUELS

In 2006, the government of Indonesia launched a plan to rapidly develop a biofuels industry and pledged to invest a total of Rp 13 trillion (US$1.4 billion) to encourage the development of the country's fledgling biofuel industry. They plan to allocate nearly 500,000 hectares of land for new biofuels development in 2007. Under the plan, Indonesia will use palm and jatropha as the primary feedstocks for biodiesel, and sugarcane and cassava as the primary feedstocks for ethanol. A national standard for biodiesel was established in 2006. Following successful tests in 2006 by PLN, Indonesia also plans to use an 80% blend of pure palm oil with diesel to replace diesel for power generation in power stations off the main grid that now use diesel. Because of the minimal processing, pure palm oil is less expensive to produce than biodiesel. Indonesia also and plans to develop a Bio Oil product to displace household uses of fossil fuels. The following table gives production targets through 2025.

Table 4: Biofuel Production Targets through 2025

Type of Energy	Unit	2010	2015	2020	2025
Biodiesel	kl	720.000	1,500.000		4,700.000
Gasohol	kl	550.000	850.000	1,500.000	
Bio oil	kl	400.000	700.000		900.000

Biodiesel: Biodiesel is made by combining vegetable oil with an alcohol through a process called transesterification, and can be made from a range of feedstocks. Pertamina began distributing a 5 % blend of biodiesel in May 2006 through 197 retail outlets in Jakarta, adding 12 retail outlets in Surabaya in August. Indonesia is the world's second-largest palm oil producer. Currently, two companies supply the biodiesel being used by Pertamina. The Indonesian company PT Bakrie Sumatera Plantations Tbk announced plans to build the country's first modern biodiesel plant in a joint venture with the construction firm PT Rekayasa Industri. The plant, projected to cost $25 million, would add capacity to produce 60,000-100,000 tons of biodiesel per year. Indonesia, which consumes 27 million kilolitres of diesel oil per year, plans to expand production to 4.7 million kilolitres by 2025. The figure below shows the Road Map for Biodiesel development. The plan does not consider other promising pathways for the production of biodiesel.

Feedstock represents 70-80% of the total cost for biodiesel. When the Indonesian biodiesel program was launched, the price of palm oil was low enough that biodiesel could be produced for less than the price of diesel. Oil prices have declined slightly and disruptions in international vegetable oil markets have pushed up prices for palm oil so that the cost to produce biodiesel at the end of 2006 is greater than the cost of diesel.

Ethanol Fuel: Ethanol can be made from a variety of sugars and starches with a fermentation and distillation process. In most countries, ethanol has been introduced to markets as a straight blend with gasoline. Most automobile manufacturers warranty their cars for blends with up to 10 % ethanol. Ethanol has some solvent characteristics that in higher concentrations can cause corrosion of certain metals and deterioration of some plastics and rubbers used in gaskets and hoses. Ford and General Motors both manufacture flexible-fuel versions of some of their vehicles that can accommodate up to 85% ethanol. However, the solvent characteristics of ethanol can cause greater problems in the ethanol distribution and storage infrastructure.

Road Map on Biodiesel

Figure x: Road map on Biodiesel

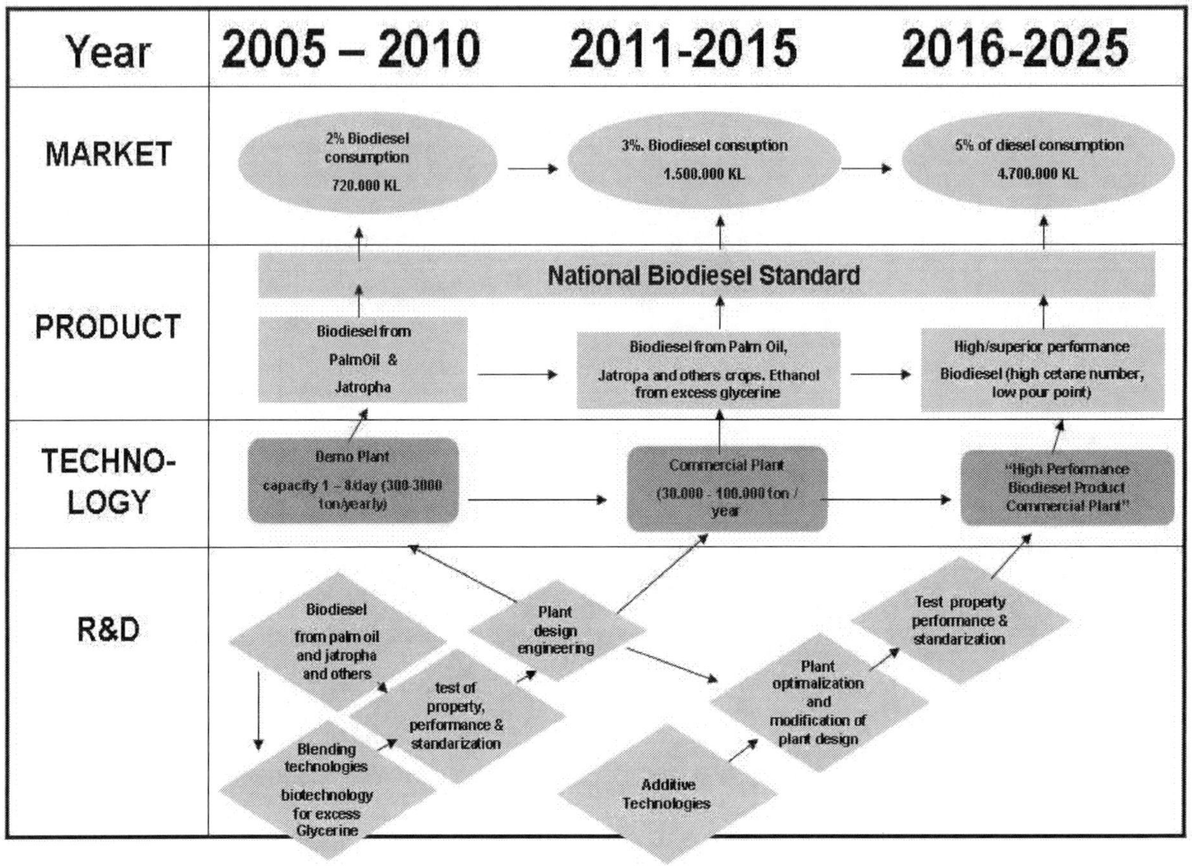

Progress with the introduction of ethanol blends in Indonesia has been slower. Pertamina has begun sales of an ethanol blend at one station in Malang. The figure below provides the Road Map for Bioethanol in Indonesia. Notably, it does not address distribution and storage issues for ethanol fuels. Although, the fermentation process has been known for thousands of years, significant improvements have been made in recent decades that reduce the costs and energy requirements of the process. New enzymatic hydrolysis processes hold the promise of producing ethanol from lignocellulosic materials. The Road Map for Indonesia does include research on lignocellulosic paths to the production of ethanol in later years.

For ethanol, feedstocks represent 50-70 % of the total cost depending on the feedstock selected. Brazil is the lowest cost ethanol producer in the world and believes it can produce ethanol to compete with crude oil at as little as \$35 per barrel. The estimate of the price at which ethanol from corn becomes competitive without subsidies in the United States is \$55 per barrel of crude.

Road map on Bioethanol

Figure x: Road map on Bioethanol

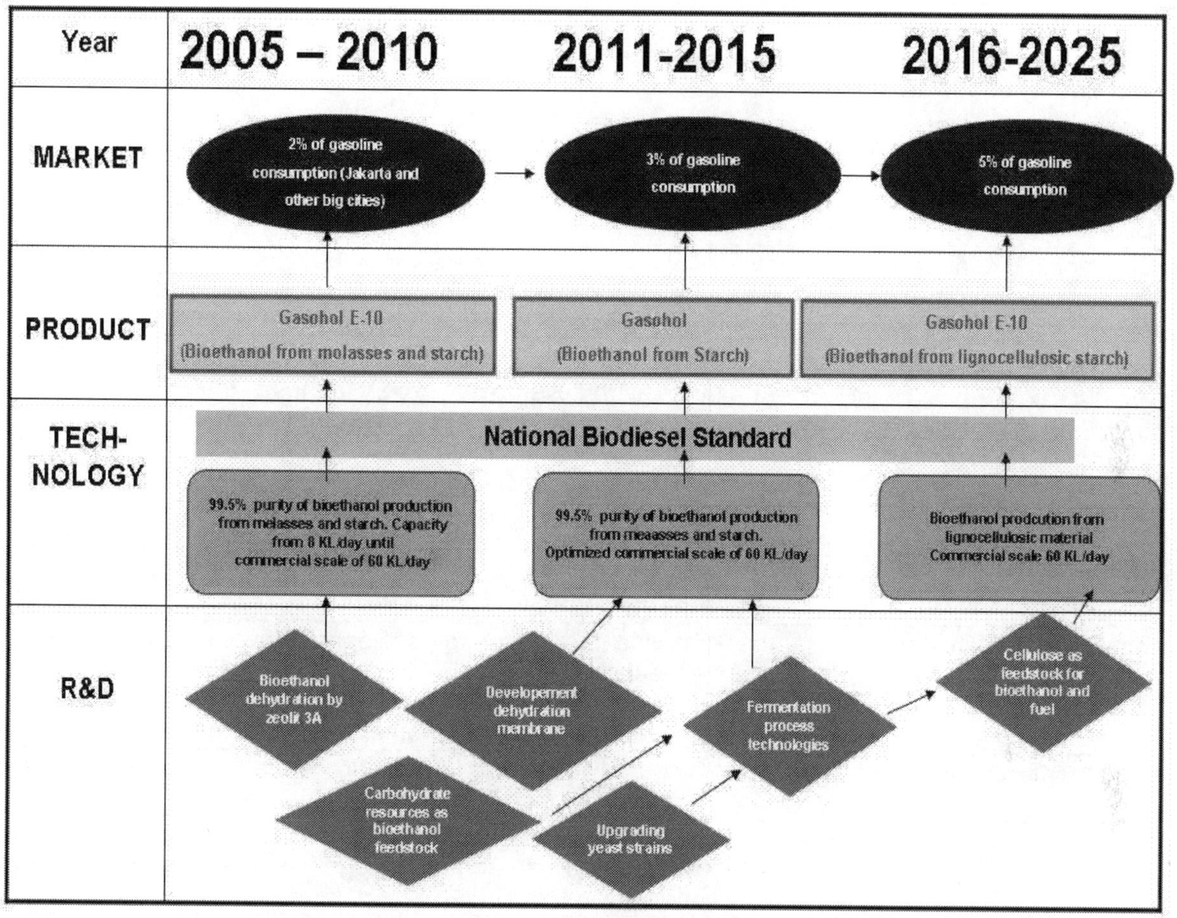

Pure Palm Oil: Pure vegetable oil can be burned directly in most diesel engines saving the cost of transesterification. Because the viscosity of pure vegetable oils increases as temperatures decrease, it is not commonly used as a direct fuel in temperate climates. Different oils have different characteristics and some (e.g. coconut oil) blend more easily with diesel fuel than others. In Indonesia's climate, it should be possible to use pure palm or jatropha oil in diesel engines with minor engine modification. PLN has been actively experimenting with the use of pure palm oil for its remote diesel generators and announced in December 2006 that it would expand the program to other sites. (Jakarta Post, 2006) It has asked regional mangers to survey their regions for potential local suppliers who may be willing to provide pure palm oil under contract to PLN.

NATIONAL POLICY FOR BIOFUELS

Introduction of biofuels can have a broader impact on national economic development than reducing fuel costs and emissions. In Indonesia, plans call for both biodiesel and bioethanol to be produced from agricultural crops. Achieving the production targets set out in the Road Maps will require investment to increase yields of existing crops and to expand planted area. It can also produce higher revenue per hectare and create jobs in rural areas where the majority of Indonesians live.

Presidential Instruction on Biofuels issued in 2006 underscores the need for a multisectoral approach to biofuels development and commercialization and illustrates the complexity of implementing a successful program. The Presidential Instruction specifies roles for eleven different ministries in biofuels development as outlined in Table x. A special budget of $11.2 million for 2006 was allocated to four ministries: Agriculture, Industry, Research and Technology and Energy and Mineral Resources.

Roles Assigned Collaborating Ministries for Biofuels Develoment

Ministry	Roles
Coordinating Minister for Economy	Coordinate the preparation of the Implementation Plan for the utilization of biofuel as a replacement for other fossil fuels (petrol and diesel in particular)
Ministry for Energy and Mineral Resources	Coordinate the compilation of policy incentives for biofuel utilization developers (except in those where the raw material to be used to make biofuel is already commercial) Compile and provide standard for biofuel quality Design a system and simple procedure to test biofuel Prepare simple licensing & business regulations for biofuel to fit in to the existing fossil fuel regulations. Coordinate the utilization of biofuel through the Coordinating Agency for National Energy (BAKOREN)
Ministry of Agriculture	Provide information to developers/growers of biofuel crops especially on critical land Facilitate supply of seeds and seedlings Push for the integration of growing and after harvest processing of biofuel
Ministry of Forestry	Provide licenses for forestry areas that are not productive and can be used for growing biofuel crops
Ministry of Industry	Increase the development of local facilities for biofuel production Push and encourage the development of biofuel processing industry
Ministry of Transport	Facilitate the increased utilization of biofuel in this sector
State Ministry of Research and Technology	Coordinate research and development of growing, processing and utilization of Biofuel
State Ministry of Cooperatives and Small Enterprise	Help and push cooperatives and small and medium enterprises to participate in growing and processing biofuel crops
State Minister of State-Owned Companies	Push state companies in agriculture and plantations to participate in growing and processing
	biofuel crops Push state companies in the industry sector to develop biofuel processing plants Push state companies in the field of engineering to develop biofuel processing technologies
Ministry of Interior	Coordinate with the Governors and Regents to facilitate the availability of land in their respective areasfor the growing of biofuel crops.
Ministry of Finance	Provide fiscal incentives for the biofuel sector, growers and biofuel processing industry

LAND CLEARANCE AND BIODIVERSITY

Increasing demand for biofuels as countries seek alternatives to diesel and gasoline will require identification and development of new feedstocks. Indonesia announced a plan in 2006 to set aside an additional 500,000 hectares of land for development of palm oil plantations. Several NGOs have expressed concern that development of biofuels could have negative impacts on the environment and on the people living in areas targeted for biofuels development. Environmental impacts include increased pressure on forests and other areas of critical biodiversity, expanded use of fertilizers and pesticides, and effluents from the biofuel processing facilities. Social impacts include pushing local people, especially the poor, off of the land they depend upon for survival and exploitative labor practices.

Recent studies by WWF, Friends of the Earth, and Greenpeace have highlighted the environmentally and socially destructive nature of the palm oil industry in South East Asia, one of the leading suppliers of feedstock oils for the biodiesel industry. While most palm oil is used in the food processing, cosmetic industries, and other non-fuel industries, anything that increases the demand could exacerbate the problem.

From a base of 120,000 hectares in 1968 there are now around 5 million hectares of land in Indonesia given over to palm, most of it owned by a few large producers. At least 19 million hectares more have been earmarked for plantations by provincial governments, while the central government has said there are 27 million hectares of 'unproductive forest land' available for development. Land clearance not only has implications for the health and sustainability of ecosystems, but also for climate change. Bringing forest, peatland or grassland under cultivation releases CO2 into the atmosphere.

RENEWABLE ENERGY DEVELOPMENT INITIATIVES

The Blue Print for the National Energy Plan 2005 – 2025 sets out targets for renewable energy development. Table shows the targets through 2025. Indonesia has successfully developed a wide range of renewable energy applications.

Targets for Renewable Energy Development through 2025

Type of Energy	Unit	2010	2015	2020	2025
Geothermal	MW	3,442	4,600	6,000	9,500
Wind	MW		25.6		255
PV	MW	25	50	70	80
Micro Hydo	MW	169	298	488	700

SOLAR PHOTOVOLTAICS APPLICATION PROJECTS

Demonstrations and Pilot Projects: The first rural electrification in Indonesia using photovoltaic cells packaged as solar home systems was in Sukatani village, Java, where 85 SHS (solar home systems), public lighting systems and 15 streetlights were installed in 1989. The program was funded by the Dutch aid agency DGIS, and involved the BPPT, the Ministry of Co-operatives, the local government and the international and local branch of the PV manufacturer "R&S" (now Shell Solar). The success of the Sukatani project led to a similar rural electrification scheme in the village of Lebak, in which a further 500 45Wp SHS were installed. By the end of 1996, government sponsored solar PV projects and private initiatives had resulted in the installation of over 1 MWp of PV capacity in rural areas.

These pilot programs were followed by the Presidential Aid Programme (BANPRES), which provided interest-free credit for 3,000 SHS. Programs to disseminate solar home systems in Indonesia have faced difficulties with cost recovery as they are often initiated by donors or government agencies rather than as a response to a market demand. Consequently, they can be perceived as entitlements rather than commercial transactions (see more detailed BANPRES description below).

PV systems were first demonstrated in Indonesia in 1979 through a water-pumping project carried out by the German Aid Agency, GTZ, with the assistance of the Indonesian government. Since 1993, the Department of Health had installed 270 medical clinics with PV-powered lighting and refrigerators, requiring around 75,000 PV modules. Eight hybrid mini-grids had also been installed in eastern Indonesia (Nusa Penida). The hybrid generators used a meter prepayment system, which aimed to match customer consumption with ability to pay.

Additional major projects included the GTZ Eldorado Sun, a project funded by the German Aid agency GTZ that developed solar pumping projects on four islands, and a project funded by the Australian Aid agency, AusAid that aimed to install 36,400 50Wp SHS.

In 1997, the GOI set a target for 50 MWp of PV by 2005, to be installed across 6,000 of the 13,000 islands of Indonesia. In the same year, the World Bank and the Global Environment Facility (GEF) launched a project to assist the Government of Indonesia (GOI) achieve its one million SHS target, providing funds for the installation of 200,000 PV systems (10 MWp) for households and commercial establishments in West Java, South Sulawesi and Lampung, Sumatra. The plan was to concentrate on 50 Wp solar home systems. Solar equipment was to be exempted from import tax and duties if equivalent products were not manufactured in Indonesia.

Unfortunately, the financial crisis hit Indonesia starting in 1997 reducing the affordability of 50 Wp systems and although the program was modified, only $0.1 million of the $20 million had been used by December 2000 and the World Bank loan was closed early. The GEF grant was reconfigured. By 2003, a total of 8,054 solar home systems had been installed.

Promoting a Domestic PV Industry: In recognition of the potential market in Indonesia, all of the major international photovoltaic power system manufacturers, including BP Solar, Kyocera, Shell Solar, Siemens and Solarex have a subsidiary or local distributor in Indonesia. They seek to maximize the use of local components and employ local labor for installations in order to keep costs down. The primary weaknesses in the Indonesian solar industry are the lack of local manufacturing facilities for high quality batteries and for photovoltaic modules.

The government actively supports development of solar PV applications. The lead institution for PV testing and standards has been the BPPT. BPPY has developed several laboratories, including equipment to test electronic control devices and a computer-based data acquisition system, as well as PV pumping and battery test facilities. BPPT closely monitors the standards and performance of systems installed under the existing SHS projects and has worked with the private PV industry to develop standards for SHS. The BPPT lab has obtained ISO17025 accreditation for testing PV components and has carried out testing for PV systems used in other countries.

Evolving Financing Mechanisms: The Indonesian government identified the lack of appropriate financing as an impediment to increased use of photovoltaic systems for rural electrification.

BANPRES was created to address this shortcoming by providing interest free loans through village level Local Cooperatives (KUDs) in 13 provinces. Purchasers were required to make an initial or down payment of 5 % of the 400 USD cost, followed by payments of approximately 3 USD per month for ten years. This was equivalent to the monthly fee if connected to the grid.

Although the program led to the purchase of around 3,300 systems, it did not succeed as a financing program. Consumers expressed the need to make the down payment by installments over several months and by 1993, BANPRES fee collection was in arrears by 40 %. A major problem has been the need to replace batteries, the high cost of which often resulted in the purchase of inferior technology and poor system performance further hampering repayments. Defenders of the program maintained that the targeted end users did not have sufficient income for repayment

Subsequent government projects focused on end-users that could pay for the systems on a consumer-financing basis. Three major local dealers - Pt. Sudimara, Pt. Kyocindo and Pt. Walet - offered short term credit (3 years) at around 18 % interest with a 30 % down payment. Typical monthly fees of around 8 USD were required. In 1995, Sudimara reported system sales at 400 USD plus 10 % VAT. Around 10 % to 15 % of sales were cash. Credit terms were 140 USD down payment and 40 monthly payments of 10 USD interest at 1.5 % a month. These were being sold through regional service centers (with a potential market of approximately 10,000 to 20,000 families per service center) in three provinces. One PV manufacturers has focused on the outer provinces, where publicly funded programs usually do not reach. The company has sold thousands of units over the past decade. In general, payment plans lasting longer than two years do not seem to work because of the problems with battery failure, at which time collection rates fall sharply.

In general, successful companies emphasize the development of strong local dealers including the provision of supplier credit for dealers. as more crucial for successful marketing and distribution. Many systems have sold for cash, proving that lack of effective dealers may be more of an immediate constraint than the absence of special financing mechanisms to SHS users. Accelerating market growth will require mechanisms to expand affordability for lower-income customers

BIOMASS FOR POWER GENERATION [1]

Biomass is the traditional and oldest energy source in Indonesia, accounting for around 35 percent of total energy consumption. Most biomass is used to meet energy needs for cooking and small rural enterprise. Many forest products and agricultural processing companies use their wastes to supply heat for process needs. Several also generate electricity with total installed power capacity of 302 MW (Agus 2006). New projects are being developed including a 30MW plant in east Kalimantan and a 10.3 MW plant in Northern Sumatera.

National studies estimate the potential electricity generation potential to be about 50 GW from the roughly 150 million tons of biomass residues produced per year equivalent to roughly 470 GJ/year. However, these studies estimate the main source of biomass energy in Indonesia to be rice residues with a technical energy potential of 150 GJ/year. Indonesian policy requires rice processing to occur close to where the rice is grown so most rice processing occurs in small decentralized plants. Rice husks have low bulk density and are expensive to transport so the technical potential from rice residues overstates the economically viable potential.

[1] Asia Pacific Energy Research Centre, "New and Renewable Energy in the APEC Region – Prospects for Electricity Generation" – Institute of Energy Economics, Japan – www.ieej.or.jpapercpdfnre_report2004.pdf.url

Other potential sources are rubber wood with 120 GJ/year, sugar residues with 78 GJ/year, palm oil residues, 67 GJ/year, and the rest with smaller than 20 GJ/year are from plywood and veneer residues, logging residues, sawn timber residues, coconut residues, and other agricultural wastes. Development of biofuels will likely expand planted areas for some crops and increase the use of residues to provide the heat and power needed for biofuel production.

Wood processing industry. Although there are over 400 sawmills producing wood residues ranging from 10,000 and 50,000 m^3/year, very little has been used to generate electricity. Most sawmills generate power with diesel generators. Plywood factories use wood wastes to generate process steam needed for dryers and hot presses so they are familiar with the operation of boilers but most do not generate power. An estimated 40 plywood factories produce less than 50,000 m3/year of wastes while 70 plants produce between 50,000 to 200,000 m3/year. The ASEAN COGEN programme supported development of a 5.5 MW wastewood power plant at T Siak Raya Timber in Pekanbaru, Sumatra. Earlier feasibility studies and demonstration projects show favorable returns on investment for biomass cogeneration. With the elimination of diesel subsidies, many of these projects will have very high rates of return.

Sugar industry. At present, there are 56 sugar mills in operation in Indonesia. with a large potential to produce excess electricity from bagasse. Three co-generation plants utilizing bagasse with a combined total capacity of 25 MW are planned in Java (18MW), Sumatra (6MW) and Sulawesi (1MW). Another 12 bagasse projects are planned with a total combined capacity of 104 MW. These plants are to be located in Sumatra (38 MW), Sulawesi (13MW), Kalimantan (4 MW) and Java (29 MW). Some 100 MW of (mostly biomass and bagasse co-generation) projects have been identified for power generation and supply to the grid, under the PSKSK (SPPA) arrangements with the utility purchasing the excess power at fixed rates.

The GOI has an active policy to relocate sugar cane fields from East Java to Kalimantan, Sumatra and Sulawesi. Setting up mill to process the newly developed sugar cane plantations outside Java offers an immediate opportunity to install new cogeneration systems, some in locations that currently depend heavily on diesel generators for power supply. In these locations, financial rates of return for investments in cogeneration will also be high.

Financing biomass cogeneration: Although financial rates of return will be high for cogeneration investments at plywood and sugar factories, it will be difficult to mobilize private capital for investment. Traditional lenders in these sectors are not familiar with energy investments. Capital requirements are high compared to the investment made in processing plant equipment and cost recovery normally occurs over longer time periods.

GEOTHERMAL ENERGY

The nearly 200 volcanoes and 100 geothermal fields make Indonesia a promising area for geothermal power generation. As of 2002, there had Geothermal development started in Indonesia as early as the 1920s with its first shallow steam wells drilled at the Kamojang Province. The field, which produced 250 kWe in 1979, has been producing 140 MWe from dry steam since 1985. Expectations for geothermal power development were high in the early 1990's when independent power contracts were awarded for eleven projects with a total committed capacity of 3,417 MW. The financial crisis in 1997-98 led to the suspension of seven geothermal projects. As of 2005, Indonesia had developed 802 MW out of an estimated potential 30,000 MW.

The primary barriers to development of Indonesia's geothermal resource has been pricing for electricity and a shifting regulatory environment. Geothermal resources were managed by Pertamina until 2000 when they were transferred to regional authorities. In October 2003, the GOI passed a new Geothermal Law No. 27/2003 to promote a more favorable business environment for geothermal. Under the law, geothermal exploration and development are conducted under a license following competitive bidding on prospective sites. They have three years to conduct exploration (with a possible two-year extension) and then two years to complete a feasibility study.

Significant progress was made to develop geothermal projects in 2005 and 2006 with constructions started on a 110 MW plant being jointly developed by Pertamina and Cheveron Texaco and commitments made for 300 MW to a joint venture between PLN and Pertamina in North Sumatera. Several smaller plans also began development. Local company Star Energy announced in November 2006 that it had found what it claims is the world's largest geothermal steam well at its Wayang Windu concession in West Java. The company said the discovery would allow them to expand the Wayang Windu geothermal plant to 400 MW by 2010. The Wayang Windu geothermal power plant started operation in early 2000. Star plans to bring the second unit online in 2008, adding 110 MW to the power plant's existing capacity of 110 MW (US Embassy 2006). The Blue Print projects geothermal capacity will grow to 9500 MW by 2025. Figure 3 shows the distribution of geothermal potential across Indonesia.

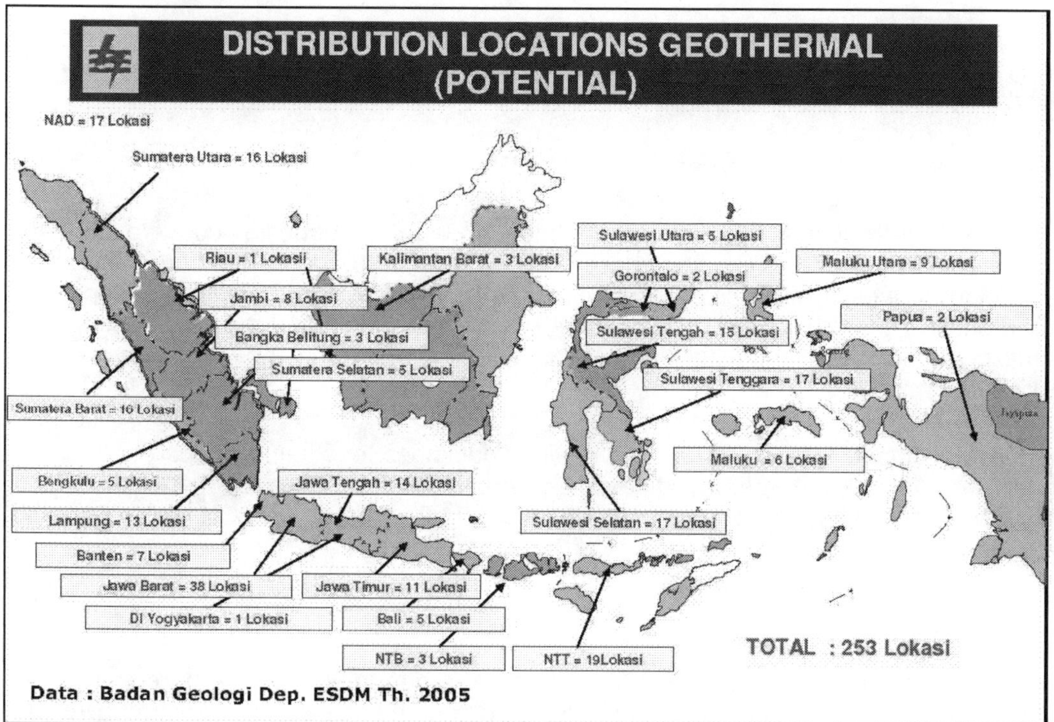

NAD = 17 Lokasi
Sumatera Utara = 16 Lokasi

DISTRIBUTION OF GEOTHERMAL POTENTIAL ACROSS INDONESIA

Pricing will continue to be the biggest challenge for geothermal projects to overcome. The surge in development in the 1990s was stimulated by electricity prices between $0.069 and $0.085. PLN is seeking to pay prices under $0.05/kWh.

In general, geothermal projects require the more capital to develop than other projects. The project developer must invest in drilling to map and evaluate the quality of the resource prior to the investment in capital to build the plant. Of course, the higher capital costs are offset by the zero fuel cost (unless you treat the resource mapping as an up-front fuel cost) but it will be hard for private developers to move forward with these projects without greater confidence in the long-term pricing for electricity.

SMALL HYDROPOWER

Indonesia has abundant hydro resources and success with development of micro. mini, small and large plants over the past decades. Hydro contributed 12.9 TWh to total electricity supply in 2005. Official data estimates the national hydro potential at about 75 GW with small hydro project accounting for only 500 MW. Because of the processes used for these estimates, the small hydro potential is likely larger than 500 MW although still a small percentage of total potential.

The current installed capacity for all hydro plants is about 4,260 MW of which small hydro makes up about 64 MW. There is a significant portfolio of small hydro projects in planning or under construction. Hydro sites under 10 MW will likely receive significant private interest following the change in the law requiring PLN to purchase power from plants up to 10 MW. Financing will be the major challenge because of the difficulty in negotiating long-term power purchase agreements from PLN.

Hydro systems with capacities below 500 kW are an important component of community-led rural electrification efforts and for meeting power needs of small rural industries.

a) German-Indonesian Mini Hydro Power Project (MHPP)

The Mini Hydro Power Project (MHPP) is a continuation of German-Indonesian cooperation in the mini-hydro sector started in 1991. The first phase of MHPP (1991-96) focused on the introduction of MHP technology to local institutions and individuals active in micro hydro project development. In the second phase (1999-2002), the scope of intervention was broadened to include policy dialogue, scaling up MHP technology packages, and improving operation and management practices applied..

The project has developed a variety of MHP technology packages together with local manufacturers in Java and Sumatra, which have been applied in more than 100 installations over the past 10 years. These schemes presently supply over 20,000 families with a clean sustainable source of energy.

b) IBEKA Mini Hydro Projects

Institut Bisnis and Ekonomi Kerakyatan (IBEKA) or People Centered Business and Economic Institute is an NGO involved in renewable energy development, particularly micro-hydropower systems. IBEKA's goal for their micro hydro development projects is to empower the community. IBEKA has assisted communities to build small hydro power plants that produce excess electricity, which is sold to the national grid.

COMMUNITY-BASED INTEGRATED RENEWABLE ENERGY PROJECTS

There have been initiatives to provide energy through RETs that integrate energy development with the overall rural development strategies in target communities. Table 7 lists several community-based projects.

Community Based Projects

Program	Activities
UNDP-GEF-SGP Community Based RET Projects	Demonstrate use of renewable energy to generate rural income Technologies: biogas, solar dryers, pico- and micro-hydro, solar lanterns,
E7 AIJ Pilot Project	Implemented from 1997-2000 Technologies: solar home systems, micro-hydro Brought electricity to 4000 people in 8 communities
YBUL Renewable Energy Project	Technologies: Micro-hydro, solar dryers
WOMINTRA Village Level Energy Projects	Technologies: Micro-hydro, solar home systems, PV for telecommunications
Yayasan Dian Desa Biomass Energy Projects	Technologies: Improved stoves to reduce indoor air pollution

PROGRESS TOWARDS TARGETS

The key barrier to scaling up clean energy projects in Indonesia continues to be the difficulties associated with achieving acceptable rates of return on investment. Demand side energy efficiency must overcome the low cost of electricity, the limited buying power of the majority of energy consumers, and the lack of fuel efficiency standards for vehicles, all of which suppress investments to reduce power and fuel consumption. Supply side energy efficiency projects in the power sector are possible and can be scaled up by working with PLN, but they must compete for capital with investments needed to expand generating capacity to meet growing demand.

In the electric power sector, project developers continue to face challenges to negotiate and sign power purchase agreements with PLN. Projects that want to sell power to the Java-Bali grid must compete with relatively low power generation costs. Projects that sell power to the many mini-grids primarily supplied by diesel generators with high power generation costs must negotiate purchase agreements with PLN.

Despite policies that attempt to make the negotiation process more transparent, project developers face the underlying reality that the price for which PLN sells power is significantly lower than the cost of production. PLN loses money on almost every kWh sold in areas off the main grid so simply expanding supply with clean energy sources increases losses and must be budgeted. Theoretically, buying power from clean energy sources that are less expensive than current diesel plants allows available subsides to support expanded rural supply. However, long term contracts require careful planning because of the implied commitment of future subsidies and are challenging to administer in a volatile fuel market based on cross-subsidies from urban consumers for a decentralized system that necessarily involves multiple layers of government. The government continues to make progress in the clarification of policies to make the process of negotiation with PLN more transparent but the process is slow.

The fossil fuel sector faces the same return-on-investment barrier but decisions in this sector are driven by national economic and environmental considerations. With declining oil resources and high international oil prices, Indonesia has increased its use of coal and plans to dramatically expand coal use over the next three years adding 10,000 MW of new capacity with coal projected to supply more than 60% of primary electricity generated by PLN by 2010. Decisions made on the environmental standards for the new fleet of coal power plants will have major implications for emissions of SOX, NOX, particulates, and other pollutants associated with coal combustion. Decisions made on the efficiency of the new fleet of coal plants will have major impact on future

CO2 emissions. The government of Indonesia plans to seek international tenders and must decide what environmental standards to apply knowing that stricter standards will increase capital costs but produce measurable health benefits.

Low domestic prices for electricity and petroleum products has led to Indonesia's high energy intensity and created a significant burden on the government budget. In 2004, about $7.9 billion dollars went to fuel subsidies equal to 14 percent of government expenditures.

GROWTH RATE IN RENEWABLES DEPLOYMENT

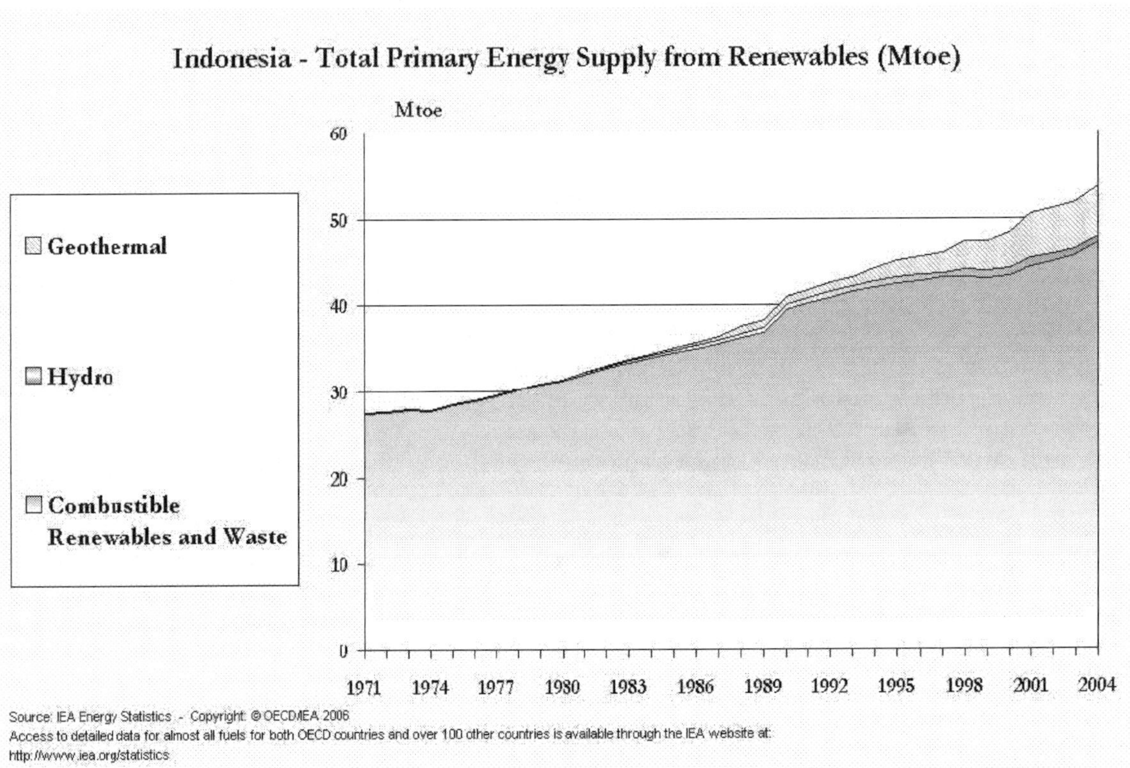

Indonesia has a wide variety of abundant energy resources including fossil and non-fossil resources and has been a net energy exporter for many years. The primary energy source fueling Indonesia's economy continues to be oil although decreasing domestic production, increasing domestic consumption and global price hikes over the past few years have reduced oil consumption and prompted efforts to develop other domestic energy resources. Coal will be the primary domestic resource used for new power generation. The second most important fuel source for Indonesia's population is biomass although most of its use occurs in the informal part of the economy. Biofuels are planned to be the primary domestic resource developed to diversify transport fuels.

In addition to the role they play in the energy sector, oil and gas together continue to lead export earnings for Indonesia increasing to $19.2 billion in 2005 from $17.6 billion in 2004 compared with $15.2 billion in 2003. Despite being an exporter of crude oil, Indonesia is now a net importer of oil. Indonesia is the only Asian member of the Organization of Petroleum Exporting Countries (OPEC) outside of the Middle East, and the only OPEC member that is a net oil importer. Its fuel production has declined over the years, owing to aging oil fields and lack of investment in new equipment. Indonesia remained the world's leading exporter of LNG in 2005, with 18.8 percent of the world market allowing it to remain a net exporter when both oil and gas are considered. Oil and gas imports increased to $17.4 billion in 2005 from $12.1 billion in 2004, compared with $8.4 billion in 2003. Coal exports have grown dramatically exceeding $5 billion in 2006.

COAL POLICY IN INDONESIA[3]

Indonesia is fortunate in having various mineral and energy resources most importantly oil, gas, coal, geothermal and hydro. In the last three decades mineral and energy resources contributed significantly to state's revenue. Mineral and energy sectors including coal are also used for domestic energy need and as raw materials for various types of industries from small industries to large industries. As most minerals deposits located at remote part of the country, this sector also contributes in accelerating regional development and in providing necessary infrastructures that can be used for other activities. Indonesian coal has competitive advantage due to its low sulfur and ash content and geographically close to major coal market in Asia and Pacific. Coal will also play a more important role on Indonesian energy mix given the fact that oil reserve is depleting. Despite some internal and temporary shortcomings, the Government is committed to maintain and enhance coal contribution to national development by continuously provide good environment for coal development.

Indonesia is fortunate in having various mineral and energy resources most importantly oil, coal and gas. Mineral and energy resources contributed significantly to national export earnings at around 70% during the early development plan and around 30% of total national earnings in 2001. Contribution of mineral and coal to national income in 2001 was nearly Rp. 50 trillion or 3.2% of total GDP. Apart from this, coal and mineral also absorb significant amount of employment or around 35 thousand people. Indonesia has around 6 billion tons of proven coal resources. As most of Indonesian coal is of low rank, coal contribution as export earnings can not be expected in the long term and must be better utilized for domestic use unless upgrading related coal technology is well developed. One of the issues facing coal development in Indonesia is the drop of investment in the last five years. This drop is mostly attributable to the internal problems exist in Indonesia such as regional autonomy, land use, and fiscal issues.

In the context of energy, Indonesia experiences high energy growth in the last two decades accounts for 9% per annum. This growth is attributed to population growth and rapid industrialization process. Oil share on energy mix is still substantial at more than 50% although this figure is much lower than it was in 1980-s at around 70%. The Government develops program to diversify oil with coal, gas and other sources of energy.

This paper outlines coal policy development in Indonesia with particular attention to its share on energy mix. It will also discuss prospect and challenges faced by coal industry and how the Government deals with this issues. This paper will also present latest energy outlook in Indonesia with particular attention to coal and in the context of energy policy development in Indonesia.

COAL SUPPLY AND DEMAND

Coal Supply

Coal in Indonesia is found in many part of the country but mostly in Kalimantan and Sumatera. In Sumatera, coal is found in South and West Sumatera whereas in Kalimantan, coal is found in South and East Kaliamantan. Other areas that produce coal include Riau, Jambi Bengkulu and Central Kalimantan. Most of Indonesian coal (58.6%) are of low quality (lignite) followed by sub bituminous coal (26.6%), bituminous (14,4%) and anthracite (0.5%). Coal resources in Indonesia is amounted at around 51 billion tons, of which 4.9 billion tons is proven reserves (Table 1). Coal production has been increasing and reached about 103 million tons in 2002. With this rate of production, coal will last in almost 50 years. Most of coal is produced by coal contractors with the rate of production grows at nearly 7 million per-annum (Table 2). Coal production is projected to increase to 171 million tons in 2010 and 214 million tons in 2020.

Coal Demand

Coal in Indonesia is consumed primarily for electricity generation and cement industry. In 1993, domestic coal consumption was only 8.4 million tons and increased considerably to 29.3 million tons in 2002 (Table 3). Coal consumption for electricity has grown almost 4 times in less than 10 years period indicating the increasing number and capacity of coal fired power plant in Indonesia. Apart from electricity, coal also used in cement industry, textile and metal industries. At current stage, the last two consumer are using small portion of coal but this amount is predicted to grow to meet the increasing demand of this industry. The prospect of briquettes for household and small industry is also bright. This prospect will even brighter if the Government is phasing out all subsidy for kerosene.

On export side, coal export increases significantly in the last ten years from 4.3 million tons in 1990 to 52.1 million tons in 2000 and 74.1 million tons in 2002 (Table 4). Indonesian coal is exported mostly to Asian market including Japan, Taiwan and South Korea. The average growth rate of export in the last five years was 5.8 million tons/annum.

Coal demand in Indonesia is projected to increase from 29 million tons in 2002 to 58.5 million tons in 2010 and 95.6 million tons in 2020. (Table 5). Electricity demand for coal will increase from 20.1 million tons in 2002 to 72 million tons in 2020. The highest growth will be expected on industries as price of coal products is likely to compete with oil product especially kerosene.

COAL POLICY DEVELOPMENT

Coal will play a more important role on energy development as well as remains a major source of income for government. This role must be maintained and enhanced for the maximum benefit of the whole people in the country. In the past, coal policy is seen only as part of mining policy or treated more on upstream basis rather than a holistic upstream (mining) and downstream (energy) link. This approach is no longer considered valid as the role of coal on energy mix increases and become strategic to meet the increasing demand of electricity in the country. The role of coal on energy mix increased from 8% in 1995 to 13.1% in 2002. The share of coal on electricity is even greater from 20% in 1985 to 34.8% in 2002. This share will continuously increase inline with the construction of new coal fired power plant and the increase on the capacity of existing coal fired power plant. Coal share on electricity is projected to increase from 34.8% in 2002 to 60% in 2020. On energy mix, coal share is projected to increase from 13.1% in 2002 to 19% in 2020. (Table 6).

The reorientation of coal policy is directed by some issues including the depleting nature of oil reserve, its increasing share on electricity, availability of coal resource and technology development that allow conversion of coal to environmentally friendly fuel substitution or as raw materials for industry.

In the context of energy sector, the Government has been restructuring its energy policy especially on oil and gas (by enacting Law No 22/2001) and electricity (Law No 20/2002). The restructuring policy aims to create transparency by applying competition and fair playing field. One of the key feature is the separation of competitive and non-competitive region. In the competitive region, all players have the same level of playing field including the state owned electricity company (PT PLN) whose status is changed to a limited liability company. Another policy taken by the Government is restructuring energy pricing. For a long time, energy price (oil fuel) and electricity tariff in Indonesia are subsidized. This policy not only burdening Indonesia's state budget but also unfair for poor people. It also makes other source of energy difficult to compete with oil fuel. These two laws have direct affect to coal development in Indonesia. The oil and gas law will provide good opportunity for coal products such as briquette to compete with kerosene as the Law promotes the rationalization of fuel price into its economic price including that of kerosene. The Oil and gas and electricity law also directed to promote energy diversification. It is expected that the Laws will provide much better climate for investors to invest including that on coal fired power plant as coal is still the least cost option.

The National Coal Policy (NCP) would match the existing laws on oil and gas and electricity sectors. The NCP is now on its final stage and is expected to be launched by the Minister in a matter of time. The NCP would also be a good input in the drafting of general mining and energy laws.

CHALLENGES OF COAL DEVELOPMENT

Coal development in Indonesia is facing some shortcomings. This shortcomings are internal and external in nature. Internally, there has been some regional autonomy issues that considered as contra-productive by many coal industries as it create high economy and inconsistency with Government's policy and regulation. This issue seems to be temporary in nature due to euphoria of reform. The Government is working closely with local government to find a better way to address the interest of local government and the industry.

Another issue is the limited coal infrastructure (roads, rails and ports) available. Under Indonesian system, coal company is responsible in providing their own infrastructures. With the increasing production and export of coal, better infrastructure facilities and capacity is needed. In this regard, the Government sees the prospect of establishing multi client system of coal transportation by coal consortium. This system will not only cheaper than if each company build its infrastructure but also make the management system of coal transportation easier and more effective and efficient. With regard to the issue of illegal mining, the Government is committed to increase law enforcement, but on the other side develop a partnership program between local traditional mining and coal mining. On the issue of environment, the Government is continuously promoting best practice on coal mining and utilization. With consistent and committed effort to better manage the industry, the Government believes that foreign investors would be more interested in invest in Indonesia.

Coal Resources and Reserves by Province (2003)

Million tons

Province	Resources	Reserves
Banten	10.0	-
Bengkulu	50.0	51.4
Jambi	48.8	-
Nangroe Aceh Darussalam	28.1	-
North Sumatera	1.5	-
West Sumatera	909.9	107.0
South Sumatera	11,324.9	1,511

Riau	2,340.4	-
East Java	0.14	-
East Kalimantan	5,772.5	1,885
Central Kalimantan	19,772.9	-
South Kalimantan	5,649.5	1,412
West Kalimantan	186.6	-
Papua	121.6	-
Total	46,3	4,9

Source: Directorate General of GMR (DGGMR), 2003

Coal Production by Companies

Million tons

Company	1990	2000	2001	2012
State Owned Company (PTBA)	4.8	10.7	10.2	9.5
Coal Contractors (PKP2B)	3.8	61.7	76.5	87.1
Mining Authorization/Cooperatives	1.1	4.3	5.7	6.8

Source: Directorate General of GMR (DGGMR), 2003

Domestic Coal Consumption, Million Tons

Consumer	1993	2001	2009
Power Plant	5.4	19,5	20.1
Cement Industry	2.6	5.1	4.7
Metallurgy	0.5	0.22	0.24
Pulp Industry		0.82	0.5
Briquette	0.008	0.031	0.024
Others		1.5	3.7
Total	8.4	24.98	29.3

Source: Directorate General of GMR (DGGMR), 2003

Coal Export by Producers, Million Tons

Producers	1990	2000	2001	2008
State-owned company (PTBA)	1.5	2.14	1.89	1.85
Coal Contractors	2.3	52.1	59.36	66.5
Mining Authorization	0.5	4.1	4.02	5.8
Total	4.3	58.46	65.27	74.17

Source: Directorate General of GMR (DGGMR)

Domestic Coal Production and Consumption Projection, 2002 – 2020

Millions Tons

Year	Production	Average Domestic Consumption Projection			
	Projection	Electricity	Cement	Other Industry	Total
2002 (R)	103.4	20.1	4.2	1.1	25.2
2005	130.5	24.7	4.5	3.4	32.6
2010	171	36.1	6.1	4.5	46.7
2015	199	54	8.5	7.6	70.1
2020	214	72	11.1	12.5	95.6

Source: Directorate General of GMR (DGGMR)

Projection of Coal Contribution on Energy and Electricity Mix, 2002-2020

Million Tons

Year	Coal for Electricity Mix		Coal for Energy Mix	
	%	Total (Million Ton)	%	Total (Million Ton)
2002 (R)	34.8	20.1	13.1	25
2005	40	24.7	15	33
2010	50	36.1	16	47
2015	55	54	18	70
2020	60	72	19	96

Source: Directorate General of GMR (DGGMR)

NUCLEAR ENERGY

Nuclear activities in Indonesia began with the establishment of the State Committee for the Investigation of Radioactivity in 1954, which was assigned to investigate the possibility of radioactive fall-out in Indonesia territory due to nuclea rweapon test in the Pacific Ocean.

Noting that the development and application of atomic energy could enhance the welfare of the people, the Government issued Government Regulation No. 65 on December 5, 1958 establishing the Atomic Energy Council and the Atomic Energy Institute. Thiswaslater followed by the enactment of Act No. 31 year 1964 regarding the Basic Stipulations on Atomic Energy and Government Regulation No. 33 year 1965 which also renamed the Atomic Energy Institute into the National Atomic Energy Agency (Badan Tenaga Atom Nasional or BATAN). However, the 5th of December has been retained as the date of anniversary of BATAN.

In 1965, the operation of the first research reactor (Triga Mark II) was inaugurated in Bandung. In order to improve mastery of nuclear science and technology, several research & development and engineering facilities were built, among others are the Nuclear Technology Research Centre of Pasar Jumat, Jakarta in 1966, and the Nuclear Technology Research Centre of GAMA, Yogyakarta in 1967, where the Kartini research reactor was then built in 1979. To further support the nuclear energy programme, research & development and engineering facilities, the 30 MW Multipurpose Research

Reactor was inaugurated in 1987. It's supporting laboratories including facilities for fuel fabrication of research and power reactors, reactor safety testing, production of radioisotope and radiopharmaceutical, management of radioactive wastes and other nuclear facilities have been built in the PUSPIPTEK science and technology research complex in Serpong.

Further development saw the enactment of Act No. 10 in 1997 on Nuclear Energy, which stipulated among others the separation of the executing function on the beneficial applications of nuclear energy (BATAN), from the regulatory function held by the Nuclear Energy Regulatory Agency (BAPETEN).

Of the key arguments put forward by many fellow Indonesians regarding Indonesia's Nuclear Energy Plan, safety concern comes first followed by the fact that Indonesia is sitting in the so-called "Ring of Fire" making it vulnerable to have nuclear based energy. The story went to the concern of suspected corruption, which may undermine the project itself and lastly argued that there are other more promising alternative resources of energy besides nuclear. In the end, the story proposed that the Government should be more adaptive in addressing the people's concerns over the respective issue.

Discussion about safety concerns will always place facts and statistics among the dominant parameters. Much of the arguments will be based on the cost and benefit analysis; availability of energy resource versus the safety of the people. The Nuclear Power Plant (NPP) against the risk of safety of the people. Why can't we have both? Why can't Indonesia have NPPs that do not undermine the issue of safety? The fact that 30 countries in the world today enjoy the energy resulted from more than 400 operating NPPs which makes 17% shares of world energy production certainly can be treated as a serious consideration.

The second concern deals with the geological fact related with the Ring of Fire. Such concern might be reconsidered with the fact that Indonesia is not the only country in the world sitting in the hot zone. Japan for example, is situated in a volcanic zone on the Pacific Ring of Fire with risks of destructive earthquakes and tsunamis, which by far posses' higher risks both in terms of quantity and quality compared to Indonesia, and yet it has 55 operating NPPs contributing 29,3% for its total production of electricity. Furthermore given the fact that Japan is one of Indonesia's main investors, surely, we can learn something than just finding an escape goat.

The third issue put forward is the classical problem of corruption, which may undermine the construction of NPP. This is not of without valid reason as there are strong general tendencies that almost all sector of developments in Indonesia are suspected of corruptive practices. However, as cultural as one may argue, Indonesia is currently fighting against corruption. The fact that the recently dissolved Corruption Criminal Acts Eradication Team has successfully saved Rp.3.946 trillion of state money, in addition to assets and money of regions amounting to Rp.4.105 billion by far can be treated as a positive sign in corruption eradication. As the efforts against corruption continues, one may assert that in a glass of water analogy the fighting is half full rather than half empty.

The last concern argued is that Government should look for another alternative of energy resources than nuclear. This is true in the long run but what needs to be underlined is that even nuclear based energy is not enough to solely support the high energy demand of a country, never mind the fact that 78,5% of France's energy is derived from nuclear. According to IAEA, in Japan for instance, the nuclear average annual growth rate for electricity production is 4.75% compared to that of thermal and hydro respectively 2,78% and 0,54%. This signifies that alternative energy resources may be considered but it would be very difficult not to rely on nuclear energy to sustain the high demand of energy in the growing economy as Indonesia.

With regard to Government's attitude towards the people on nuclear energy plan. One may argue that the gist of democracy is among others that people can voice their interest and opinion towards any Government's policy without fear of suppression or intimidation. Such condition arguably does exist in Indonesia, for example, in the last rally against the Government plan to start using nuclear energy by 2016. As democracy works both ways, both the Government and the people must respect each other respectively. Whilst the people always claim that the Government is being ignorant to the people's voice, the people to some extent also fail to notice the notion that such Government plan is formulated for the benefit of the people. In conjunction to the benefit for the people, by building NPP arguably Indonesia will not only resolve its issue of energy demand but also acquire and be exposed to the world-class technology of nuclear energy. Certainly if this understanding is being affectionately socialized in proper manners, the Indonesian people in general will understand and therefore accept the plan.

Conclusively, it would be very difficult to contest the notion of growing economy without including the necessities of supporting variables thereof. The acquiring and balancing of these supporting variables are of course not an easy task. The concern of safety of the people against the high demand of energy to sustain the ever growing of economy of Indonesia, let alone the facts that Indonesia is a country with a population of more than 200 million people and situated in one of the most geographic and vibrant economies of the world. Nevertheless if we are to grow then we need to ensure that the supporting variables are available and ready to use. As complex as it may be to choose, Indonesia needs not to choose as the answer does not necessarily electing between the two, rather a compromise; a safe nuclear energy plan for Indonesia.

REGULATORY FRAMEWORK

The Indonesian Nuclear Energy Law, separates fully the promotional and regulatory functions in accordance with the article 3 and 4, and it establishes the regulatory body for control of the application of all nuclear energy through regulations, licensing and inspection. The regulatory body, Badan Pengawas Tenaga Nuklir (BAPETEN) . or Nuclear Energy Regulatory Agency ,

was then established by Presidential Decree No 76 / 1998 in May 1998 which was revised by Presidential Decree No103 / 2001 .

Utilization of nuclear energy . as stipulated in Nuclear Energy Law is defined as any activity related to nuclear energy utilization that includes research, development, mining, fabrication, manufacturing, production, transportation, storage, transfer, export, import, decommisioning and radioactive waste management to enhance people's welfare.

The basic principles of nuclear energy regulation practices in Indonesia set out in the Nuclear Energy Law provide that the control of nuclear energy application is aimed to : Assure the welfare, security and peace of the people;

Assure the safety and health of workers and the public and protection of environment; Maintain law enforcement in implementation of any nuclear energy application; Increase legal awareness of the user to promote the creation of safety culture in nuclear field; Prevent unauthorised changes in the purpose for which nuclear materials are used; and Promote and assure the maintenance of worker discipline in the application of nuclear energy.

Any activity related to the aplication of nuclear energy is required to be conducted in manner which observes safety, security and peace, and protects the health of workers and the public and the environment.

BAPETEN is established as a governmental agency, under and directly responsible to the President of the Republic of Indonesia. The BAPETEN has the following functions :

The rulemaking of national policy in the field of the control of nuclear energy utilization; The planning of national programme in the field of the control of nuclear enrgy utilization; The guidance and rulemaking in the implementation of the nuclear safety, radiation safety, and safeguards assesments; The implementation of licensing and inspection at the developments and operation of nuclear reactor, nuclear instalation, nuclear material facility, radiation source, and development of nuclear preparedness; The implementation of cooperation in the field of the control of nuclear energy utilization with the government agencies or other organizations either internally or externally to the Government of Indonesia; The implementation of safeguards and state's system of accounting for and control of nuclear material (SSAC)

The implementation of the guidance and counseling for effort that related to the safety and health of worker and the people, and effort of environmental conservation;

In executing the three main activities in controlling of utilization of nuclear energy , Bapeten requires to execute activity of review and assessment and research and development on nuclear safety and radiation safety as scientific justification in determining the safety requirement which is put in regulation, licensing and inspection. Through the technical and scientific activities, all activity on controlling nuclear energy have been executed on the basis of correct scientific assessment for assure that all utilization of nuclear energy is peaceful to worker, society and environment.

IMPORTANT ENERGY POLICY DEVELOPMENTS

ENERGY BRIDGE; THE FUTURE IN ENERGY TRANSPORT

Energy Bridge is an offshore Liquefied Natural Gas regasification and delivery system which involves the use of purpose-built vessels for the transportation and vaporization of LNG through specially designed offshore and near shore receiving facilities.

Energy Bridge Regasification Vessel or EBRV, are purpose-built LNG tankers which incoporate onboard equipment for LNG vaporization and delivery of high pressure natural gas. These tankers load in the manner as standard tankers, but are able to discharge the cargo in three ways; as liquid in conventional receiving terminals, as gas through the EBRV's high pressure gas connection to a buoy moored in the hull of the ship, and as gas through a high pressure gas manifold located forward of the EBRV's LNG manifolds.

To accomplish its task, an EBRV consists of five major components, namely; High pressure cryogenic pumps (to bring the LNG from the cargo tanks up to high pressure prior to vaporization), Shell and tube vaporizers (to convert LNG back to gasoueus natural gas), Oversized boilers (to provide the steam necessary to power and sustain vessel operations along with the shipboard regasification process), Buoy compartment (to accommodate the sub-sea buoy system serving as a moor for the vessel and to deliver natural gas to a seabed pipeline), and reinforced LNG cargo tanks and internal pump towers (to withstand loads encountered throughout all loading levels).

EBRVs may be operated in several ways; closed loop, open loop, and combined. In closed loop, steam from the EBRV's boilers is used to heat water circulated through the shell-and-tube vaporizers in the regasification plant, which results in minimal seawater usage.

Open loop mode draws sea water through EBRV's sea chests, which is used as a heat source, passed through the shell-and-tube vaporizers which vaporizes the LNG. During the process, seawater temperature is lowered by 13F which is why this system is not suitable for water temperatures below 45F. In combined mode, seawater at temperatures between 45F and 58F may be used when heated by steam from the EBRV's boilers for LNG vaporization.

There are two ways to distribute gas from an EBRV, both offshore via a gateway system, and dockside via a gas port system. In the gateway system, the EBRV connects to a series of undersea buoys which are securely attached to the sea floor by anchors, chains, and wire ropes. Once connected, the gateway serves as a mooring for the EBRV while it discharges natural gas through a flexible riser connected to a subsea pipeline tied in into a a downstream delivery infrastructure. The gateway system is commonly located 10 miles from land.

The gas port system utilizes a shore mounted high pressure gas arm that will connect to the EBRV's high pressure gas manifold. The natural gas will be delivered from the EBRV at downstream pipeline pressure, which allows an EBRV to function as a flexible LNG receiving terminal. A gas port system may be designed for "across-the-dock" transfer of LNG so that a conventional LNG carrier may be able to deliver directly into the EBRV to ensure uninterrupted flow of natural gas.

GOVERNMENT IMPLEMENTS MANDATORY BIOFUEL USE STARTING JANUARY 2009

Starting January 1st 2009, users and fuel distributors are obligated to use biofuel, with products such as biodiesel, bioethanol, and biokerosene. This mandatory use is stipulated in Minister of Energy and Mineral Resources Decree no 32/2008 regarding a three month transition period consisting of several phases which began in October 2008.

"This mandatory use is a way of increasing biofuel usage for the transport, industry, and power sectors" explains the Minister of Energy and Mineral Resources, Purnomo Yusgiantoro on Monday (13/10). Aside from supporting the energy diversification program, this step is also expected to contribute in reducing fuel subsidy costs.

According to the decree, the Director General for Oil and Gas, Evita H. Legowo explains that as of January 2009, the government will mandate the use of biodiesel at a minimum of 1% for the transportation sector (both PSO and non PSO), while the industrial and commercial sector is targeted at 2,5% and powerplant at 0,25%.

The government has mandated the use of bioethanol at a minimum of 1% for the PSO transportation sector, 5% for the non PSO transportation sector, and another 5% for the industry and commercial sector. Biofuel as a source of alternative energy is targeted to fulfil 0.25% of the power plant fuel needs, while low and medium rpm operating machines of the industrial and sea transport is each targeted at 1% starting January 2010.

These numbers are set by the the government in order to implement the targeted usage of Biofuel susge of 5% by 2025" explains Evtta H. Legowo. This biofuel will be domestically supplied, and not imported.

Both Purnomo Yusgiantoro and Evita Legowo commented that the biofuel development program for the three year biofuel development program has been carried out very well, with several challenges and obstacles that must be faced together. "There are of course some plus and minuses, for which all of these challenges must be faced together" concludes the minister.

USE OF BIOFUEL FOR POWERPLANTS

In order to accelerate the use of biofuel, the Minister of Energy and Mineral Resources have issued Ministerial Decree No. 32/2008 regarding the supply, use, and sale of Biofuel as an alternative form of energy. One example of this ministerial decree is the regulation on mandatory use of biofuel in the transportation, industry, commercial, and power sector until 2025, which is to be carried out in several phases. For powerplants, these phases include the use of biodiesel amounting to 20% and biofuel amounting to 10% towards the total needs of powerplants.

The Department of Energy and Mineral Resources through its Directorate General for Electricity and Energy Usage, in coordination with Neaue Maschinbeau Halberstadt GmBH (NMBH) have completed the converter installation needed for biofuel conversion of a 10 MW Diesel Power Plant powerplant in Riau. NMBH is a German private company which has carried out intensive research for conversion development and biofuel powerplant.

By adding a converter, a diesel fueled powerplant will be able to fueled by biofuel, which in this case is Crude Palm Oil (CPO). Using this technology, Crude Palm Oil which is in abundant in the local area will be able to be directly used as fuel. Use of this converter in a 10 MW powerplant will be the first in Asia.

The suceess of this program will help boost Indonesia's effort in maximizing the use of biofuel, as more than 1.600 of PLN's powerplant is still powered by diesel. There are several benefits to be taken from biofuel use, including efficient fuel costs, as the current price of CPO is half of that of diesel price. Biofuel will also potentially reduce air pollution because it is based on renewable energy.

BIOFUEL WILL BE AVAILABLE IN PETROL STATIONS AROUND INDONESIA BY 2010

PT Pertamina has targeted that biofuel products such as Biopertamax, Biopremium, and Biosolar, will be available in petrol stations around Indonesia by 2010.

By the end of 2008, Pertamina has targeted Biofuel distribution in West Java, East Java, and Bali. By 2009, this product will be distributed in all regions of Java, Sumatra, part of Kalimantan and in Makassar, South Sulawesi. This is explained by the Pertamina Managing Director, Arie H. Soemarno during the launching of Pertamina Biosolar for industry and the expansion of biofuel sales in petrol stations around Indonesia on Tuesday (11/11).

As a national company with 51 years of experience in the oil and gas sector, Pertamina has learned that technical difficulties is not a major obstacle in the distribution of fuel. By distributing biofuel, Pertamina has contributed to the central government's plan for nature conservation.

The current biofuel sold by Pertamina will be able to be used for motor vehicles and the industrial sector. The use of biofuel for motor vehicles will not need any form of engine modification, while at the same time increasing engine life because of its cleaning attributes and being environmentally friendly.

Natural gas and coal will play a grater role in fulfilling Indonesia's energy demand due to the decrease of crude oil production. In planning for the 2009 proposed state budget (RAPBN), the Department of Energy and Mineral Resources have assumed a gas lifting of 7.526 BBTUD or

1.384 MBOEPD. Coal production is expected to be at 250 million tonnes or equivalent to 2.875 MBOEPD. Electricity subsidy is estimated to be at Rp. 60,42 trillion.

These assumptions on gas lifting and coal production is part of the materials discussed during the meeting between the Minister of Energy and Mineral Resources, Purnomo Yusgiantoro with Commission VII of DPR RI. on Monday (8/9) in Jakarta. Also present were officials from echelon I and II of the Department of Energy and Mineral Resources. This meeting was attended by most members of Commission VII DPR RI.

The majority of coal production which will reach 250 million tonnes will be marked for export, nevertheless the coal allocation for domestic use have also seen a sharp increase. This sharp increase is needed to fulfill the demand of the 10.000 MW powerplant acceleration program which will be operable at the time. For the following years, coal production is expected to keep increasing, while export levels will be relatively unchanged.

Natural gas production will exceed 10.000 MMSCFD with export allocation level relatively steady compared with the previous year. Part of the excess in production will be allocated for domestic consumption. For the following years, gas allocation for domestic needs will keep increasing, while on the other hand export will keep declining.

Indonesia's energy supply is changing. After years of dependency on oil and natural gas, the role of coal in energy proportions have steadily increased. Since the year 2007, the role of coal have exceeded that of crude oil, and this role of coal is projected to keep increasing due to the declining production of oil and natural gas.

ELECTRICITY SUPPLY INCREASES

According to data from The Energy and Mineral Resources (ESDM) Department, the projected development of electricity supply will increase in 2009. To support this development, power plant capacity is expected to increase 0,2% to a total of 6,7% per year or equivalent to 35.265 MW.

With the rapid development of electricity supply, Indonesia's electrification ratio will also increase by 1% compared with 2007 to a total of 65,34%. The majority of areas that are within reach of electricity is located in the Western provinces of Indonesia where there is a large power plant capacity. One prime example is Jakarta with an electrification ratio of 100%, this province is covered by a power plant capacity of 23.302 MW.

1500 INDUSTRIES READY TO IMPLEMENT 5 MINISTER'S DECREE

The 5 Minister's Decree (SKB) regarding work schedule diversion to optimize power usage has gained support among the industrial sector. Data from PT PLN has shown that there have been 1500 industries based Java and Bali that have signed an agreement with PT PLN to switch their working hours to Saturday and Sunday.

"By shifting the industries working hour to Saturday and Sunday, we can achieve a power load diversion of 150 MW from Monday to Friday" explains PLN's Director for Java – Bali, Murtaqi Syamsuddin during the decree's implementation in Tangerang, Banten on Friday (1/8).

PT PLN's industrial consumer (I3) in Java and Bali amounts to 6.856 customers with a total power usage of around 6000 MW. From those figures, the number of customers that operates on Monday – Friday equals to 3.052 customers, so it is expected that more industries will move their operating hours to Saturday and Sunday.

In implementing the 5 Minister's decree, PT PLN have set 12 groups of industrial customers which must move their operating schedule to Saturday and Sunday per province. These clusters are set according to the total power load of 150 – 180 MW per day.

PLN have coordinated with each of the industrial customers, regent heads, mayors regarding the socialization of the decree. The industries that are ready to divert the power load have begun to shift their working hours starting July 31st 2008, according to the agreed schedule.

GOVERNMENT EXPLANATION ON GOVERNMENT OF INDONESIA DECREE REGARDING THE REDUCTION OF FUEL SUBSIDY AND OTHER RELATED POLICIES

The global price of crude oil has been increasing during the past year. Last year the global price of oil is at USD 80/barrel, currently this price is above USD 130/barrel. This has caused an increase in the amount of fuel subsidy to an extent that it is no longer sustainable. Given that the price of crude oil is at USD 120/barel for 2008, the fuel subsidy will exceed Rp 200 billion, while according to UU No 16/2008 regarding the 2008 national budget which have been approved by the house of representatives, the maximum budget allocated for fuel subsidy is only Rp 135,1 billion.

With the increasing fuel subsidy, the government's ability to fund programs which are oriented to the improvement of lives for the poor is dramatically reduced. These programs include education, health facilities, National Program for Community Empowerment, small business credit facilities, and the development of infrastructure.

On the other hand, the fuel subsidy is mainly consumed by those who are not targeted by the program. As much as 40 percent of high income families benefit from 70 percent of the subsidy, while 40 percent of the lowest income families only benefit only 15 percent.

Increase of Fuel Price as the Final Alternative

The government have attempted to manage and minimize the effect of rising global oil price towards the community. Steps such as reducing government spending, increase of tax income, efficiency of PLN and Pertamina, conversion and economic spending of subsidized fuel will continue to be taken.

Despite all of these efforts, they are not enough to control the effect of rising global oil prices, therefore the government have been forced to implement the option of fuel price increase.

The price of fuel in Indonesia is currently one of the lowest in the Asian region (lower than East Timor, Cambodia, and Bangladesh as lower income countries). After the 28,7 % increase, our fuel prices is still far lower than those of East Timor, Phillipines, Thailand, and Singapore.

Compensation Program

The government is fully aware that the increasing fuel price will also affect living costs and increase the burden of the public. With this in mind, the government have prepared programs to reduce the burden of low income societies before implementing new fuel price. These programs include direct cash assistance (Bantuan Langsung Tunai), towards 19,1 million families, for a total amount of Rp 14,1 billion for the 2008 fiscal year.

Alongside this program, the government has set other compensation programs, which include:

• Food Sustainability Program:

a. o Low Priced Rice (Raskin) is kept at Rp 1600/kg, which was planned for Rp 1900/kg.
b. o Time period of Low Priced Rice distribution towards 19,1 million households has been extended from 10 months to 12 months.
c. o Fund support for children's education of government employees category I/II, Tamtama TNI/Polri.
d. o Subsidy increase for low scale credit facilities.

After careful consideration and preparation of BLT distribution, the government through The Minister of Energy and Mineral Resources Regulation No.16/2008, have increased the price of subsidized premium fuel, diesel, and kerosene, effective as of May 24 2008, 00.00 WIB (refer to attachment).

Besides, the government will also continue to implement the economic use of subsidized fuel through the control card program. This program was launched in Semarang, Central Java, and the smart card program will be put to trial this September.

All of these policies are aimed to strengthen and stimulate the national economy, while also improve the social fairness for all citizens of the Republic of Indonesia.

OPPORTUNITY FOR OIL AND GAS SUPPORTING INDUSTRIES

Opportunity for oil and gas supporting industries to develop is huge given the fact that oil and gas sector in Indonesia will continue to grow. The industries is encouraged to explore all the opportunities to take part in the oil and gas industry development in Indonesia.

In his speech on the closing of the workshop of the oil and gas supporting industries in Batam (3/4), Minister Purnomo said that the government will provide fair opportunity to all players in this sector as well as develop necessary infrastructures to support the oil and gas industry development across the country. Minister Purnomo also said that the government will continue to develop the role of oil and gas operator and supporting industries in the oil and gas upstream and downstream services. He also added that Indonesia has already proposed the energy service classification on WTO meeting to enhance the competiveness of local energy services.

The workshop which was attended by around 200 participants agreed to adopt what so called as Batam Declaration aimed at at enhancing national capability on, inter alia, capital, technology , human resources and managerial skill. This will be done through the finalization of drafting the Blueprint of National Capability Enhancement, formulating policy and strategy on oila dn gas supporting businesses as well formulating the policy and regulation on oil and gas goods and services procurement. In addition to that, the government will also identify the capability of oil and

gas supporting industries and the requirement of oil operation as well as enhancing coordination among government institution related to national goods and services empowerment.

RESTRUCTURING AND REFORM AGENDA

In the 1990s, restructuring process has been carried out in various ways. Reform process was undertaken by implementing Power Sector Restructuring Policy in 1998 consisting a set of program to put the foundation for future development of Indonesia's power sector in place. Key activities of the policy are: redefining government role; strengthening regulatory framework; establishing IRB; restoring financial stability by gradually increasing the tariff; introducing competition; and rationalizing and expanding private participation. The introduction of IPP was also one of the ways to secure national power supply through public-private participation.

Restructuring energy sector, then, entered the new era in early 2000s after promulgating three new laws in oil and gas, electricity, and geothermal as mentioned in the previous chapter.

However, the subsequent actions are still to be awaited. Delay in finishing up the related policy and regulation has cost the industry with delaying of infrastructure development and calamity. Controversy of LNG terminal development plan is one of the examples. It is not known who does what and who have decided. All of this shows that formulating the implementation and operational regulations for the laws, by and large, is not an easy task.

DECENTRALIZATION AND ENVIRONMENTAL CONCERNS.

Another issue in national level is the decentralization of the governance to the *Kabupaten* and *Kota* (city) level (Law No 22 and 25/1999). Decentralization process starting at 1999 has been one of the main challenges to energy security. Decentralization also means "headache" for the execution of most of energy development plans. Various additional tax and regulation imposed abruptly by some local governments have increased cost for the survey, exploration, exploitation, and transportation of energy products.

Illegal mining and smuggling practices have also been the main concerns. It is predicted that almost one third of coal industry in South Kalimantan came from illegal mining through both legal and illegal ports[7]. Like other utilities, regulated prices in energy sector have been set below long-run marginal cost for decades, leading to excessively high demand, weakened incentives for efficient use, and unnecessarily environmental damage. In additions, energy product such as petroleum and electricity are of poor quality and must be improved all the time, particularly in phasing out the lead from gasoline and stabilizing the electricity supply. Poor air quality in some big cities will be a constant threat for people's health.

Other important issue is Public Service Obligation (PSO). This is very much important to curb inequality to energy supply among people. Ocampo (2004) underlined in Recent 19 World Energy Congress in Sydney that limited access to modern energy services will undermine the living condition and productivity of one third of world populations, cause hardships, and worsen disparity. PSO is somehow not yet comprehensively discussed in many forums in Indonesia.

PSO is critical due to the fact that the cost to provide services in rural area is much higher than urban. World Bank (2004) in its recent study on infrastructure calculates that connection cost for electricity in rural is about 33% higher than that of urban areas. So far, demand and supply response are restricted to large consumers. In such high demanded public utilities sector and limited national budget, then the policy on PSO is somewhat unclear. Basically, the main concern in PSO is to provide accessibility and affordability of energy services to all people including who live in rural and isolated areas.

GOVERNANCE PRACTICES

Last but not least is KKN . Indonesia is still ranked as one of the highest KKN practice country in the world. In the transition period, rent seeking behavior and lack of good corporate and good governance practices in such huge investment and business activities in globalization era may thwart programs in regulating the sector. Therefore, discussing development issue especially regulation in energy sector cannot just put KKN issue away. Although some basic regulations for sector have been in place, the potential negative impact of KKN on energy sector still remains.

Energy regulation is a complex subject. For Indonesia, it covers several cross cutting issues, namely: social, economy, environment, and decentralization. Wim Kok (2004) notes that affordability and security of energy supply are some of the main topics in the social issue. Contribution to welfare, employment generation, and maximum contribution to government income have been among the most important economy issues for decades. Meanwhile, the issues on environment and ecology consists of least emissions per kWh generated, least emissions per ton fuel of refineries, least emissions per ton.km or car.km, and high percentage of renewable energy. To put simply, the purpose of the energy sector in Indonesia as a developing country is, then, to contribute to the national development by providing energy services to as many people as possible, as cleanly as possible, and as cheaply as possible.

Regulation is needed since energy is fundamental to society. "Societies collapse when the energy flow is suddenly impeded", Rifkin (2002) wrote in one of his latest books. We do not want Indonesia to be commonly or fully dependent on other country's investment and technology, nor depends on multinational company as it happened in Azerbaijan. Kleveman (2003) reminds the rest of the world to avoid the condition of what once a former British Petroleum (BP) spokesman told him that Azerbaijan would collapse if BP pulls out from Baku field. Therefore, better regulation and continuous effort to increase self-reliance in energy sector is very critical.

INDEPENDENCE

The global trend of restructuring and reform in energy sector has pushed regulatory issues to the fore, among them the role of regulatory bodies or agencies. In the developed market, these agencies have a long history. Nowadays, creating such agency or empowering the existing regulator has become a central end of reforms around the globe. Some governments are reluctant to give up political control over regulatory decisions. Many issues, however, remain debatable, particularly the notion of independence.

Independence can be interpreted in any different ways. It makes establishing an independent regulatory body difficult. It is even more challenging in country with a limited tradition of independent public organizations and lack of regulatory experience such as Indonesia.

Smith (1997) assess that formal safeguards for the independence requires, among others: (1) providing regulator with a separate legal directive, free from ministerial influence; (2) setting down professional criteria for filling the positions; (3) involving both executive and legislative branches in the recruitment process; (4) assigning the member of the agency for fixed terms and protecting them from arbitrary removal; (5) exempting the agency from civil service salary rule that make it difficult to attract and keep hold of well-qualified staff; (6) providing agency with a unswerving source of financial support, usually allocated levies on regulated firms or consumers; (7) and staggering terms so that they do not coincide with election cycle.

6 Part of the sentence was found frequently in many books and paper, especially in World Bank publications.

However, being independent is not enough. It needs to be brought together with measures to make regulator accountable for its actions. Checks and balances are definitely required to ensure that regulator does not stray from its mandate, engage in KKN practices, or become entirely inefficient. Fitting independence and accountability in proper balance is quite difficult, but the following measures have been adopted by many countries: (1) commanding rigorous transparency in decision-making process; (2) ruling out conflicts of interest; (3) providing helpful arrangement of appealing the agency's decisions; (4) providing for scrutiny of the budget; (5) subjecting the regulator's conduct and efficiency to inspection by public or independent auditors and other public watchdogs; (6) allowing regulator's removal from office in case of proven misconduct or incapacity.

Regulation in infrastructure and utilities, including energy, usually has three main objectives: (1) to protect consumers from abuse by dominant market power company; (2) to support investment by protecting industry from arbitrary action by government; and (3) to promote economic efficiency. According to Warrick Smith (1997), regulating energy sector is naturally also complicated by three factors. *First*, price for energy is usually political. Energy history is abounding with examples of justifiable price increases being withheld at the cost of investors and the long-term concern of consumers. *Next*, investors are typically aware of these pressures and of the vulnerability of their huge, long-term, and immobile investment. Unless a government has made a credible assurance to rules that make certain an opportunity to have reasonable returns, private investment will not flow in. *Third*, the long-term nature of most investments in energy sector makes creating credible commitments difficult. Highly specific rules can provide assurance to investors and lower the cost of capital. But they make it difficult to change regulation to unforeseen developments, including changes in technology and market conditions.

One of the most fragile aspects in the new regulatory era is the relationship with ministry office. It is suggested that the ministry is responsible for policy and the agency for regulation. However this division is unsupportive in practice, because the line between concepts is blurry. Consequently, qualified personals with well-built background and proficiency on the subject are strongly recommended to fill the position in regulatory body.

The enactments of new Oil and Gas Law No. 22/2001 and Electricity Law No. 20/2002 have significantly brought changes in the sectors. Responding to the new challenge of an open market, new Oil and Gas Law restructures oil and gas business. Hence, it has also brought the monopoly of Pertamina to an end. In upstream sector, the authority of mining right is now returned to the government after conducted by Pertamina in the past. Any state-owned and local government owned companies, cooperatives, and private company might now participate in oil and gas exploration and exploitation. Thus, oil and gas upstream activities will be conducted in more open and transparent way.

Liberalizing down stream business as one of the keywords to deregulation, including refining, transportation, storage, distribution and marketing the products, will also be conducted in an open and more transparent way. All players are free to enter the market, including state and regional owned companies, cooperatives, and private companies. In additions, there is a clear demarcation between upstream and downstream players. This is expected to give more room for market mechanism. In additions, through GR 31/2003 Pertamina has been transferred into a limited liability company (Persero) and has no longer its regulatory function.

Based on the new law, nowadays the development and regulatory aspects of oil and gas sector are administered and managed by the following: (a) Directorate General of Oil and Gas (DG Migas) as the main policy making agency; (b) BP Migas as the executing body; (c) and BPH Migas as the regulating agency .

The role of DG Migas is now limited to supervision and control function for both upstream and downstream business activities. Upstream activities include exploration and exploitation, while downstream covers processing, transportation, storage, and distribution/commerce. In addition, DG Migas is also responsible to conduct all of government activities and to formulate the policy measures in oil and gas business.

BP Migas, on the other hand, has to control and manage the execution of upstream oil and gas activities based on Cooperation Contract (CC). Some of BP Migas task includes: providing input to Minister on preparation of operational territory; preparing the CC; giving approval of business plan and budget proposal; and monitoring the implementation of the CC. In the meantime, BPH Migas is responsible, among others, for the provision and distribution of BBM and natural gas through the pipelines to all people in the country, price and tariff setting including toll fee, and maintaining the national oil fuel reserves.

In power sector, the new Electricity Law No 20/2002 accommodates long-term future market trend for an open market, based on ideal condition for multi buyer-multi seller system. Electricity supply business is to be unbundled into subsystem such as generation, transmission, distribution, exchange and operation market, and last mile or retail sale. Other supporting businesses, for example consulting services, construction, testing, operation and maintenance, research and development, and training center facilities will be carried out by separate entity. However, electricity transmission and distribution will still be preserved for state-owed companies.

In terms of regulation, as mentioned in previous section, EMSB has been established through Government Regulation No. 53/2003. EMSB is an Independent Regulatory Body to regulate and monitor electricity business for competitive area. It's main role are, among others, creating market rules; monitoring fair competition; protecting public interest; enforcing regulations; and facilitating dispute resolution. EMSB is directly responsible to the President. In the meantime, the Directorate General of Electricity and Energy Utilization (DGEEU) still plays the key government role for over all policy and regulation in the power sector. The recruitment and selection process for EMSB's personals is now underway.

For a successful practice in regulatory framework, all personals of existing regulator (the DGs) and new regulators - BP Migas, BPH Migas, and EMSB - have also to keep informed with the latest development in energy sector. Not only do they have to master the current issues, but they also have to be able to cope with the trend and long-term vision in international market. These are at most important for the "new comers". Appendix 1 illustrates summary of the latest development in the sector and the long-term anticipation for the future.

BASIC REGULATIONS

Table gives a highlight of current situation of regulatory framework in Indonesia's Energy industry. Having had a set of new regulations in energy sector and successfully establishing separate regulation agencies for oil, gas, and electricity businesses have made Indonesia's energy sector reform on the right tracks. On one side, government officer has frequently expressed their satisfaction upon the achievement and the performance of the sector. They also show their confidence for a prospective future of Indonesian industries. On the other side, some experts, observers, and investors, however, cannot cover up their deep concern for some business practices that might leading to a setback.

At this moment, judicial review process for the new Oil and Gas Law No 22/2001 is underway in national Constitution Court to ensure better benefit for all Indonesian. In regulatory framework, transition period and lack of socialization may threat the implementation of independent regulator. The following are some of conditions that have put regulator somehow "under attack".

Existing Condition of Policy and Regulation in Energy Sector

SECTOR	Blue Print	Law	Regulator/ Executing Body	Competition	Main SOEs
OIL Upstream Downstream	None* None	Available Available	Available Available	Semi Limited	PT. Pertamina PT. Pertamina
GAS	In Progress	Available	Available	Very limited	PT. PGN/ Pertamina
ELECTRICITY	Available	Available	In Progress	N o n e	PT.PLN
GEOTHERMAL	None	Available	None	Very limited	PT. Pertamina/ PT. PLN

**) To some it means no need for Blue Print or Master Plan in Upstream. Source: Enhanced from Eddy Satriya (2003)*

Firstly, government political wills to finance the newly born regulator in transition period is unclear. This makes the agency difficult to run the office effectively and in an efficient manner.

Budget allocation process still has to follow regular process in Ministry of Finance, meaning that their salary cannot escape from the regular standard of civil servant. This condition violates formal safeguard for the independent requirement mentioned in Section 3. Relatively low formal salary is pretty much true for BPH Migas, while BP Migas made it successfully to more comfortable level by benchmarking at least to Pertamina's salary.

Secondly, even though related laws and government regulations have stated a clear job description, it is inevitable that some gray areas emanate in the transition regulatory framework, especially in oil and gas sector. Kurtubi (2004) clearly said that BP Migas has already out of its track in managing and marketing the gas and national LNG industry to the investors, leading to a weak bargaining position. My experience teaches that in transition, there always exists an overlapping of authorities and budget power dispute. Somehow this situation provides new opportunity for new players and existing one in the industry.

Next, in terms of independence, it seems that new regulators had difficulties in freeing up the institution from government intervention in two ways. First, at the early stage regulators depends heavily on government for budget supports. It might put them in difficult position whenever they have to make decision on contracts, negotiation, or choosing the winner since PT. Pertamina's share belongs to government. Second, some personals from government and regulator also have position as board of commissioner of several state-owned companies in energy sector. In short, conflict of interest has not easily been ruled out as recommended for a better regulation.

Fourthly, It was not an easy task for the Independent Regulator and Executing Body to recruit skilled staff members to fill the position in the new agencies. It is for people knowledge that many of top executives in BP Migas and BPH Migas came from the "migration" of existing regulators in ministry and operators. This condition will need time to proceed for a real independence agency. Qualified staff member with strong background in energy sector is not merely enough. Being an independent regulator demands also other supporting disciplines such as economics, law, and social adequacy.

Then, incumbent phenomenon is other crucial aspect in regulation. Most of the case, competition in transition faces hard situation regarding the incumbency. Incumbent phenomenon relates to resistance of the early player in the market. In the case that the player is a state-owned company, regulator usually has to work hard to provide a fair path for competition. Incumbent naturally is unwilling to give up their market share or tends to put additional efforts to keep their "slice" bigger. Recently, PT. Pertamina has issued a specific letter to all Gas Station Owners in Jakarta, West Java, and Banten provinces in order to protect their market. Basically, Pertamina gave warning to the owner for not to sale competitor products such as lubricants, fuel treatment and other additive products in their stations. Any violation will end up with sanction to the owners.[8] Similarly, in the future, I envisage that PT. PLN would not be that easy to giving up their market share to competitors.

Finally, slow progress in finalizing implementation regulations for all new laws in energy sector have resulted in the delay of energy sector reform. Non-existence of Gas Master Plan also worsens the regulating aspects especially in downstream business of oil and gas. Up to now, DG Migas have just completed three Government Regulations out of seven as mandated in oil and gas sector. Mean while in electricity, government has to complete at least six Government Regulations in electricity business and two regulations in engineering section.

At the moment, government completed one Government Regulation on EMSB (GR No. 53/2003). Intensive support from related agencies and funding assistance from the National Budget would expedite the process of drafting and finalizing regulations. Quoted from *Koran Tempo* (2004) last August, State Secretary has returned two drafts of Government Regulations for both upstream and downstream oil and gas activities proposed by MEMR due to potential dispute on land certification and the matter of utilizing Pertamina's infrastructures.

As mentioned earlier, regulatory framework is a cross-sector issue. This situation leaves the Ministry and Regulators to always seek solutions for the unsolved and to be more prepared for the following challenges.

The first is the hardship on decentralization practices may still put some barrier for regulating the sector. In line with this issue is the misconception of some province or *Kabupaten* to explore their region due to the environmental and ecological issues.

Other is dualism in supervising state-owned company (SMSOE and MEMR). In the past, all state-owned companies in energy sector were responsible mainly to MEMR. Today, they also have to report in detail especially the financial aspects and development plan to SMSOE. This dualism is not solely happened in energy sector, but also to other sectors.

The third is that the necessity to gain maximum profit mandated by SMSOE may hamper the provision of energy supply to all people through Public Service Obligation. Article 66 Law No. 19/2003 concerning State-Owned Companies stipulates that PSO may only be imposed to the State-Owned Companies after getting approval from the Minister or Stakeholder Meeting (*RUPS*). One of most crucial aspect is the uncertainty of the provision of oil supply and related products across the country, since Pertamina's obligation will be ended by November 2005.

The fourth is poor coordination and too much overlapping in energy sector management. Too many offices took part in one specific issue, but too few action and result. The scarcity of gas supply to Asean Aceh Fertilizer factory in gas-rich province Aceh proves this. This observable fact also happens to other specific energy issues such as renewable energy, *blue-sky* program, and other technical and environmental aspects.

Next is the early stage of BP Migas, BPH Migas and EMSB might have a possible overlapping job description with KPPU. KPPU stands for "Komisi Pengawasan dan Persaingan Usaha.", or National Commission for Antimonopoly Practices.

Especially in clearing up dispute on competition issues such as dominant market abuse and other monopolistic practices. KPPU is established following of the enactment of Antimonopoly Law No 5/1999.

Finally is that worsening political turmoil in Central Asia contributes to the unprecedented oil price to a level of US$ 50.0 per barrel. At certain points all of these might amplify the complexity in regulating the sector. Moreover, recent geopolitics development does not only involve Central Asia and U.S. multi national companies in oil and gas, but in any way may also spill it over to Indonesia as one of OPEC member and the biggest Moslem country in the world.

In general, Government of Indonesia has been on the right track in conducting energy sector reform. Issuing three consecutive laws after economic crisis and political turmoil in 1998 have shown how all related stakeholders put regulation issues at the highest priority. This step has been followed by establishing separate agencies taking care of regulation issues, namely BP Migas as an executive body for upstream oil and gas business, BPH Migas to regulate downstream oil and gas business, and a brand new EMSB to supervise electricity market for competitive areas.

However, these regulatory agencies still suffer from problems such as unclear financing policy, lack incentive for the independence, existence of gray area, strong incumbency, and limited qualified staff member to fill position in the agency. In the future, decentralization process, slow progress in publishing related regulations, dualism in supervising state-owned operators, unclear Public/Universal Service Obligation policy, and lack of good corporate and good governance practices might also jeopardize the existence and the performance of the agency.

Therefore, in order to keep energy sector reform unharmed and to improve the performance of the independent regulator, it is recommended to refocus job description, reassess the agencies' independency, accelerate market competition, and prioritize the finalization of related government regulation and other decrees.

LONG TERM PERSPECTIVE

Ten Years Ago	TODAY	10 Years From Now	20 Years From Now

Fairly efficient world crude oil and petroleum product market had formed. Natural gas contracts in Europe and Asia were at least partially linked to spot prices in petroleum product markets. The US gas market was in the midst of deregulation The first experiments with competitive electricity market started.	Crude oil and petroleum markets continue to expand. Gas-to-gas competition is a reality in the US and UK. The EU gas directive has been passed and the first open access regimes are being introduced, e.g. in Holland. Competitive electricity market have been or are being introduced in all major countries of North and South America, the UK, Australia, New Zealand and Europe.	The policy-induced barriers among energy sectors such as gas, power, and petroleum products are disappearing. Techniques like slim-hole drilling and intelligent drills reduce the cost of oil production. New super cars become commercially available using only one third to half the energy needed today. New techniques such as high-pressure pipelines and floating LNG facilities maintain the competitiveness of natural gas. Gas and power compete as the cheapest way to transport energy as high-voltage direct current transmission and some super-conductive transmission lines lower the cost of electricity transport. Micro-turbine and some form of fuel cells become competitive rendering distributed generation a reality in some uses. Better means of IT based electronic communication and control systems reduce energy losses and waste. Despite the low costs of traditional energy sources, some renewable forms of energy are profitable in a number of niche market, including in developing country. Competition within and between energy sub-markets stimulates the development of new technology and organizational solutions.	State-owned oil companies are a rarity. Oil equivalent to today's Brent quality costs about US$ 10/bbl (in money of 1998) as technological improvement mean that the exploitation of heavy oil shales and tar sand is economic at this price. Few countries maintain special petroleum tax regimes. The distinctions between oil, gas, and power companies have disappeared. New types of energy companies form, partly organized around customer solutions. The real price of energy has continue to drop. Energy efficiency is increasing rapidly as new building techniques and IT based control systems are introduced on massive scale world wide.

ENERGY EFFICIENCY AND ALTERNATIVE SOURCES

RENEWABLE ENERGY DEVELOPMENT

NATIONAL ENERGY POLICY AND REGULATIONS ON RENEWABLE ENERGY

In 2007, the Government of Indonesia issued the Energy law, which consists of two main policies on energy development as follow:

a. Energy Conservation

b. Energy Diversification

Until the time the National Energy Council (DEN) establishes a new National Energy Policy (KEN), the National Energy Policy of 2006 is in force. The aim of this policy is to:

 Achieve energy elasticity to GDP of less than one by year 2025

 Realise an optimum primary energy consumption mix in 2025, with shares as follows:

– oil,—to become less than 20%

– natural gas—to become greater than 30%

– coal—to become greater than 33%

– biofuels—to become greater than 5%

– renewable energy and other energy including nuclear—to become greater than 10%

– liquefied coal—to become greater than 2%.

The details of the energy programs and targets of the National Energy Policy are elaborated in the *Blue Print – National Energy Management 2005 to 2025* (ESDM, 2006).

Indonesia's 2006 energy policy expects the combined share of renewable energy and nuclear in the overall energy mix in 2025 to exceed 17%. The policy places special emphasis on enhancing the share of biofuels. Renewable energy and other energy including nuclear (as in the list above) is expected to be made up of at least a 5% share from geothermal and a combined share of biomass, hydropower, solar, wind and nuclear power to comprise the remainder of the 10% by 2025.

CURRENT RENEWABLE ENERGY DEVELOPMENT

1 GEOTHERMAL ENERGY

Indonesia is a country with an abundant potential resources of geothermal energy that spread along the belt of active volcanoes from Sumatra, Java, Bali, Nusa Tenggara, Sulawesi, Maluku and West Papua. The data from the Geological Agency, the Ministry of Energy and Mineral

Resources indicates that until 2012 had identified 299 locations with total potential capacity of 28 GWe for further development as shown in Table 4 below.

Indonesia's Geothermal Potential, 2012

RESOURCES (MWe)	%	RESERVE (MWe)	%			
Speculative	Hypothetical	42.37%	Possible	Probable	Proven	57.63%
7,247	4,886		13,391	823	2,288	
12,133		16,502				
28,635						

Until mid 2013, 58 Geothermal Working Areas (GWAs) had been determined by the government of which 19 GWAs were the existing GWAs that had been established before the enactment of Law No. 27 Year 2003 regarding Geothermal

However, the current indirect utilization of geothermal energy for electricity is only reached 1,341 MW or 5% of Indonesia's total potential. On the other hand, direct utilization of geothermal energy (for non-electricity purposes) has not been yet optimizing.

Based on the Presidential Regulation No. 5 Year 2006 regarding the National Energy Policy (KEN), the government has set a target of 9,500 MW electricity by 2025 or equal to 5% of total primary energy, as shown in Figure 10 below.

In order to accelerate the geothermal utilization, the government issued the 10,000 MW Fast Track Program Phase 2 (FTP2), in which geothermal is targeted to contribute 4,965 MW of electricity.

In order to support the program, at least USD 15,000 million of investment is needed. This large amount of fund can not be borne entirely by the government. Thus stakeholder's participation and support from partners both within and outside the country, are strongly needed.

There are 2 ways for new investors for becoming a geothermal developers as shown in figure 5, as follows:

1. Preliminary Survey Assignment;

2. Obtaining Geothermal Business Permit (IUP), which awarded through GWA tender mechanism.

In 2012, total steam production of geothermal energy was about 69,882,712 tonnes and total electricity production of geothermal energy was about 9,291,090 MWh as shown in Table 5.

Steam and Electricity Production 2012

GWA	STEAM PROD (TONNES)	ELECTRICITY PROD (MWh)
Kamojang	10,878,385	1,455,258
Lahendong	3,261,670	446,809

Sibayak	160,361	15,960
Ulubelu	1,393,112	206,988
Salak	25,797,305	2,981,160
Darajat	14,174,881	2,189,644
Wayang Windu	13,169,817	1,845,674
Dieng	1,047,181	149,597
Total	69,882,712	9,291,090

2 Hydro Energy

Currently, Indonesia's total hydro energy potential resources was 75,670 MW which spreads across in many locations with the total installed capacity of 4,200 MW or 5.6% of the total hydro energy potential, as shown in Table 6.

In the regions with hydro potential, the central and local governments manage to build micro-hydro power plants (MHP) and after the projects completed it handed over to local people for management, operation and maintenance of MHP as efforts to provide electricity access to the community and to encourage community participation in managing its own energy.

Potential and Installed Capacity of Hydro Energy

No.	Island	Large Scale (>10 MW)	Small Scale (<10 MW)	Total (MW)			
		Potential	Installed Capacity	Potential	Installed Capacity	Potential	Installed Capacity
1	Sumatera	16,100.00	1,154.00	281.76	83.44	16,381.76	1,237.44
2	Java	12,050.00	2,012.50	222.02	212.32	12,272.02	2,224.82
3	Kalimantan	5,999.50	30.00	277.75	31.27	6,277.25	61.27
4	Sulawesi	14,550.00	352.00	167.56	118.05	14,717.56	470.05
5	Bali-Nusa Tenggara	4,900.00	0.00	31.64	12.25	4,931.64	12.25
6	Maluku-Papua	21,057.00	23.00	32.78	4.67	21,089.78	27.67
Total	74,656.50	3,571.50	1,013.51	462.00	75,670.01	4,033.50	

However, there are some challenges currently faced by the government in the development of hydro energy, as follows:

- Despite the good quality of local production turbines, the capacity of turbines are still below 1 MW. Therefore, it still needs support in research and development, as well as local manufacturing capabilities to produce turbines with capacity more than 500 kW;

- Hydro energy resources is generally located in forested areas where there are no adequate existing transportation infrastructure, thus its causing high investment cost;

- MHP often utilizes the resources of the state-controlled forests or indigenous people. MHP development is often hampered by the difficulty of obtaining permits from various parties. Therefore, it needs a mediation (clearing house); and

- MHP sustainability depends on the preservation of water resources, therefore the development of MHP should be protected by policies and regulations on water resources use and spatial planning to ensure water sustainability and availability.

3 Solar Energy

Located in equator line, Indonesia has an abundant resources of solar energy which is suitable to use in small islands, remote areas, and border areas where there are no existing electricity installation. Very small scale electricity system in remote areas is currently planning to be supplied by solar power plants (SPP) in collaboration with the local government.

There are three alternatives of solar project implementation scheme, which are 1) Solar Photovoltaic (PV) or Solar Home System (SHS), 2) Centralized PV power plant, and 3) Hybrid System (using solar PV together with other existing resources, especially diesel power).

In 2009, Indonesia's total solar photovoltaic/solar home system installed capacity was 13.49 kW or increased by 55.6% from 2008 as shown in Table 7. The total installed capacity of solar photovoltaic/solar home system is still expected to increase in the near future due to the government's electrification ratio program in small islands, remote areas and border areas will use solar photovoltaic/solar home system to electrify community.

Installed Capacity of SPP/SHS

No.	Island	Capacity/Year (kW)				
		2005	2006	2007	2008	2009
1	Sumatera	0.33	0.78	1.69	2.65	4.28
2	Java	0.33	0.40	0.53	0.78	1.15
3	Kalimantan	0.16	0.38	0.71	1.11	1.93
4	Sulawesi	0.12	0.64	1.37	1.98	2.85
5	Bali, Nusa Tenggara	0.12	0.35	0.62	1.00	1.41
6	Maluku, Papua	0.16	0.36	0.71	1.15	1.87
Total	1.22	2.91	5.63	8.67	13.49	

However, there are some challenges currently faced by the government in the development of solar energy, as follows:

- A strategy to encourage the capability of local solar panel industries increased and expanded as well as other components in developing SPP; and

- Changing people's image of SPP reliability. SPP devices are already widely trading in the market with vary of the quality that it has not been yet controlled by the government at the moment. So that there are many solar panels with low quality are selling in the market and making disappointed to consumers to use, thus damaging the image of SPP reliability.

In order to encourage the participation foreign companies in developing the solar industry in Indonesia as well as to boost utilization of solar energy, currently the government is finalizing the draft of the Ministerial Regulation regarding feed-in-tariff (FIT) in the field of solar energy. Through this FIT scheme, more incentives will be given to both local and foreign investors who can maximize the local content in their products. Based on the Presidential Regulation No. 54 Year

2010 regarding Procurement of Goods/Services in the Government, the minimum local content that has been set by the government is 40%.

Regarding PV development in Indonesia, there are 3 government programs that could be implemented, namely:

a. Rural PV: the aim is to increase electricity accessibility for rural community, especially in remote areas, small islands, and border areas using Solar Home System and Centralized PV System (off grid).

b. Urban PV: the aim is to provide electricity for own used to middle and high class households, real estates, office and commercial buildings, hotels and resorts, industries and other (off grid and on grid).

c. On grid PV: the aim is to offer business opportunity to Independent Power Producers (IPPs) in developing Solar Power Plant and selling the electricity to PLN.

4 Wind Energy

Since Indonesia is located on the equator line thus by nature the wind energy potential is relatively small. However, there are some areas have a good wind speeds as an effect of nozzle in the narrowing area between two islands or in the mountain slopes between two adjacent mountains.

On the other hand, in order to deal with intermittent characteristic of wind energy, application of wind power generation system with a special design is required for Indonesia.

Noting this situation and condition, the development of wind energy in Indonesia will be focused on or concentrated in the areas with wind potential, and the development of wind power plants in remote areas, small islands, and border areas is prioritized using hybrid systems together with existing diesel power plants.

In 2010, Indonesia's total installed capacity of wind power plant was 1,962.36 kW or increased by 5% from 2009 as shown in Table 8.

Installed Capacity of Wind Power Plant

No.	Island	Capacity/Year (kW)					
		2005	2006	2007	2008	2009	2010
1	Sumatera	1.50	81.50	81.50	81.50	81.50	85.58
2	Java	285.75	285.75	285.75	285.75	285.75	300.04
3	Kalimantan	0.00	0.00	0.00	0.00	0.00	0.00
4	Sulawesi	148.70	228.70	388.70	588.70	588.70	618.14
5	Bali, Nusa Tenggara	591.05	591.05	911.05	911.05	911.05	956.60
6	Maluku, Papua	2.00	2.00	2.00	2.00	2.00	2.00
Total	1,029.00	1,189.00	1,669.00	1,869.00	1,869.00	1,962.36	

However, there are some challenges currently faced by the government in the development of wind energy, as follows:

 There is not comprehensive data and maps of wind energy potential available.

 Investment cost of wind power plant is still relatively high (energy prices are still high) compared to other conventional power plant investments.

 There is no significant incentives available for industries in developing wind energy utilization.

5 Bioenergy

As agricultural based country, Indonesia has an abundant resources of bioenergy. The biggest bioenergy potential are coming from agro-industries and municipal wastes. It is estimated that the potential of bioenergy is about 49,810 MW. Similar with other renewable energy, bioenergy is also categorized as a clean energy, which not only could increase the national energy security but also give a good contribution to the environment.

The development of bioenergy in Indonesia is a part of green energy development under the program of the National Action of Greenhouse-gas Mitigation (*Program Rencana Aksi Nasional Gas Rumah Kaca*–RAN GRK). Bioenergy is also considered as local energy sources that considered as the most proper energy to improve the modern energy access to people in rural areas. Therefore, bioenergy development has become one of the priority energy development in Indonesia in order to increase the national energy security. Under the Presidential Regulation No. 5 Year 2006, bioenergy is expected to have a significant contribution in the national primary energy mix, with minimum share to 5% by 2025 or even the minimum share of bioenergy targets set higher to 8.9% by 2025 under the government's new vision, namely Vision 25/25.

In Indonesia, products of bioenergy are developed in the form of biofuel (liquid), biogas, bio-solid (bio-briquettes, pellets) and bio-electricity, as shown in Figure 12 below.

Even though Indonesia has a big potency of bioenergy, the utilization of bioenergy is still very low compared to its potency. It is estimated that the total bioenergy installed capacity has reached about 1,618.40 MW.

As mentioned above, the products of bioenergy are included bio-briquettes and pellets. Bio-briquettes and pellets are not a priority to be developed by the government currently, since its utilization can be replaced by the traditional biomass or simply by agricultural wastes. However, bio-briquettes and pellets are developed by some industries from some agro-industry wastes, and these products are exported to some countries.

A very different situation can be seen in the solid-biomass, the development of biogas currently is done in intensively manner, both for the domestic utilization (cooking and light), and also the electricity production. Biogas technology is considered as the grass-root technology where Indonesia is able to produce the technology. In Indonesia, the biogas development is carried out through 3 (three) following approaches, namely:

a. Non-Commercial (under government investment)

This scheme is addressed to increase the access on modern energy to rural people, especially for those who live in isolated areas. This approach also has aim to introduce the bioenergy technology to stimulate people's interest to utilize biogas system. Currently, under the Ministry of Energy and Mineral Resources budget, there are two programs using this scheme, namely the Biogas-Energy Self-sufficient Village (ESSV) program and the Special Allocation Fund Program

(*Program Dana Alokasi Khusus*). The non-commercial approach for the biogas development is also conducted by other ministries, such as the Ministry of Agriculture, the Ministry of Cooperative and Small-Medium Enterprises, the Ministry of Environment, and the State Ministry for Development of Disadvantaged Regions.

b. Semi Commercial (under application of partial subsidies)

This approach has a mechanism where subsidy is applied to stimulate participation of people to use the biogas system and the involvement of private sectors and financial institutions. Currently, Indonesia has such program which is implemented under a bilateral cooperation between Indonesia and the Netherlands, namely the Domestic Biogas Program (*Program Biogas Rumah-Program BIRU*). The chosen area to implement this program is the area where there is credit available to the consumers/biogas users provided by financial institutions. The program provides Rp 2 million to each beneficiary to construct the biogas unit. Remaining cost of the

biogas unit construction is paid by the beneficiary, by cash or through the credit from the financial institution who involved this program.

c. Commercial (under Private Investment or Public-Private Partnership)

The scheme for commercial approached is under business-to-business (B to B) in where usually biogas is produced in large scales to produce electricity. What the government can do to encourage involvement of private sector in this scheme is to provide proper regulations and to improve the investment climate.

Further, Indonesia has also a priority to develop the biofuel (liquid), which consists of biodiesel, bioethanol and bio-oil. In 2008, the government issued the Ministerial Regulation No. 32 Year 2008 regarding Mandatory of Biofuel Utilization. This regulation mandates the utilization of biofuel in the transportation, industry as well as power plant sectors. Based on this regulation, the minimum of utilization of biodiesel is expected to become 20% and bioethanol to become 15% by 2025, as shown in Table 9 below.

MANDATORY ON BIOFUEL UTILIZATION

(Ministerial Regulation No. 32 Year 2008)

BIODIESEL (Minimum)

SECTOR	Year					
	2008	2009	2010	2015	2020	2025
Transportation, PSO*)	1%	1%	2.5%	5%	10%	20%
Transportation, Non PSO	-	1%	3%	7%	10%	20%
Industry	2.5%	2.5%	5%	10%	15%	20%
Electricity Generation	0.1%	0.25%	1%	10%	15%	20%
BIOETHANOL (Minimum)						
SECTOR	Year					
	2008	2009	2010	2015	2020	2025
Transportation, PSO	3%	1%	3%	5%	10%	15%
Transportation, Non PSO	5%	5%	7%	10%	12%	15%
Industry	-	5%	7%	10%	12%	15%

PURE PLANT OIL (Minimum)							
SECTOR	Year						
	2008	2009	2010	2015	2020	2025	
Industry and transportation (low & medium speed engine)	Industry	-	-	1%	3%	5%	10%
	Marine	-	-	1%	3%	5%	10%
Electricity Generation	-	0.25%	1%	5%	7%	10%	

**) PSO fuel: Public Service Obligation fuel means subsidized fuel*

Indonesia has various sources of raw material for both biodiesel and bioethanol. Biodiesel can be gained through the process of oil from diverse plants, such as from palm, coconut, *jarak pagar* (*Jatropha Curcas*), *nyamplung* (*Callophylum*), and *kemiri sunan* (*Aleurites Trisperma*). However, the current raw material for the Indonesian's biodiesel production is crude palm oil (CPO), in where

Indonesia is the biggest producer of the CPO in the worlds. In 2011, the production of CPO was reached 23 million ton or increased about 5% compare 2010's production (Kemtan, 2013).

Similar with biodiesel, bioethanol also can be produced from various raw materials, such as from mollases from sugar industry, cassava, and sorghum.

Biodiesel industry is more advanced nowadays in Indonesia compared to bioethanol industry. It is estimated that Indonesia produces about 21 million ton per year of biodiesel. Currently, there are 23 biodiesel producers with the total installed capacity of around 4.5 million KL/year. For the bioethanol industry, it is identified that there are only 7 bioethanol producers which have commercial business licenses with the total installed capacity of about 286,686 KL/year. Some small industries have been identified in producing mitanol (kerosene from bioethanol) as a fuel for domestic purposes, for example for cooking and lighting. However, since mitanol is produced in small-scale industry, data and information related to this is limited.

Currently, under the mandatory regulation, the implementation of biodiesel mixed fuel in PSO transportation sector has achieved to 7.5%, in which 7.5% of biodiesel was mixed into the PSO diesel fuels, called as B-7.5. These fuels are distributed by Pertamina (State Owned Oil Company) as the off-taker. Furthermore, the government has a plan to mix 2.5% of biodiesel into the non-PSO diesel fuels, and it will be sold by other oil companies, such as Shell and Petronas.

In the national market, biofuel is sold based on the Market Price Index, which is regulated under the Ministerial Decree. In order to support the implementation of the mandatory regulation, the government provides subsidy but only for the biofuel which is mixed in PSO-fuel for transportation. Subsidy is defined as the difference between the Market Indeks Price of biofuel and the reference price of conventional fuels (solar/gasoline), which is currently taken from Means of Platss Singapore (MOPS). The figure of subsidy can be seen in Figure 13 below.

In 2011, the subsidy was set by the government at the maximum of Rp 2,000,- per litre, both for biodiesel and bioethanol. In order to accelerate the utilization of biofuel, the government was allocated subsidy about Rp 2,937 billion in 2012, of which Rp 2,083 billion for biodiesel with quota to 694 thousand KL and Rp 854 billion for bioethanol with quota to 244 thousand KL.

With this figure, in 2012, the subsidy for biodiesel was became Rp 3,000,- per litre or increased Rp1,000,- per litre compared to 2011's subsidy and bioethanol became Rp. 3,500,- per litre or increased Rp1,500,- per litre compared to 2011's subsidy.

In 2012, there were 669 thousands KL of biofuels from blending biofuels and fossil fuels in Indonesia or increased almost doubling from 358 thousands KL in 2011. As of January 2013, the utilization of biodiesel was reached around 59 thousands KL.

A very different situation can be seen in the Indonesian bioethanol utilization, due to constraints on pricing, it is causing reluctances to supply bioethanol in Indonesia and as a result there was no utilization of bioethanol since 2010. To overcome this obstacle, the government under the Ministry of Energy and Mineral Resources has submitted a new proposal on biofuels price index revisions to the Ministry of Finance; however it has not yet been approved.

The Government of Indonesia always recaps and reviews progressing toward the achivement of government program and policy on biogas and biofuels development as an evaluation and annual monitoring. Until the end of year 2012, the government had been implemented several programs for accelerating biogas utilization. One of the programs was Desa Mandiri Energi (DME), which was utilizing the biogas for rural areas. The other program was cooperation between the government and Hivos (Non-Government agency from the Dutch Goverment) for providing biogas unit for households. Table 10 shows the several achievements of biogas program in Indonesia up to year 2012.

Achivement for Biogas Program in Indonesia until the End of Year 2012

NO	IMPLEMENTOR	SPECIFICATION AND CAPACITY	TOTAL
1	Directorate General of New Renewable Energy and Energy Conservation, MEMR (Energy Rural Program/Program DME for Biogas Communal and Household)	□ Household Biogas Digester: capacity of 4 - 6 M3 , □ Communal: capacity of 20 - 30 M3, and □ Waste of Tofu Industry: capacity of 40 - 136 M3	□ Capacity of 6 M3 : 682 units, □ Capacity of 20 M3 : 54 units □ Capacity of 40 M3 : 1 unit □ Capacity of 90 M3 : 4 units □ Capacity of 136 M3 : 1 unit
2	Indonesia Domestic Biogas/ Program BIRU (Cooperation Between GOI-HIVOS)	Target 8,000 of biogas units for household (capacity of 4 - 6 M3)	Realization: 7,432 units
3	PT Navigat Organic Energy Indonesia	Waste Power Plant from Municipal Solid Waste (Biogas), Capacity of 26 MW (COD in Sept. 2012 with total 8 MW of Biogas Landfill), on Grid	1 unit with capacity of 26 MW of electricity
4	PTPN V, Kebun Tandun	Waste Power Plant from Palm Oil Mill Effluent (POME), Capacity of 1 MW (COD in March 2012), off grid	1 unit with capacity of 1 MW of electricity

		for own using in palm plantation	

For biofuel, both biodiesel and bioethanol, the realization of biodiesel utilization in Indonesia until the end of year 2012 was only 544,963 kL or 38.93% from 1,400,000 KL of the mandatory

regulation target, as shown in Figure 14 below. Therefore, to increase the utilization of biofuel and to achieve the mandatory regulation target, the government has increased the subsidy for biofuel in 2013.

OTHER NEW AND RENEWABLE ENERGY

1 Nuclear

Based on calculation conducted by PLN, nuclear power plant (NPP) cannot compete with other types of power plant, such as a of 1,000 MW supercritical coal power plant. The Fukushima Daichi NPP accident in March 2011, where thousands of residents who live near the plant must be evacuated to a safe area, has led to an escalation of opposition to develop nuclear energy for power generation. The decision to build a nuclear power plant is not based solely on economic and energy availability considerations, but also other considerations such as political, security, social, cultural and environmental. With those multi-dimensional aspects, the nuclear power plant construction program in Indonesia can only be decided by the government.

2 Liquefied Coal and Gasified Coal

As stated in the Government Regulation No. 5 Year 2006 regarding the National Energy Policy, the share of coal in primary energy consumption is expected to become greater than 33% in 2025. In Indonesia, coal has been projected as the main energy source to replace oil fuels until 2025, mainly for power generation. Therefore, in order to reduce CO2 emission from coal fired power plant, it is necessary developing an environmental friendly power generation through clean coal technology implementation such as coal gasification and coal liquefaction.

Currently, Indonesia has reserves of about 60 billion tons of coal which found it throughout of Indonesia. Nearly 85% of the reserves are young coal (lignite) or in other words, low quality coal. It is because contains of 30% water content and has a low calorie with a low sale value. Lignite, also known as brown coal, in which could be developed as an alternative to petroleum.

In Indonesia, there are 2 potential locations for constructing a liquefied coal factory, one is in Banyuasin in South Sumatra, which has reserves of 2.9 billion tons of coal and the other is in Berau, East Kalimantan, which has reserves of 3 billion tons of coal. In addition, the large number of coal resources in Indonesia caused coal reserves can be utilized both as for gas fuel or power generation through coal gasification process.

3 Coal Bed Methane (CBM)

CBM resources in Indonesia reached 453.3 Trillion Cubic Feet (TCF) spreads in 11 areas of the coal basin at various locations in Indonesia, namely Sumatra, Java, Kalimantan and Sulawesi. Indonesia's CBM potential of 453.3 TCF includes proven reserves of 112.47 TCF and potential reserves of 57.60 TCF.

In 2015 Indonesia will be predicted to produce CBM up to 500 million cubic feet per day (MMSCFD) and this production can be increased to 900 MMSCFD in 2020. CBM production in Indonesia is expected to reach 1,500 MMSCFD by 2025.

4 Ocean Energy

Indonesia is a country in the world that has the largest maritime territory, with approximately two-thirds of its total area. Indonesia has the second longest beach in the world after Canada, where the length of the beach is about 80,000 km and the width of the ocean is about 52 million km2.

Indonesia's ocean area along the southern coast of Java to Lombok is a location that has a large wave energy potential ranging from 10 to 20 kW per meter waves. Some studies have concluded that wave energy at some point in Indonesia could reach 70 kW per meter waves in some locations. The west coast of southern Sumatra island and the south coast of western Java have ocean wave energy potential about 40 kW per meter waves.

There are several alternative technologies that could be chosen for utilizing ocean wave energy. One of them that will be developed by Indonesia in the south coast of Java island is Tapered Channel Technology (Tapchan). However, until now the utilization of wave energy in Indonesia is still in the research stage with small capacity developed by the research and development institutions (BPPT-the Agency for the Assessment and Application Technology, PLN) and other educational institutions.

Indonesia has also tidal energy potential in some areas such as in Bagansiapiapi, Riau where the tidal high reaches 7 feet, in Palu Bay, Central Sulawesi where its geological structure enables the tidal phenomenon, in the Gulf of Bima in Sumbawa, Nusa Tenggara Barat, in West Kalimantan, in Papua, and in the south coast of Java where the tidal can reach more than 5 feet of high.

Not much different from tidal energy, ocean thermal energy in Indonesia also recently reached in the stage of research (potential data collection and feasibility study) conducted by several research institutions. As for ocean currents, current utilization for power generation has reached the implementation phase (pilot project) by T-files team from Bandung Institute of Technology (ITB) and Dr. Erwandi from Indonesia Hydrodynamics Laboratory, the Agency for The Assessment and Application of Technology (BPPT) and the Ministry of Marine Affairs and Fisheries.

STANDARDIZATION

1 Standardization of Bioenergy Products

Like in any other countries, the development of bioenergy in Indonesia also considers the standards and safety aspects in order to protect the consumers from low quality products and to increase the competitiveness of the bioenergy producers. The development of standard is also one effort to make conducive business environment to the bioenergy producers with the guidelines. Some the following standards on bioenergy have been issued by the government, namely:

a. SNI 7826 : 2012 - Biogas producer unit with digester fixed dome type from concrete

b. SNI 7182 : 2012 - Biodiesel

c. SNI 7390 : 2012 - Denaturized bioethanol for gasohol

Some standards are under preparation, namely:

a. RSNI on Basic security system of hydrogen fuel

b. Biomass stove performance standard

2 Standardization of Hydro Power Plant

The development of hydro power plant in Indonesia also considers the standards and safety aspects in order to protect the consumers from low quality products and to increase the competitiveness of the hydro power plant industries. The development of standard is also one effort to make conducive business environment to the hydro power plant industries with the guidelines. Some the following standards on hydro power plant are under preparation, namely:

a. RSNI on Capacity design and lay-out of micro hydro power generation system with capacity up to 25 kW.

b. RSNI on Technical specifications of cross-flow turbines with capacity up to 35 kW.

3 Standardization of Solar Power Plant

The development of solar power plant in Indonesia also considers the standards and safety aspects in order to protect the consumers from low quality products and to increase the competitiveness of the solar power industries. The development of standard is also one effort to make conducive business environment to the solar power plant industries with the guidelines. Some the following standards on solar power plant are under preparation, namely:

a. RSNI on Characteristic parameters of stand-alone photovoltaic (PV) system

b. RSNI on Crystalline silicon terrestrial photovoltaic (PV) modules - design qualification and approval type

4 Standardization of Wind Power Plant

The development of wind power plant in Indonesia also considers the standards and safety aspects in order to protect the consumers from low quality products and to increase the competitiveness of the wind power industries. The development of standard is also one effort to make conducive business environment to the wind power plant industries with the guidelines. Some the following standards on wind power plant are under preparation, namely:

1. RSNI on Testing performance of electrical power produced by wind turbine

2. RSNI on Plan requirements of small scale wind turbine

CLEAN ENERGY DEVELOPMENT

The Indonesian Government has commitment to support the development of clean energy, which is not only dedicated to maintain the environmental friendly, but also to secure the sustainability of energy supply for the nation. The development of new and renewable energy in Indonesia is a major part of clean energy development, a part from clean energy technology and energy efficiency implementation. With the growing concerns on climate change and reducing GHG emission from energy sector, new and renewable energy development obtains enourmous support in the clean energy development.

As a part of clean energy development, new and renewable energy also provides co-benefits to the country, as it supports the sustainable development both in local and national level.

1 NATIONAL MITIGATION ACTIONS

In 2010, the President of Republic of Indonesia announced the Indonesia's voluntary commitment to reduce national GHG emission to 26% in 2020 from the business-as-usual scenario and it can be further reduce to 41% with the international support. This voluntary commitment is part of Indonesia's active participation to implement the Bali Action Plan that called for the developing countries to contribute in the global GHG emission reduction.

To achieve the voluntary commitment, the Presidential Regulation No. 61 Year 2011 regarding the National Action Plan on Greenhouse Gasses Emission Reduction (RAN-GRK) has been issued by the government. The RAN-GRK is not only enacted the national targets, but also mandated national mitigation actions to achieve the targets, instructed the implementation and evaluation of the actions plans by line ministries/institutions, and guided the development of the Local Action Plan on Greenhouse Gases Reduction (RAD-GRK) by the local governments.

The core activities (defines as the activities that directly reduce the GHG emission) are conducted in 5 main mitigation sectors:

* Agriculture;
* Forestry and peat land;
* Energy and trasnportation;
* Industry; and
* Waste management.

As defined in Table 11 below.

Core Activities of the RAN-GRK

Sector	Emission Reduction (Giga Ton CO2e)	otal	Mitigation Actions	
	26%	15% (Total 41%)		
Forestry and Peatland	0.672	0.367	1.039	□ Control on forest and peatland burning; □ Management of water supply & network system; □ Forest and land

				rehabilitation including Industrial forest and community forest; □ Removal of illegal logging; □ Preventing deforestation; □ Community emporwerment
Agriculture	0.008	0.003	0.011	□ Introduction of low emission rice variety; □ Eficiency of irrigation water; □ Usage of organic fertilizer
Energy and Transportation	0.038	0.018	0.056	□ Biofuel utilization; □ Utilization of high efficiency engine; □ Increasing quality of public transport and roads; □ Demand side management; □ Energy efficiency; □ Renewable energy

As stipulated in the Presidential Regulation No. 61 Year 2011, new renewable energy development, such as renewable energy-based power generation and bioenergy development (biogas and biofuel), is not only measured for its energy supplied, but also monitored and reported for its contribution on GHG emission reduction in national level.

As a part of the RAN-GRK, the Presidential Regulation also stipulated that the local governments must participate in reducing greenhouse gas emission in each respective provinces/regencies through the development of Local Action Plan on Greenhouse Gases Reduction (RAD-GRK). The RAD-GRK is defined as the workplan to implement activities that directly and indirectly reduce the GHGs emission reduction according to local development target and taking into account the RAN-GRK. The RAD-GRK will be legally binding policy for 1 (one) year after the issuance of the Presidential Regulation by Governor Decree (approximately by September 2012). The RAD-GRK will be reported to the Bappenas (the Ministry of National Development Planning) and the Minister of Interior.

In December 2012, the Bappenas and the Minister of Interior officially launched the 33 RAD-GRK developed and issued by the local governors in Indonesia.

2 Indonesia Carbon Market: Policy and Status

Indonesia participation in carbon market officially started in 2004, after the ratification of Kyoto Protocol under Law No. 17 Year 2004. The entry of Indonesia as part of the Kyoto Protocol enabled Indonesian project proponent to be involved in the flexible mechanism, namely Clean Development Mechanism (CDM). In order to answer the inquiry on the national approval of the CDM projects from the United Nations Framework Convention on Climate Change (UNFCCC), the Ministry of Environment had issued the Ministerial Decree No. 206 Year 2005 regarding the National Commission on CDM, which is mandating several representative of the ministries as the Committee members and appointing the Ministry of Environment as the authorities' institutions to manage the approval process. The 2005 decree was renewed by the Ministerial Decree No. 522 Year 2009, where the management of the approval process has been moved to the National Council on Climate Change (NCCC).

The first project of CDM conducted by the Indonesian project proponent was the solar cooker project in Province of Aceh, registered in the Executive Board of CDM in year 2006. Up to now, the National Committee on CDM has been reported that there are 240 CDM projects in Indonesia, in which 81 projects registered to the EB of CDM, including 4 Programmatics of CDM. Most of the projects are coming from energy sector related to energy supply, energy distribution and transmission, and energy efficiency in demand side. In Indonesia, proposing the energy projects as the CDM projects is perceived getting financial and political supports to the projects development. However, since the CDM windows are closed for the Indonesian stakeholders, with the EU—ETS as the main buyer of CER restricting the supply to the Least Developed Countries (LDCs).

Apart from CDM, the National Committee on CDM also processes and facilitates the national approval of voluntary carbon market. For anticipating new market mechanisms under the Post-Kyoto regimes, the National Council on Climate Change will explore and participate in international, regional and bilateral carbon market initiatives, such as new CDM, REDD+, crediting NAMAs, regional market, domestic market, joint credit mechanism (i.e with the Japanese Government), framework of various approaches (FVA), and others.

Those various initiatives and mechanisms on carbon markets can be classified into several carbon markets development strategies as follows:

- Multilateral carbon market: currently being negotiation; robust, and slightly complicated; environmental integrity and sustainable development criteria are not clearly defined yet;

- Bilateral and regional carbon market: directly between Indonesia and some developed countries; International carbon offset; Japan has already expressed their interests and it would be started soon; and

- Domestic carbon market: can be mandatory or voluntary basis; national boundary; simple and robust; can be used for NAMAs.

As for domestic carbon market, currently the NCCC develops the Nusantara Carbon Scheme (SKN) as the national appropriate carbon crediting mechanism. There are several considerations for the development of SKN, as follows:

1) as an alternative for the financial support in domestic mitigation actions;

2) as a catalyst for other new market mechanisms;

3) to maintain the momentum of Indonesia carbon market development; and

4) to ensure environmental integrity and sustainable development in Indonesia mitigation actions.

Currently, the NCCC has been developed at least 6 methodologies to the projects that could be divided into the afforestation and reforestation sector, biomass boiler utilization, composting, high-efficient boiler, biomass cookstove and mini/micro hydro power plant.

POLICY ON INCENTIVES AND FISCAL

In order to encourage and attract the participation of foreign and domestic investors as well as developers in developing new and renewable energy in Indonesia, the government has provided some initiatives on incentive and fiscal, as follows:

1. Increasing tariff/electricity selling price of new and renewable energy through feed in tariff (FIT)in which PLN is obliged to purchase renewable energy at a predetermined price (MEMR Regulation No. 04 Year 2012 (for Small and Medium Renewable Power Plant) and MEMR Regulation No. 17 Year 2013 (for Solar Photovoltaic Power Plant));

2. Giving fiscal incentive such as provision of corporate income tax relief or reduction facility (Regulation of the Minister of Finance (PMK) No. 130/PMK.011/2011). The government also gives tax breaks and customs (PMK No. 21/PMK.11/2010).

3. Incentive of funding;

4. Incentive of market support; and

5. Ease of licensing/finishing of overlapping the land.

INDONESIA'S RENEWABLE ENERGY POTENTIAL

Indonesia has a large new and renewable energy potential, which includes 450 MW of mini/micro hydro power, 50 GW of Biomass, 4,80 KWh/m2/day solar power, 3-6 M/sec wind power, and 3 GW of nuclear energy. This new new and renewable energy potential is explained by the Director General of Electricity and Energy Conservation at the Focus Group Discussion on the Supply and Demand of New and Renewable Resources held by Pusdatin ESDM.

Currently the development of renewable energy is regulated by Presidential Decree No.5 / 2006 regarding the national energy policy. This decree states that the contribution of new and renewable energy in the 2025 national primary energy mix is estimated at 17%, consisting of 5% biofuel, 5% geothermal power, biomass, nuclear, hydro, and wind, and also liquefied coal at 2%. The government will take measures to to add the capacity of Micro Hydro power plants to 2,846 MW by 2025, Biomass of 180 MW by 2020, wind power (Bayu power plant) of 0,97 GW by 2025, solar of 0,87 GW by 2024, and nuclear power of 4,2 GW by 2024. The total investment needed for this development of new and renewable energy sources up to the year 2025 is projected at US$ 13,197 million.

The steps needed for biomass development involves waste recycling from the farming and forestry industry as a source of energy which is integrated with its respective industry, integrating biomass development with the local economic development, boost the fabrication of biomass energy convertion technology including its supporting sectors, and increasing research and development of waste recycling, including city waste, for energy.

Efforts to develop wind energy includes its development for both electricity and non-electricity usage (as a pump for irrigation and clean water), development of small scale (10 KW) and middle scale (50-100 MW) wind energy technology and supporting factories to mass produce small and middle scale SKEA

Development of solar energy includes the use of solar powered power plants in the rural and city areas, supporting the commercialization of solar powered power plants by maximizing the private sector role, development of domestic solar powered power plants industry, and the creation of an efficient funding system through the involvement of the banking sector.

To develop nuclear energy, the government needs to take socialization measures to gain the support of the public and coordinate with several countries to master its technology.

In developing microhydro, the government needs to integrate the microhydro powered power plants with the local economy, maximizing the irrigation potential for the power plant and developing the domestic microhydro industry, and developing several partnership and funding patterns.

To support the development of new and renewable energy, the government has issued several rules and regulations, consisting of Presidential Decree No. 5 / 2006, on the National Energy Policy, Law No. 30 / 2007 on Energy, Law No. 15/1985 on electricity, Government Regulation No. 10 / 1989 which is renewed by Government Regulation No. 03 / 2005 and No.26 / 2006 regarding the supply and usage of electricity, Ministerial Regulation No. 002 / 2006 on the commercialization of middle scale renewable energy power plants, and Minister of Energy and Mineral Resources Decree No.1122k/30/MEM/2002 on the spread of small scale power plants. Currently the government is composing a law on new and renewable energy which includes the supply and usage of new and renewable energy along with its incentives.

EFFICIENCY IMPROVEMENT GOALS

The National Energy Conservation Master Plan (2005)—(*Rencana Induk Konservasi Energi Nasional,* or RIKEN*)* states that Indonesia's goal is to decrease energy intensity by an average of 1% per year to 2025.

The National Energy Management Blueprint (2006)—(*Blueprint Pengelolaan Energi Nasional*, or PEN)1 states Indonesia's goal of achieving 41% reduction of total primary energy supply (TPES) in 2025 against TPES forecast in the business as usual case, though measures of RIKEN in energy efficiency and conservation (EE&C).

The National Energy Policy (2006)2 states that Indonesia's goal is to achieve energy elasticity of less than 1 in 2025. Note: energy elasticity is defined, in this case, as the rate of change of total primary energy supply, over the rate of change of GDP.

SECTORAL ENERGY EFFICIENCY IMPROVEMENT GOALS

RIKEN identified energy saving potential in the sectors, as follows:

Industry sector (for select industries)—15% to 30%

Commercial building sector—electricity savings of 25%

Residential sector—10% to 30%.

ACTION PLANS FOR PROMOTING ENERGY EFFICIENCY

a) Objectives

The objective of Indonesia's energy conservation program is —to conserve natural energy resources and increase energy supply resilience to support sustainable development□.3

b) Applicable sectors

Industry, commercial building, households, and vehicles; beginning with those buildings and vehicles of government departments and agencies, provincial governments, and state-owned enterprises.

c) Outline

Energy programs:

Mandatory energy conservation of government office buildings: Government departments and agencies and regional governments are mandated to implement best-practice energy saving measures as explained in the government's guidelines and directives on energy saving in government buildings, and are required to report their monthly energy use in buildings to the National Team on Energy and Water Efficiency, every six months4.

State-owned energy service company (ESCO): The state-owned ESCO (established in 1986) is expected to take a leading role in providing energy conservation related services, particularly to industry. The government expects its ESCO to maintain forefront expertise in the field of energy efficiency and conservation in Indonesia, and to encourage a greater role for private-sector ESCOs in the future.5

Public—Private Partnership Program on Energy Conservation: The Partnership Program on Energy Conservation is a government-funded energy audit program that is available to industries and commercial buildings. Participating industries and commercial buildings are required to implement the recommended energy saving measures identified in the energy audit.

The *Energy Conservation Clearinghouse* was created for the purpose of data and information exchange on energy efficiency and conservation, particularly for the industry sector and commercial buildings.

Energy benchmark and best practice guide for specific industrial energy use, and energy use in commercial buildings.

Energy Labelling

Indonesia's energy labelling program began in 1999. A dual energy rating system was considered for electrical appliances, initially for refrigerators. The energy labelling system design shows: (1) information about the kWh per year energy consumption of a product and its relative position on a line from the lowest to highest case of kWh per year consumption of similar products in the market (Indonesia), and (2) an energy consumption star rating—of four stars—that shows the product's energy efficiency rank, relative to similar products in the market (Indonesia) at the time of assessment.6

This energy labelling system was discontinued, however, to be replaced by a new energy labelling system and design.

A new energy labelling system is currently being developed. The design will provide information on: (1) the absolute energy efficiency or performance of a product; and (2) an energy efficiency star rating of four stars. The star rating is to be assigned by an independent and accredited test facility that tested the product. The new energy label design is shown (right). It shows an example for the case of an energy label for compact fluorescent lamps—CFLs; the energy label provides information on the lumens produced per watt.

BRESL: To remove barriers in implementing energy standards and labelling (ES&L), Indonesia is currently participating in a UNDP-GEF project: Barrier removal to the cost effective development and implementation of energy efficiency standards and labelling project—BRESL. The program involves six developing economies of Asia. BRESL has five major programs in promoting ES&L. The programs are: (1) policy making, (2) capacity building, (3) manufacture support, (4) regional cooperation, and (5) pilot projects.7

Energy efficient lighting program in the residential sector: The lighting program in the residential sector is primarily a demand-side management (DSM) programme, in addition to energy savings. There are two lighting programs. They are: (1) the Caring Program (*Program Perduli*)—a program of the state owned electricity company—PLN and (2) Brightness Program (*Program Terang*)—a government program. The programs provide subsidised, and in certain cases, free CFLs to eligible households.

Energy Awards: Indonesia is an active participant in the ASEAN Energy Award program, specifically the Best Practice Competition for Energy Efficient Buildings and Best Practice Competition for Energy Management in Buildings and Industries. Indonesia has won several awards in these programs.

d) Financial resources and budget allocation

An annual government budget is allocated for energy conservation programs and R&D. The government budget for the Energy Conservation Partnership Program—energy audit was 4 billion IDR in FY2009, 20 billion IDR each of FY2010 and FY2011.

e) Method for monitoring and measuring effects of action plans

Energy consumption data is obtained on a regular basis by the Central Statistical Agency—(*Badan Pusat Statistik*, or BPS). Specific data are collected through various government programmes on EE&C. The Partnership Program of (the former) Directorate General of Electricity and Energy Utilization provides data on energy intensity and energy saving potential in industry and commercial buildings. As of 2009, 292 industries and commercial buildings had been audited by the program. Data on energy use in buildings of government departments and agencies and regional governments is obtained regularly. Voluntary reporting within the activities of the Energy Conservation Clearinghouse provides further information and data on the effects of measures.

f) Expected results

Indonesia's energy conservation program expects to realise the goal of energy savings identified in RIKEN, which are based on studies on energy saving potential and energy audits.

g) Future tasks

Continuing the energy conservation program such as: implementation and completion of energy conservation policy, conduct activities related to public awareness, training, energy conservation partnership program, preparing the certification of energy managers and energy auditors and enhance international cooperation.

INSTITUTIONAL STRUCTURE

Under the Energy Law, energy policies are formulated by the National Energy Council (*Dewan Energy Nasional,* or DEN), established in 2008. DEN consists of stakeholders of energy that includes seven ministers and high-ranking government officials, and eight stakeholder members from industry, academia, technology experts, representative of environmental concerns, and consumers.

Presidential Decree No. 43/1991 mandates relevant government ministries and agencies to issue coordinated government rulings and programs within their respective jurisdictions and regulatory roles, to promote and encourage energy conservation8. This Decree is directed to establish cross-sector regulations to provide incentives and disincentives to conserve energy.

The Ministry of Energy and Mineral Resources is the focal-point of national energy conservation and energy efficiency programs. The regional governments are responsible for implementing energy efficiency and conservation programs within their jurisdiction in the regions.

a) Name of organisation

The Ministry of Energy and Mineral Resources (MEMR), Directorate General of New Renewable Energy and Energy Conservation, Directorate of Energy Conservation

b) Status of organisation

Government

c) Role and responsibility

Formulation of energy conservation policies; implementation of energy conservation policies; establish norms, standards, process, and criteria regarding energy conservation; provide technical training and evaluation of energy conservation programmes.

d) Covered sectors

All economic sectors of industry, transport, commercial, and the residential sector

e) Established dates

August 2010.

f) Number of staff

The number of staff of the Directorate of Energy Conservation is to be established (information at time of writing, January 2011).

INFORMATION DISSEMINATION

a) Information collection and dissemination

The Clearing House of Energy Conservation (CHEC) is the centre for data and information related to energy efficiency and conservation activities. The operation of CHEC is currently limited, its capacity as a data and information centre is being up-rated. The Government of Indonesia received bilateral assistance from the Danish International Development Agency (DANIDA) in the creation of CHEC.

b) Awareness-raising

The ‗National Energy Efficiency Movement' of the Ministry of Energy and Mineral Resources promotes energy conservation awareness through seminars and workshops, talk shows, public advertisements, brochures and leaflets; it is directed to households, specific industries and transport. The state-owned electricity company PLN promotes energy conservation in electricity use. Other institutions also promote awareness, including the Agency for the Assessment and Application of Technology (BPPT).

c) Capacity-building

Indonesia is instituting mandatory training and accreditation for energy managers and energy assessors. In addition, training is given to government officials responsible for mandatory energy saving and reporting of energy use in government office buildings. The Education and Training Centre for Electricity and New Renewable Energy of the Ministry of Energy and Mineral Resources (*Pusdiklat Ketenagalistrikan dan Energi Baru Terbarukan,* or KEBT) actively organises training on energy efficiency and energy conservation, new and renewable energy technologies, and in energy planning and modelling. The centre will be responsible for training of energy managers and energy assessors.

RESEARCH AND DEVELOPMENT IN ENERGY EFFICIENCY

PLN Electricity R&D Centre (PLN-LITBANG) conducts research and development related to the power industry and provides testing services for certain electric appliances, electric lighting, including compact fluorescent lamps (CFLs). The Agency for the Assessment and Application of Technology (BPPT) had developed an energy audit mobile unit, for energy auditing and assessment of energy saving potential in industrial energy use and energy use in commercial buildings.

MEASURES FOR ENERGY EFFICIENCY IMPROVEMENTS

GOVERNMENT LAWS, DECREES, ACTS

a) Name

Law No. 30/2007 regarding Energy (The Energy Law)

b) Purpose

The Energy Law is the legally binding legislation regarding energy, including energy conservation.

c) Applicable sectors

All sectors of the economy, and government departments and agencies, and regional governments

d) Outline

The Energy Law elucidates principles regarding the utilisation of energy resources and final energy use, security of supply, energy conservation, protection of the environment with regard to energy use, pricing of energy, and international cooperation.

The Energy Law defines the outline of the National Energy Policy (*Kebijakan Energi Nasional,* or KEN); the roles and responsibilities of the government and regional governments in planning, policy and regulation; energy development priorities; energy research and development; and the role of enterprises.

Under the Energy Law, the National Energy Policy will address the sufficiency of energy to meet the economy's needs, energy development priorities, utilisation of indigenous energy resources, and energy reserves.

e) Financial resources and budget allocation

The Government allocates budget for its energy efficiency and conservation programmes.

f) Expected results

Achieve significant energy saving levels identified in the new National Energy Policy (KEN) and in the new National Energy Conservation Master Plan (*Rencana Induk Konservasi Energi Nasional,* or RIKEN).

REGULATORY MEASURES

On 16 November 2009, the government issued Governmental Regulation No. 70/2009 regarding Energy Conservation.

Regulatory measures include:

the formulation of a National Energy Conservation Master Plan (RIKEN), which is to be updated every five years, or annually, as required

the mandatory assignment of an energy manager, energy auditing, and the implementation of an energy conservation program for users of final energy of 6000 toe (tonnes of oil equivalent) or more

mandatory energy-efficiency standards and energy labelling

the implementation of government incentives, including tax exemptions and fiscal incentives for imports of energy-saving equipment and appliances, and special low interest rates for investments in energy conservation

the implementation of government disincentives, including written notices to comply, public announcements of noncompliance, monetary fines, and reductions in energy supply for noncompliance.

In 2011, Indonesia issued set of implementing regulations, such as:

Presidential Instruction No. 13 year 2011 on Energy and Water Saving

Presidential Regulation No. 61 year 2011 on National Action Plan on GHGs Emission Reduction and Presidential Regulation No. 71 year 2011 on National GHGs Emission Inventories.

Ministry of Energy and Mineral Resources Regulation No 6/2011 on Sign-tagging of save energy for CFL

Director General of Renewable Energy and Energy Conservation Regulation No. 1287.K/06/DJE/2011 concerning Codes of Conduct on Swabalast Lamp.

Regulations on energy conservation that were issued prior to the Energy Law that may still apply or provisionally apply include:

Presidential Instruction No. 9/1982 on Energy Conservation (in government departments and agencies, and state owned enterprise office buildings and official vehicles).

Presidential Decree No. 43/1991 on Energy Conservation.

This Presidential Decree calls for inter-ministerial coordination on policies and programs on energy conservation that includes, policy on investment, funding of energy conservation programs and pricing of energy in relation to achieving energy conservation goals. The contents of this regulation appear in Government Regulation No. 70/2009.

Ministerial Decree No. 100.k/48/M.PE/1995 National Energy Conservation Master Plan (RIKEN) and revision in 2005. RIKEN was revised in 2005. RIKEN is an economy-wide plan on energy conservation.

Ministerial Decree No. 0002/2004 regarding Development Policy on Renewable Energy and Energy Conservation - *The Green Energy Policy*. The Green Energy Policy is an economy-wide policy.

Presidential Instruction No. 10/2005 regarding Energy Saving (for government and regional government office buildings).

Ministerial Regulation No. 0031/2005 regarding Process of Energy Saving, which is the guidelines of implementation of Presidential Instruction No. 10/2005.

Presidential Regulation No. 5/2006 regarding National Energy Policy.

Blueprint National Energy Management 2008 – (Blueprint Pengelolaan Energi Nasional – Blueprint PEN) revises the National Energy Policy of Presidential Regulation No. 5/2006. Blueprint PEN elaborates on the energy policy, including on energy conservation.

Presidential Instruction No. 2/2008 regarding Conservation of Energy and Water as revised version of Presidential Decree No. 10/2005 on Energy Efficiency. Under the Instruction, government agencies should report energy and water use twice a year.

Ministerial Regulation No. 13/2010 regarding Enactment of Mandatory Competency Standard for Energy Manager in Industry.

Ministerial Regulation No. 14/2010 regarding Enactment of Mandatory Competency Standard for Building Managers in the Field of Management.

MINIMUM ENERGY PERFORMANCE STANDARDS AND LABELLING

a) Name

Indonesia has minimum energy performance standards (MEPS) for select electrical appliances based on the Indonesia National Standard (*Standar Nasional Indonesia,* or SNI) and other technical standards on energy performance testing standards (EPTS) for electrical appliances.

b) Purpose

Purpose of these standards is to specify technical requirements with regard to energy efficiency and to safety, and for purpose of energy labelling.

c) Applicable sectors

Applicable to residential and commercial sectors: appliances, lighting and equipment.

Table 1: MEPS and EPTS

	Product	EPTS
1.	Ballast (magnetic)	SNI IEC 60929-2009
2.	Fluorescent lamps	SNI IEC 60901-2009
3.	Incandescent lamps	SNI IEC 60432-1-2009
4.	Room air conditioners—split type	ISO 5151
5.	Room air conditioners—window	ISO 5151
6.	Household refrigerators	SNI IEC 15502-2009
7.	Clothes washers	SNI IEC 60456-2009
8.	Electric irons	SNI IEC 60311-2009
9.	Vacuum cleaner	SNI IEC 60312-2009

d) Outline

SNI standard on electrical appliances and equipments in general are drafted and registered under the strict system and guidelines of the National Standardization Agency (*Badan Standardisasi Nasional*, or BSN). Additional energy standards on electrical appliances are being developed.

BUILDING ENERGY CODES

By Government Regulation No. 36/2005, under Law No. 28/2002 regarding Buildings, all buildings must comply with existing standards. Indonesia has four energy standards (SNI) for buildings, the standards cover: (1) the building envelope, (2) air conditioning, (3) lighting, and (4) building energy auditing. Energy building standards have yet to be mandated. However, voluntarily energy conservation and efficiency measures in commercial buildings are widely implemented.

a) Name

SNI for buildings

b) Purpose

The building energy codes are designed to improve building energy performance.

c) Applicable sectors

Applicable sectors: residential and commercial buildings

d) Outline

The standards outline:

building envelope: design criteria, design procedures, and energy efficiency standards

air conditioning systems: technical calculation, selection, measurement assessment, and energy efficiency standards

lighting systems: lighting guidelines for optimal and efficient operation

energy audit procedure: energy audit procedures for offices, hotels, shopping centres, hospitals, apartments and residences.

The standards/codes provide recommendations that take into account productivity, comfort and cost effectiveness.

Table 2: SNI for Buildings

1.	SNI 03-6389-2000	Energy conservation for building envelope of building (Konservasi energy selubung bangunan pada bangunan gedung)
2.	SNI 03-6390-2000	Energy conservation for air conditioning systems in building (Konservasi energy system tata udara pada bangunan gedung)
3.	SNI 03-6197-2000	Energy conservation for lighting systems in building structures (Konservasi energy system pencahayaan pada bangunan sedung)
4.	SNI 03-6196-2000	Energy auditing procedure for building (Prosedur audit energy pada bangunan gedung)

e) Financial resources and budget allocation

The Government provides funding for the Partnership Programme, while follow-up of the programme and voluntary EE&C measures are self and commercially financed.

f) Expected results

The standards are expected initiate construction of more energy efficient buildings and improved overall energy efficiency of existing buildings (through retrofit).

FUEL EFFICIENCY STANDARDS

Indonesia currently does not have minimum fuel efficiency standards; however, fuel efficiency standards are expected to be implemented in the near future, as they were confirmed at COP-15 in December 2009.

Current emissions standards are equivalent to Euro II compliance, implemented in 2006. Indonesia expects to advance to Euro IV-equivalent emission standards by 2012. The state-owned oil company Pertamina is working on plans to upgrade their refineries to produce Euro IV compliant gasoline. The refinery upgrading projects are expected to be completed during 2014-16.

2.2. Voluntary Measures

Voluntary energy efficiency and conservation measures are being implemented by industry and commercial buildings through commercial financing. Certain energy intensive industries such as the fertiliser, cement, pulp and paper and steel industries; and certain commercial buildings have implemented EE&C measures including installation of automated energy management.

FINANCIAL MEASURES TAKEN BY THE GOVERNMENT

1. Tax Scheme

The government currently does not have a tax scheme, such as tax deductions, in relation to investments in energy efficiency and conservation.

2. Low-Interest Loans

The government currently does not have low-interest loans for investments in energy efficiency and conservation measures, devices and equipment, to reduce energy use and conserve energy.

3. Subsidies and Budgetary Measures

Government subsidies and budgetary measures are provided for energy conservation programs such as the (1) partnership program on energy conservation in energy auditing, (2) the lighting program—for eligible households in relation to demand-side management (DSM) programs and saving energy, (3) BRESL, and (4) other programs such as for information dissemination.

4. Other Incentives

However, in accordance with the action plan (Governmental Regulation No. 70/2009), the government is expected to introduce government incentives that include tax exemption and fiscal incentives on imports of energy saving equipment and appliances, and special low interest rates on investments in energy conservation in the near future. The Indonesia government through the Ministry of Industry has provided funding for the industry sector in order to improve the efficiency of plants and the Ministry of Energy and Mineral Resources are proposing similar financing through the government budget to promote and accelerate energy efficiency achievement.

ENERGY PRICING

The government seeks to gradually remove fuel and electricity subsidies, and to bring their retail price to reflect the cost of supply.

Government subsidy for gasoline RON 88 octane and diesel oil, which are most consumed in transport sectors are expected to be reduced this year. There is remaining government subsidy in gasoline RON 88 octane, automotive diesel oil for transport, some part of kerosene for households, LPG in the government kerosene-to-LPG conversion program for households; and electricity price for small households and small businesses.

In renewable development, just recently, the Government has issued the Ministry of Energy and Mineral Resources Regulation number 4 year of 2012 concerning Renewable Energy Based as Power Plant that mandated a Feed-in Tariff on Biomass, Biogas, and Municipal Solid Waste Based Power Plants. Based on the regulation, the PT. PLN Persero as the utility is obligated to buy the electricity from biomass, biogas and municipal solid waste based power plant up to 10 MW capacity with feed-in tariffs as follows:

If interconnected to the medium voltage the feed-in tariff:

Biomass Rp. 975/kWh9 x F

Biogas (non municipal solid waste) Rp. 975/kWh x F

Municipal Solid Waste (Zero Waste Technology) Rp. 1050/kWh

Municipal Solid Waste (Landfill Gas) Rp. 850/kWh

If interconnected to the low voltage the feed-in tariff:

Biomass Rp. 1325/kWh x F

Biogas (non municipal solid waste) Rp. 1325/kWh x F

Municipal Solid Waste (Zero Waste Technology) Rp. 1398/kWh

Municipal Solid Waste (Landfill Gas) Rp. 1198/kWh

Which F is incentives factor for:

- Java, Bali and Sumatera region F = 1

- Kalimantan, Sulawesi, NTB and NTT region F = 1.2

- Maluku and Papua region F = 1.3

In addition, the Ministry of Finance issued Regulation No. 130 year of 2011 concerning Tax Exemption or Reduction for Renewable Development.

In the electricity development, the Government just recently also issued the Government Regulation no. 14 year of 2012 concerning Electricity Business Activities. This regulation mandated among others the business regulation of electricity business for public use and own use, business license and electricity tariff.

OTHER EFFORTS FOR ENERGY EFFICIENCY IMPROVEMENTS

1. Cooperation with Non-Government Organisations

Currently, most non-government organisations (NGOs) that are working in the field of energy are involved in small scale new and renewable energy development, their programmes are nonetheless contributing to conservation of fossil energy reserves, through use of locally available energy resources.

2. Cooperation through Bilateral, Regional and Multilateral Schemes

Ongoing cooperation in energy efficiency and conservation are: (1) Indonesia-JICA (Japan): Study on Energy Conservation and Efficiency Improvement in the Republic of Indonesia; (2) Indonesia—Denmark: Energy Efficiency in Industrial, Commercial, and Public Sector (EINCOPS); (3) Indonesia—UNDP/GEF: Barrier Removal to the Cost-Effective Development and Implementation of Energy Efficiency Standards and Labelling (BRESL); (4) Indonesia—the Netherlands: Energy Efficiency Improvement in Industry; (5) Indonesia—ASEAN: Promotion of Energy Efficiency and Conservation; (6) Indonesia-UNIDO: Promoting Energy Efficiency in the Industries through System Optimization and Energy Management Standard.

3. Other Cooperation/Efforts for Energy Efficiency Improvements

Indonesia has reviewed the APEC—Energy Working Group (EWG) Peer Review on Energy Efficiency in 2011.

REFERENCES

Ministry of Energy and Mineral Resources (DESDM), Blueprint *Pengelolaan Energi Nasional 2006-2025*, in accordance with Presidential Regulation No. 5/2006,http://esdm.go.id/.

Directorate General of Electricity and Energy Utilization (DJLPE),http://djlpe.esdm.go.id.

Clearinghouse *Energi Terbarukan & Konservasi Energy, Konservasi Energi*, http://energiterbarukan.net/.

Badan Pembina Hukum Nasional (BPHN), Pusat Jaringan Dokumentasi dan Informasi Hukum (JDIH) Nasional.

Presidential Instruction (Instruksi Presiden - INPRES) No. 9/1982 regarding Energy Conservation; Presidential Decree (Keputusan Presiden –KEPRES) No. 43/1991 regarding Energy Conservation; Presidential Instruction (INPRES) No. 10/2005 regarding Energy Savings, Presidential Instruction (INPRES) No. 2/2008 regarding Energy and Water Efficiency, http://bphn.go.id/ .

CLASP, Energy Efficiency Standards & Labelling Information Clearinghouse, Standards & Labelling Program Summary Worldwide, Refrigerators – Indonesia, www.clasponline.org/.

Han Wei (2009), Barrier removal to the cost-effective development and implementation of energy efficiency standards and labelling (BRESL), UNDP-GEF (2009).

RENEWABLE ENERGY MARKET BUSINESS OPPORTUNITIES[1]

Few countries offer the renewable energy potential of Indonesia. The country is home to 40 percent of the world's known geothermal resource and offers opportunities in wind, solar, biomass, and hydropower. By 2025, Indonesia aims to install 6.7 gigawatts (GW) of new renewable energy capacity by increasing the proportion of renewables from 7 percent to 15 percent of total energy production.[1]

Indonesia has a thriving economy at the intersection of the Pacific and Indian oceans, between Asia and Australia. It is home to 240 million people and is the world's fourth-largest country in terms of population. Its territory covers approximately 17,500 islands. [2] (See Box 1 for more facts.) Eighty percent of the population lives on the islands of Java and Bali; the rest is scattered widely among the country's 6,000 other inhabited islands. Indonesia's economy grew 4.3 percent in 2009, in line with the country's 3–6 percent growth rate since the Asian financial crisis of the late 1990s.[3] Future growth is likely to accelerate as the world recovers from the global recession of 2008–2009.

Although Indonesia lacks strong policy incentives for renewable energy, it provides companies with a stable political system and a growing economy in which to do business. Projects range from large-scale power plants with power purchase agreements (PPAs) negotiated with PLN, the state-owned utility company, to development projects coordinated by multilateral development banks or local governments. Competition often comes from foreign development agencies, which feed projects directly to their domestic industries through development aid and export loans.[4]

U.S. companies have been active in Indonesia for years and have developed significant working relationships with the Indonesian government and private sector. For companies that are new to exporting, working with an established U.S. partner can be an effective means of entering the market. For companies interested in joint partnerships, Indonesian firms are often willing to cooperate, but few have the capacity or capital required to develop large-scale renewable energy projects.

The government of Indonesia realizes that foreign direct investment (FDI) is critical to economic growth and seeks to create an investment climate that is capable of attracting foreign investors. Coupled with a sound fiscal policy and strong domestic consumption, this commitment has increased foreign interest in the Indonesian market. However, investors still face significant challenges, which include import restrictions, insufficient infrastructure, and regulatory uncertainty. As a result, FDI has thus far not kept pace with domestic production.

[1] U.S. Department of CommerceInternational Trade Administration

STATUS OF ENERGY MARKET

Since 2000, Indonesia has transitioned from a robust energy exporter to an importing nation that, for the first time, is concerned with rising production costs, energy subsidies, and climate change. The transition has spurred a new commitment to renewable energy, domestic production, and environmental sustainability. In 2008, fossil fuels provided 93 percent of the economy's total energy capacity (29.5 GW). But unlike previous years when excess energy was exported to neighboring markets, aging wells and limited investment forced Indonesia to import oil and to eventually remove itself from the Organization of the Petroleum Exporting Countries (OPEC). Similarly, Indonesia was once the world's leading exporter of natural gas, but now ranks 56th because investment restrictions and contract uncertainty caused new production to go undeveloped.

To appease local energy consumers, the government has maintained high-energy subsidies for the country's energy consumers. When supply was high and demand low, the subsidies stimulated economic growth with low, stable energy prices. But as demand began to outpace supply, the subsidies became a burden on the nation's fiscal budget and a deterrent to investments in cleaner, renewable resources. In 2008, subsidies exceeded $20 billion. As oil prices declined, subsides decreased to $10 billion in 2009 but are expected to increase again in 2010. In addition, Indonesia's fuel subsidies have the negative consequence of making growth itself an unaffordable expense to the government of Indonesia, which must pay more money for every new unit of electricity sold—money that is not spent on economic growth needs or social programs.

In response to the fiscal challenges imposed by the country's energy subsidies, the government announced a "Crash Program" to produce 20,000 megawatts (MW) of additional energy in 2004. Phase I of the program was confined to coal-fired electricity that was sourced primarily from China.[5] Phase II includes a preference for renewable energy production and includes a guarantee for "off-take" PPAs by PLN.

Phase II will likely include a greater emphasis on independent power producers. Unfortunately, without changes to policy and regulation, it is unlikely that Phase II will garner a significant influx of foreign investment. Many of Indonesia's state-owned enterprises will continue to dominate the large-scale power generation sector for several years. The U.S. Department of Commerce, however, expects a significant opportunity for foreign equipment and services suppliers to provide technology or expertise for large renewable energy projects.

After the Crash Program, the government of Indonesia expects a 56 percent increase in overall energy investments by 2014.[6] Investments will likely leverage private investment with public money to focus on increasing energy reliability and to meet the government's 90 percent electrification target by 2020. To further international investment in the renewable energy sector, the GOI recently announced a $0.097 per kWh feed-in tariff for geothermal energy and the creation of a new $400 million fund to help with the development of the country's geothermal resources.

EXISTING RENEWABLE ENERGY MARKET

Today, renewable energy accounts for a small but growing portion of Indonesia's electricity portfolio. Most renewable energy comes from the hydropower and geothermal industries, but growth in other sectors is likely. Presidential Decree No. 5 mandates an increase in renewable energy production from 7 percent to 15 percent of generating capacity by 2025. To accomplish that goal, 6.7 GW of new renewable energy capacity must be installed in the next 15 years based on current growth projections.[7] Geothermal and biomass have been slated for the most growth, but opportunities exist in every renewable energy technology (see Figure 1).

Surprisingly, Indonesia continues to import fossil fuels to cover production deficiencies instead of fully utilizing its already installed renewable energy capacity (see Figure 2). Expanding the production of existing resources (that is, already operating geothermal plants or hydropower dams) could displace some fossil fuel imports, by lowering the cost of energy subsidies and creating additional demand for renewable energy technology and expertise.

Since 2008, the government has offered tax incentives for foreign investment, including investors in renewable energy. Incentives include a 30 percent net income tax reduction for six years, free repatriation of investments and profits, and dispute settlement.[8] Geothermal companies are afforded additional incentives. The 2003 Geothermal Law established long-term licenses for land use (more than 30 years) and a regulated price for geothermal energy. However, PLN's status as the sole legal provider of electricity through Indonesia's power grid complicates the incentive for foreign developers and often limits the profitability of projects.

In addition, Indonesia's National Energy Policy of 2006 amended several policies and regulations to support renewable energy deployment. The law codified the targets for renewable energy production by 2025 set by Presidential Decree No. 5 and strengthened the position of cooperatives, private companies, and community organizations in PPA negotiations.

GEOTHERMAL POWER MARKET

Indonesia is home to 40 percent of the world's known geothermal resource (an estimated 27,510 MW), which is more than any other country. However, only 1,052 MW (a little more than 3 percent) has been developed to date.[9] The government has mandated that geothermal energy provide at least 9,500 MW (5 percent) of the nation's electricity by 2025.[10] Phase II of the Crash Program will provide most of the growth, with an estimated investment opportunity of $10 billion according to the Indonesian Investment Coordinating Board.[11]

Already several U.S. companies have invested in the Indonesian geothermal sector—a trend that will likely intensify as the policy incentives for geothermal energy continue to improve. U.S. government support for technical capacity-building, resource assessments, and trade promotion are helping spur this development. U.S. government agencies will continue to work with the government of Indonesia to maintain a competitive investment climate for the U.S. geothermal industry.

BIOMASS POWER MARKET

Indonesia's potential for biomass power is likewise substantial but is currently undeveloped. Although 49,810 MW of production is possible, only 443 MW have been commercially developed. By 2025, the country has targeted 810 MW of biomass power, an increase of 83 percent, but that amount is still far less than the potential contribution.[12] Rice residues, sugar, rubber, and palm oil all provide biomass electricity, but the most promising commercial application for biomass is likely cogeneration in agribusinesses. Large-scale biomass projects would likely require new infrastructure to gather and deliver what is now considered a waste item.

Because of its extensive biomass resources, Indonesia is also considered a potential center for biofuel production, particularly biodiesel manufactured from palm oil. Palm oil, however, is rarely used in local biofuel development. The international price for palm oil and the higher value placed on food has meant that most palm oil is exported or used in food production. Additional restrictions to development include a lack of incentives and subsidies for fossil fuels; incomplete data on land ownership; and a lack of biofuel processing ability.

As a result, biofuel development in Indonesia remains sluggish compared to that of other countries in the region. Nevertheless, several companies have begun using fast-growing crops, such as cassava, jatropha, or sweet sorghum, for biofuel production The Ministry of Agriculture is preparing additional land for growing these high-yield feedstocks to meet the country's biofuel production goals of 5.57 million kiloliters of biodiesel and 3.77 million kiloliters of bioethanol. The introduction of these crops will likely accelerate the production of biofuels and could become a key investment opportunity for U.S. companies.

HYDROPOWER MARKET

As an island nation, Indonesia is well suited for hydropower from conventional hydro and emerging ocean energy technologies. PLN estimates that Indonesia has 75,670 MW of large hydropower potential. But unlike other countries in the region, Indonesia has not built a large hydropower facility and it currently only produces 4,264 MW of hydropower-based electricity.[13]

Estimates indicate that another 500 MW of mini- or micro-hydropower is also available for development. A 2007 grant from the United Nations Development Program allowed Indonesia to create an Integrated Microhydro Development Program to accelerate hydropower development. The program focused on removing investment barriers and fostering technical capacity, but it has only developed 17 percent of the nation's mini-hydropower potential, leaving future growth in the sector likely.[14]

With thousands of miles of coastline, the potential for ocean energy is significant at an estimated 10–35 MW per kilometer of coastline. Currently, only one demonstration project has been developed—an ocean current system in the Lombok Strait. But future growth in the industry is also likely as technologies become further commercialized.

WIND POWER MARKET

Indonesia's potential for wind energy is limited because of the lack of wind along the equator. The country's windiest regions tend to be the less populated, eastern islands, which lack a transmission infrastructure capable of sustaining large wind farms. Wind power opportunities are thus limited to small or medium-sized projects requiring lower wind speeds.[15]

To date, only a few small-scale wind farms have been attempted, and they account for only 1.1 MW of installed capacity.[16] Offshore wind is more likely to provide investment opportunities for U.S. companies due to Indonesia's lengthy coastlines and consistent ocean breezes.

SOLAR POWER MARKET

Indonesia offers significant solar power resources (4.8 kilowatt-hours per square meter per day [kWh/m2/day]), but the country has yet to develop a strong market. To date, Indonesia has installed 12.1 MW of solar power, which is mostly from roof-mounted solar photovoltaic (PV) cells in urban areas.[17]

Additional solar power development is hampered by the lack of household buy-in and a dearth of maintenance personnel trained at installing solar cells. However, there is little incentive for a service infrastructure to develop, and the market's growth appears limited. Without interconnection standards that allow consumers to sell excess electricity back to PLN, it is unlikely that the Indonesian solar power market will ever rival the markets in other countries in the region.

The GOI has however shown a desire to attract foreign solar cell manufacturers to Indonesia like the Philippines and Malaysia have done. This would not only help create jobs for local Indonesians, but also drive an interest in solar power writ large that could eventually lead to export opportunities for U.S. companies.

DISTRIBUTED GENERATION

Despite the government's efforts to promote large power plants, recognition is growing that smaller distributed generation can best electrify the country. One-fifth of Indonesia's population lives outside of the main population centers. Many residents therefore lack access to grid-connected electricity and depend on outdated and costly diesel generators. In fact, 35 percent of Indonesia's population lacks access to any electricity.

Small-scale renewable energy resources can help meet the rural electrification goals of President Susilo Bambang Yudhoyono and can be implemented without access to the nation's inefficient electricity grid. Small renewable energy projects can also create jobs for rural Indonesians, who often turn to deforestation as a means of survival.

Drivers of Renewable Energy Development

The government has begun promoting renewable energy to meet the country's robust economic growth targets. By reducing the country's dependence on imports and by increasing production of renewable energy, the government believes it can catalyze economic growth. Such a policy would deflect money back into the economy and make future growth more sustainable. Several other factors have also led to the country's new interest in renewable energy.

ENERGY SECURITY

As a net importer of oil and gas, Indonesia is particularly vulnerable to the vagaries of world energy markets. Without a substantial increase in stable, domestically produced renewable energy, Indonesia will need to import additional fossil fuels. Renewable energy, however, can reduce dependence on foreign fuel sources and promote stable energy markets.

RURAL DEVELOPMENT

Rural development is a priority for the government, which has set a bold electrification target of 90 percent by 2020, which is significantly above the country's current electrification rate of 65 percent. Meeting the challenge will require an average of 1.3 million new electricity connections each year.[18] According to the Ministry for the Development of Disadvantaged Regions, more than 10,000 Indonesian villages lack access to electricity; thousands more are forced to use expensive diesel generators.[19] Since 2005, the government has encouraged renewable energy as a solution to rural electrification problems and has urged local communities to forgo diesel generation. The *2005–2025 National Energy Policy Blueprint* specifically states that renewable energy technologies should be used to meet the country's rural electrification goals.

CLIMATE CHANGE

Indonesia is the third-largest emitter of greenhouse gases, mostly because of deforestation. The government is aware of the problem, and it understands Indonesia's unique vulnerability to climate change as an island nation whose capital city, Jakarta, sits below sea level. Indonesia has thus taken a stronger position toward carbon dioxide emissions than have most developing nations. At the October 2009 meeting of the Group of 20, Yudhoyono announced that Indonesia was crafting a policy that would unilaterally reduce its emissions by 26 percent by 2020 and would reduce its emissions by 41 percent with international support. To meet the goal, Indonesia will heavily invest in renewable energy and recommit to stopping deforestation.[20]

SUSTAINABILITY

Indonesia faces significant environmental challenges: deforestation, water and air pollution, and forest fires. The government views renewable energy as part of a suite of potential technologies that could help with those challenges.

Barriers to Renewable Energy Development

Although Indonesia's renewable energy resources are substantial and the government now seems committed to increasing the deployment of renewable energy, several barriers must be overcome for the country to reach its full potential.

Most important, Indonesia must address its pricing regime for renewable energy. The lack of financial incentives, combined with heavy subsidies for energy consumption, has forestalled the development of a cleaner energy infrastructure. Without incentives to promote renewable energy, most companies will invest in different markets with more favorable policy environments.

A more immediate problem is the country's "Negative Investment List," which restricts foreign investment in power plants producing less than 10 MW. Indonesia is currently revising its 2007 Investment Law, but no final announcement has been made. If Indonesia does not allow participation by foreign companies—which often have the technical know-how and capital needed to develop renewable energy projects—it is unlikely the country will meet its targets for economic growth and renewable energy deployment.

Indonesia also suffers from a lack of available financing. Although several well-capitalized private-sector banks exist, few are willing to lend money to technologies that are new or emerging. A sovereign debt guarantee could incentivize lending, promoting deployment and investment.

The lack of an after-sales service infrastructure also makes renewable energy projects difficult to implement. Local populations are often unwilling to accept a project that could stall or break down for lengthy periods, especially when the project requires spare parts that are difficult to find or technicians who are often unavailable. Moreover, the lack of a strong transmission infrastructure makes grid-connected renewable energy projects difficult to implement if they are located far from end users.

The government itself is sometimes a barrier to renewable energy deployment. It often struggles to implement policy initiatives, which causes bottlenecks for innovative technologies like renewable energy. To avoid the problem, local governments have been given the authority to administer tenders for projects, but they often lack the capacity to do so transparently. Several companies have complained that tenders are worded poorly and amended after the tender is announced. The U.S. Government is working with local Indonesian officials to promote better tendering practices.

If better data existed, the lack of tendering capacity would be less of a concern, but Indonesia suffers from a chronic lack of quality data, particularly in the geothermal industry. The state's planning agency is investigating ways to verify data through third-party valuation, but it has yet to make any public commitments. A standard PPA would solve many challenges and would allow concessions to be developed without negotiating with PLN.

INVESTMENT CLIMATE

Revenue generation in Indonesia is described in Box 2. Funding for renewable energy projects is available from a variety of sources, including the United States, the Indonesian government, or multilateral institutions aimed at either development or environmental projects.

The United States Export–Import Bank can provide funding to Indonesian buyers looking to purchase U.S. technology, and it is seeking to increase lending through a new credit facility. The Overseas Private Investment Corporation can provide project finance in addition to equity investments and political risk insurance. In addition, the U.S. Agency for International Development and the U.S. Trade and Development Agency can help with development-related projects and technical capacity building.

The government of Indonesia offers funding through a sovereign wealth fund designed to increase foreign investment in renewable energy. The fund is managed by the Ministry of Finance and is a self-sustaining investment vehicle. The World Bank, International Monetary Fund, Asian Development Bank, Clean Development Mechanism, and other multilateral funds can also offer funding and financing

Past Commercial Engagement

Commercial engagement with Indonesia on renewable energy has been an important aspect of bilateral negotiations between the U.S. and Indonesian governments. The Energy Policy Dialogue organized by the U.S. Department of Energy has been ongoing since 2005. The new Comprehensive Partnership between the United States and Indonesia will be formalized in 2010; a tenet of the partnership will be renewable energy cooperation.

In May 2010, Secretary of Commerce Gary Locke led a clean energy trade mission to Indonesia, promoting increased clean energy cooperation and the importance of addressing market access issues. In addition, the U.S. Department of Commerce led a similar clean energy policy mission to Indonesia in July 2009 that involved seven U.S. government agencies for two days of bilateral meetings and a daylong policy summit. The U.S. Trade Development Agency and the U.S. Department of State also hosted an ASEAN Clean Energy Trade and Investment Forum in Manila in April 2010, focusing on many of the same issues in a regional context.

IMPORTANT LAWS AND REGULATIONS

BASIC POLICIES AND REGULATIONS

On 10 August 2007, Indonesia enacted Law No. 30 Year 2007 regarding Energy. This Energy Law contains two main principles of the National Energy Policy, which are the energy diversification and conservation energy. The Law mandates the utilization of energy resources on the priority of renewable energy sources. It also defines the roles and responsibilities of the Government and Local Government to prioritize the renewable energy sources.

Further, the Energy Law mandates the formation of the National Energy Council (*Dewan Energi Nasional*, DEN). The National Energy Council has some tasks as follow:

- drafts the National Energy Policy (KEN); and
- endorses the National Energy Master Plan (*Rencana Umum Energi Nasional*, RUEN); and
- declares measures to resolve conditions of energy crisis and energy emergency; and
- provides oversight on the implementation of energy policies that are cross-sectoral.

The assembly of the DEN members is chaired by the President. As an institution, the DEN is headed by the minister responsible for energy affairs. The DEN has 15 members: Seven ministers and high-ranking government officials responsible for the supply, transportation, distribution and use of energy; and eight stakeholder members from industry, academia, expert groups, environmental groups, and consumer groups. The selection and appointment of members of the DEN was finalized in late 2008.

Both the National Energy Policy and the National Energy Master Plan are now still being formulated, which are expected to address some issues related to the sufficiency of energy to meet the economy's needs, energy development priorities, utilization of indigenous energy resources, and energy reserves.

Regarding the utilization of energy resources and final energy use, security of supply, energy conservation, protection of the environment with regard to energy use, pricing of energy, and international cooperation, the Energy Law defines the outline of the National Energy Policy (*Kebijakan Energi Nasional,* or KEN); the roles and responsibilities of the government and regional governments in planning, policy and regulation; energy development priorities; energy research and development; and the role of enterprises.

The National Energy Master Plan (RUEN) implements the KEN. By law, the RUEN is drafted by the government, namely the Ministry of Energy and Mineral Resources, in a process that involves the related ministries and other government institutions, state-owned companies in the energy sector, and regional governments as well as academia and other energy stakeholders, and has due regard to input from the public.

The DEN finalised the draft of the National Energy Policy in March 2011, a document which would need to be discussed with the parliament (the DPR) before being enacted by the government. Thus, this new national energy policy would replace the existing the National Energy Policy that was established by the Presidential Regulation No. 5 Year 2006.

Over the past decade, Indonesia has reformed its energy sector through a series of new laws: the Oil and Gas Law (Law No. 22 Year 2001), the Geothermal Energy Law (Law No. 27 Year 2003), the Mineral and Coal Mining Law (Law No. 4 Year 2009), and the Electricity Law (Law No. 30 Year 2009).

These laws were established to promote an increased role for enterprises in the energy supply chain, in terms of fair competition on an equal playing field (as an alternative to a monopolistic industry), direct contracts between energy producers and buyers, and a transparent regulatory framework.

An advanced reformation in the electricity sector, which would have to establish the possibility of direct competition in power generation, transmission and distribution through Law No. 20 Year 2002 (currently annulled), was rejected by the Constitutional Court in 2004.

1 THE LAW ON OIL AND GAS

Indonesia's oil and gas industry is currently undergoing regulatory changes. The industry was reformed in 2001 under the Oil and Gas Law (Law No. 21 Year 2001). The regulatory bodies known as BP MIGAS and BPH MIGAS were created to address oil upstream and downstream activities, respectively. Exploration and production activities were conducted based on a fiscal contractual system that relies mainly on production sharing contracts (PSCs) between government and private investors, which may include foreign and domestic companies, as well as the government-owned oil company (Pertamina).

However, on 13 November 2012, the Constitutional Court declared that the existence of BP MIGAS was in conflict with the Constitution of 1945 and ordered its dissolution. At the time of this writing, the government is drafting a new Oil and Gas Law that will determine a new industry structure and until this law can be enacted, an Interim Working Unit for Upstream Oil and Gas Business Activities (SKSPMIGAS) has been established under the Ministry of Energy

and Mineral Resources (MEMR) to undertake all BP MIGAS roles and responsibilities. Furthermore, on 14 January 2013, the government issued the Presidential Regulation No. 9 Year 2013 as the umbrella for the establishment of the Working Unit for Upstream Oil and Gas Business Activities (SKKMIGAS) with its tasks in managing the upstream oil and gas business in Indonesia.

BPH MIGAS has supervisory and regulatory functions in the downstream oil and gas sector with the aims of ensuring availability and distribution of fuel throughout Indonesia, and the promotion of gas utilization in the domestic market through fair and transparent market competition.

The enactment of the Oil and Gas Law required that the state-owned oil company, Pertamina, relinquish its governmental roles to the new regulatory bodies BP MIGAS (now handed over to SKKMIGAS) and BPH MIGAS, and mandated the termination of Pertamina's monopoly in upstream oil and gas activities.

2 THE LAW ON MINING

On 16 December 2008, the parliament passed a new law on minerals and coal mining to replace Law No. 11 Year 1967, which had been in place for 41 years. The new law was enacted by the government on 12 January 2009 as Law No. 4 Year 2009 regarding Mineral and Coal Mining. The new Mining Law basically ended the concession of work areas by contracts of work (COW) and by work agreements for coal mining enterprises, *Perjanjian Karya Perusahaan Pertambangan Batubara* (PKP2B). Concessions are now based on permits from the central and regional governments. However, as the principle and objective of exertion in mining, in article 5 said that the government have an obligation to make policy about priory of minerals for domestic purposes/Domestic Market Obligation (DMO).

Prior to the new law, the government arguably had less regulatory control over its concessions. For example, any changes to concession terms needed to be agreed to by both the government and the investor. By instituting permits, the government expects to be better positioned to promote investments and to regulate mining.

The law creates greater opportunity for smaller investments in mining and gives regional governments a greater role in regulating the industry, along with revenue from mining. The Mining Law called for regulations on:

- concession areas and concession periods (for exploration permits) and production limits (for production permits) in mining for metals, non-metals and specific non-metals;
- a requirement that prospective investors submit post-mining and reclamation plans before applying for a permit;
- an obligation on permit holders to build smelters;
- an obligation on foreign companies to divest shares to the government, or state-owned enterprises and private companies registered in Indonesia;
- payment of taxes, fees and allocation of profits; and
- reclamation and post-mining costs.

The set of Government Regulations with regard to the Mining Law was completed in 2010 and these are now operational.

3 THE LAW ON ELECTRICITY

On 23 September 2009, the government enacted Law No. 30 Year 2009 regarding Electricity. This new Electricity Law replaced Law No. 15 Year 1985, which the Constitutional Court had reinstated in December 2004 as a provisional law, upon annulment of Law No. 20 Year 2002.

A notable difference between Law No. 30 Year 2009 and Law No. 15 Year 1985 is the absence of a Holder of Electricity Business Authority (*Pemegang Kuasa Usaha Ketenagalistrikan*, PKUK). Under Law No. 15 Year 1985, the government had appointed the state-owned electricity company, PLN, as the sole PKUK and so had made it responsible for providing electricity to all parts of Indonesia.

Under the new Electricity Law, the electricity industry will be made up of electricity business entities that are title holders of electricity supply business licences, *Izin Usaha Penyediaan Tenaga Listrik* (IUPTL). The IUPTL could either be in integrated electricity supply, power generation, transmission, distribution or retailing of electricity. Indonesia's electricity systems would retain vertically integrated configurations; however, these could comprise several licensed systems—such as PLN's numerous power systems, provincial government owned systems (to be established, where necessary), and private sector power systems, each operating within their respective business areas Licence holders of specific electricity supply types (such as the IPPs, as licence holders in power generation for supply of electricity to the public) would participate in the vertically integrated systems.

By law, the government and regional governments would regulate the electricity industry within their respective jurisdictions and through electricity regulatory authorities. The Electricity Law allows appointment of an electricity supply business area, cross-border electricity trading, as well as electricity tariffs to be differentiated by region for one certain business area (to allow for different costs of supply). Under the previous Electricity Law, Indonesia had a uniform electricity tariff regime and applied cross-subsidies between regions. At the time of writing, there was no ruling as to whether PLN will implement tariff differentiation over its extensive power systems across Indonesia.

As mandated by Law No. 30 Year 2009, the MEMR has issued three Government Regulations (GR), namely GR No. 14 Year 2012 regarding electricity supply businesses activity, GR No. 42 Year 2012 regarding the buying and selling of electricity across Indonesia's borders, and GR No. 62 Year 2012 regarding electricity support businesses.

The Ministry of Energy and Mineral Resources (MEMR) Regulation No. 22 Year 2012 regarding the Obligation of PLN to Purchase Electricity from Geothermal Power Plants and the Standard Purchase Price for Geothermal Power by PLN has been issued to replace the MEMR Regulation No. 2 Year 2011 and to set out how PLN purchases geothermal power. This regulation aims to accelerate the development of geothermal power as mandated by the Government Regulation No. 59 Year 2007 regarding Geothermal Business Activities as amended by the Government Regulation No. 70 Year 2010, particularly geothermal projects under the FTP2. The regulation sets the purchase price for geothermal power plant at 10 - 18.5 US cents/kWh depends on region and type of voltage transmission where geothermal power plants are connected to system. PLN will purchase geothermal power at that price without negotiable.

4 THE LAW ON GEOTHERMAL

According to Law No. 27 Year 2003 regarding Geothermal Energy, there are 5 (five) stages in utilizing geothermal energy. The stages are preliminary survey, exploration, feasibility studies, exploitation and utilization. The government could only conduct the two early stages, preliminary survey and exploration whereas private party (business entity) could conduct all the five stages.

A business entity could initiate geothermal activities once obtaining Geothermal Business Permit (IUP) through Geothermal Working Area (GWA) tender. The authority to issue the permit is depending on the location where the GWA is located. If the GWA is located within a regent/city, IUP is issued by regent/city government. For GWA is located at across several regent/city areas, IUP is granted by provincial government. The central government (minister) is only giving IUP for GWA which lies across several provinces.

IUP period consist of:

a. An exploration period valid of 3 (three) years which can be extended 2 (two) times for a period of one year each.

b. A feasibility study valid for a maximum of 2 (two) years since the exploration time period is ended.

c. An exploitation period valid for a maximum period of 30 years after exploration has ceased and can be extended. Exploitation period may be extended at most of 20 (twenty)years for each extension

INDONESIA ELECTRICITY LAW

LAW OF THE REPUBLIC OF INDONESIA NUMBER 30 OF 2009 CONCERNING ELECTRICITY WITH THE BLESSING OF GOD ALMIGHTY THE PRESIDENT OF THE REPUBLIC OF INDONESIA,

Considering:

a. that national development aims to achieve just and prosperous society in a manner materially and spiritually distributive under Pancasila and the 1945 Constitution of the Republic of Indonesia;

b. that with electricity having important and strategic role in the achievement of national development objectives, power supply business is therefore controlled by the state and the supplies thereof needs to be kept increased to reflect the progress in development to enable the availability of power in adequate quantity, fair distribution, and quality;

c. that power supplies are labor and technology intensive and in line with the principles of regional autonomy and democratization within the life style of the society, the nation, and the state, the role of the regional governments and the people in the supply of power requires increase;

d. that however useful, electric power also poses risk such that supplies and beneficial use of it must be subject to the electricity safety requirements;

e. that Law Number 15 of 1985 concerning Electricity no longer reflects the demanding times and changes in the society life such that replacement with a new law is required;

f. that in consideration of what item a, item b, item c, item d, and item e intend, it is necessary to make Law concerning Electricity;

Bearing in Mind: Article 5 section (1), Article 18, Article 20, and Article 33 of the 1945 Constitution of the State of the Republic of Indonesia; With the Joint Consent of THE HOUSE OF REPRESENTATIVES OF THE REPUBLIC OF INDONESIA and THE PRESIDENT OF THE REPUBLIC OF INDONESIA DECIDE: To Stipulate :

CHAPTER I GENERAL PROVISIONS

Article 1

In this Law:

1. Electricity" means any matter that concerns power supply and beneficial use as well as power support business.

2. "Electric power" means any form of secondary energy that is generated, transmitted, and distributed for any purposes, not including electricity in use for communication, electronics, or signal.

3. "Power supply business" means power procurement through power generation, transmission, distribution and sale to consumers.

4. "Power generation" means any activity through which to produce power.

5. "Power transmission" means delivery of power from generation to distribution system or to consumers, or cross-system delivery of power.

6. "Power distribution" means delivery of power from transmission system or from generation to consumers.

7. "Consumer" means any person or entity that purchases power from power supply license holders.

8. "Electric power sale business" means any activity through which to sell power to consumers.

9. "General power plan" means any development plan for power supply system, including power generation, transmission, and distribution as necessary to meet power demand.

10. "Power supply license" means a license to conduct public power supply business.

11. "Operating license" means a license under which to provide captive power supplies.

12. "Business area" means an area the Government determines as a place at which power supply business is conducted by power distributing and/or selling entities.

13. "Land damages" means any payment in compensation for release or transfer of land title, including the buildings, plants and/or other objects related to land.

14. "Compensation" means an award of a sum of monies to land title holders, including the buildings, plants and/or other objects related to land for the benefit of indirect use of the land for electricity constructions without the land title being released or transferred.

15. "The Central Government," hereinafter called "the Government," means the President of the Republic of Indonesia that holds the power in the government of the state of the Republic of Indonesia as intended by the 1945 Constitution of the State of the Republic of Indonesia.

16. "The Regional government" governors, the regents, or the regional instrumentalities that elements of the administrators of Administration. means the mayors and act as the the Regional

17. "Minister" means the minister in charge of electricity affairs.

18. "Any person" means any individual or entity of either legal entity or nonlegal entity.

CHAPTER II PRINCIPLES AND PURPOSES

Article 2

(1) Electricity development shall embrace the principles of:

a. benefit;

b. just efficiency;

c. sustainability;

d. optimum economics with the beneficial use of energy resources;

e. self-reliance;

f. sound business;

g. security and safety;

h. conservation of environmental functions;

i. regional autonomy

(2) Electricity development shall aim to guarantee the availability of power at adequate quantity, good quality, and reasonable price to improve public welfare and prosperity in a just and distributive manner as well as to realize sustainable development

CHAPTER III CONTROL AND COMMERCIALIZATION

Part One Control

Article 3 (1) Power supplies shall be under the control of the state, which supplies shall be provided by the Government and the regional governments under the principle of regional autonomy. (2) To provide power supplies as intended by section (1), the competent Government and regional governments shall establish policies, regulation, and supervision and manage/conduct power supply business. Part Two Commercialization

Article 4

(1) The conduct of power supply business by the Government and the regional governments shall be authorized to state-owned entities and region owned entities.

(2) Private entities, cooperatives, and self-reliant communities may participate in power supply business.

(3) With respect to power supplies as intended by Article 3 section (1), the Government and the regional governments shall set up funds for:

a. indigent community groups;

b. the construction of power supply facilities in less-developed regions;

c. the construction of power in remote and contiguous areas; and

d. the construction of rural electricity.

CHAPTER IV AUTHORITY OF MANAGEMENT

Article 5

(1) The authority of the Government in the field of electricity shall include:

a. adopt national electricity policy;

b. make electricity laws and regulations;

c. prepare guidelines, standards, and the criteria with respect to electricity;

d. set guidelines to consumer power tariffs;

e. develop a general national electricity plan;

f. determine business areas;

g. issue cross-border licenses;

h. electricity trading issue power supply business licenses to entities:

1. whose business overlaps the provinces;

2. in state-owned-entity form; and

3. that sell power and/or rent electrical grids to Government-appointed power supply license holders;

i. issue operating licenses for installation facilities overlapping the provinces;

j. set consumer power tariffs for Governmentappointed power supply license holders;

k. give approval for power selling and electrical grid rent prices to Governmentappointed power supply license holders;

l. give approval for excess power sale to Government-appointed operating license holders;

m. issue power service licenses to state-own entities or foreign investors/with majority of shares being owned by foreign investors;

n. issue electrical grid utilization licenses in the interest of telecommunications, multimedia, and informatics to grids owned by Government-appointed power supply license holders or Government-appointed operating license holders;

o. direct and supervise entities engaged in the field of electricity whose licenses issued by the Government;

p. appoint electricity inspector(s);

q. establish functional positions for electricity inspectors at every level of the governments; and

r. impose administrative sanctions on entities whose licenses issued by the Government.

(2) The authority of the province governments in the field of electricity shall include:

a. make provincial regulations in the field of electricity;

b. develop a general provincial electricity plan;

c. issue power supply licenses to entities whose business areas overlap the districts/cities;

d. issue operating licenses for installation facilities overlapping the districts/cities;

e. set consumer power tariffs for province government-appointed power supply license holders;

f. give approval for power selling and electrical grid rent prices to entities that sells power and/or rents electrical grids to entities whose license issued by the province governments;

g. give approval for excess power sale to operating license holders whose licenses issued by the province governments;

h. issue electrical grid utilization licenses in the interest of telecommunications, multimedia, and informatics to grids owned by province government-appointed power supply license holders or province government-appointed operating license holders;

i. direct and supervise entities engaged in the field of electricity whose licenses issued by the province governments;

j. appoint provincial electricity inspector(s); and

k. impose administrative sanctions on entities whose licenses issued by the province governments.

(3) The authority of the district/city governments in the field of electricity shall include:

a. make district/city regulations in the field of electricity;

b. develop a general district/city electricity plan;

c. issue power supply licenses to entities whose business areas within the districts/cities;

d. issue operating licenses for installation facilities within the districts/cities;

e. set consumer power tariffs for district/city government-appointed power supply license holders;

f. give approval for power selling and electrical grid rent prices to entities that sells power and/or rents electrical grids to entities whose licenses are issued by thedistrict/city governments;

g. issue power service licenses to entities whose majority of shares is owned by domestic investors;

h. give approval for excess power sale to operating license holders whose licenses issued by the district/city governments;

i. issue electrical grid utilization licenses in the interest of telecommunications, multimedia, and informatics to grids owned by district/city government-appointed power supply license holders or district/city government-appointed operating license holders;

j. direct and supervise entities engaged in the field of electricity whose licenses issued by the district/city governments;

k. appoint district/city electricity inspector(s); and

l. impose administrative sanctions on entities whose licenses issued by the district/city governments.

CHAPTER V BENEFICIAL USE OF PRIMARY ENERGY SOURCES

Article 6

(1) Primary energy sources domestically available and/or imported/coming from abroad must be beneficially used in an optimum manner within the national energy policy to enable guarantee of the sustainable power supplies.

(2) Beneficial use of primary energy sources as intended by section (1) must be made while giving precedence over new energy and renewable energy sources.

(3) Beneficial use of primary energy sources domestically available as intended by section (1) shall be given precedence in the interest of national electricity.

CHAPTER VI GENERAL ELECTRICITY PLAN

Article 7

(1) General electricity plan shall be prepared under the national energy policy and adopted by the Government upon consultation with the House of Representatives of the Republic of Indonesia.

(2) General electricity plan as intended by section (1) shall be prepared in which the regional governments are involved.

(3) Regional general electricity plan shall be prepared under the national general electricity plan and adopted by the regional governments upon consultation with the Regional House of Representatives. (4) Guidelines to preparation of general electricity plan as intended by section (1) and section (3) shall be issued by the Minister.

CHAPTER VII ELECTRICITY BUSINESS

Part One General

Article 8 Electricity business shall include:

a. power supply business; and

b. power support business.

Part Two Power Supply Business Article 9 Power supply business as intended by Article 8 shall include:

a. public power supply business; and

b. captive power supply business.

Article 10 (1) Public power supply business as intended by Article 9 item (a) shall include the following types of business:

a. power generation;

b. power transmission;

c. power distribution; and/or

d. power sale.

(2) Public power supply business as intended by section (1) may be conducted in an integrated manner. (3) Public power supply business as intended by section (2) shall be conducted by 1 (one) entity within 1 (one) business area.

(4) Limited business areas as intended by section (3) shall also apply to public power supply business limited to only power distribution and/or power sale. (5) Business areas as intended by section (3) and section (4) shall be determined by the Government.

Article 11

(1) Public power supply business as intended by Article 10 section (1) shall be conducted by state-owned entities, region-owned entities, private entities, cooperatives, and self-reliant communities engaged in the field of power supplies.

(2) State-owned entities as intended by section (1) shall receive first priority to conduct public power supply business.

(3) The competent Government or regional governments shall allow opportunities to region owned entities, private entities, or cooperatives to conduct integrated power supply business in areas where power service is not yet provided.

(4) Where no region-owned entities, private entities, or cooperatives are able to supply power in those areas, the Government must commission a state-own entity(ies) to supply power. Article 12 Captive power supply business as intended by Article 9 item (b) shall include:

a. power generation;

b. power generation and power distribution; or

c. power generation, power transmission, and power distribution.

Article 13 Captive power supply business as intended by Article 12 may be conducted by government agencies, the regional governments, state-owned entities, region-owned entities, private entities, cooperatives, sole proprietorships, and other institutions/entities. Article 14 Ancillary provisions on power supply business as intended by Article 9 through Article 13 shall be governed by Regulation of the Government. Part Three Power Support Business Article 15 Power support business as intended by Article 8 item (b) shall include:

a. power service business; and

b. power industry business.

Article 16 (1) Power service business as intended by Article 15 item (a) shall include:

a. power supply installation consultancy;

b. power supply installation construction and performance;

c. power installation checking and testing;

d. power installation operation;

e. power installation maintenance;

f. research and development;

g. education and training;

h. power tools and equipment laboratory testing;

i. power tools and equipment certification;

j. electricity technician competency certification; or

k. other service business directly associated with power supplies.

(2) Power service business as intended by section (1) shall be conducted by state-owned entities, region-owned entities, private entities, and cooperatives with certification, classification, and qualification under provisions of laws and regulations.

(3) In the conduct of power service business, state owned entities, region-owned entities, private entities, and cooperatives must give precedence over domestic products and potential.

(4) Ancillary provisions on certification, classification, and qualification of power service business shall be governed by Regulation of the Government.

Article 17 (1) Power industry business as intended by Article 15 item (b) shall include:

a. power tools industry business; and/or

b. power equipment industry business.

(2) Power industry business as intended by section (1) shall be conducted by state-owned entities, region-owned entities, private entities, and cooperatives. (3) In the conduct of power industry business, state owned entities, region-owned entities, private entities, and cooperatives must give precedence over domestic products and potential. (4) Power industry business activities as intended by section (1) shall be performed under provisions of laws and regulations.

CHAPTER VIII LICENSING

Part One General

Article 18

Power supply business and power support business as intended by Article 8 shall be conducted upon receipt of business licenses. Part Two Power Supply Licenses and Operating Licenses Article 19 (1) A license under which to supply power shall include:

a. power supply license; and

b. operating license.

(2) Any person to engage in public power supply must hold a power supply license. Article 20 A power supply license as intended by Article 19 section (1) item (a) shall be issued subject to the types of business as intended by Article 10 section (1). Article 21 The competent Government or regional governments shall issue power supply licenses. Article 22 Operating licenses as intended by Article 19 section (1) item (b) shall be issued to power generators with specified capacity as governed by Regulation of the Minister.

Article 23

(1) Operating licenses as intended by Article 22 shall be issued by the competent Government or regional governments.

(2) Operating licenses as intended by section (1) shall be issued upon meeting the administrative, administratif, technical, and environmental requirements.

(3) Operating license holders may sell excess power for the benefit of public interest upon receipt of approval from the competent Government or regional governments.

Article 24

Ancillary provisions on power supply licenses and operating licenses shall be governed by Regulation of the Government.

Part Three Power Support Licenses

Article 25

(1) Power service business as intended by Article 15 item (a) and Article 16 section (2) shall be conducted upon receipt of power service licenses from the competent Government or regional governments.

(2) Power service licenses and power industry licenses shall be issued under the provisions of laws and regulations.

Article 26

Ancillary provisions on power service licenses shall be governed by Regulation of the Government.

Part Four Rights and Obligations of Power Supply License Holders

Article 27

(1) In the conduct of power supply business as intended by Article 10 section (1) power supply license holders shall in the public interest be entitled to:

a. cross rivers or lakes either above or below the surface;

b. cross sea either above or below the surface;

c. cross public roads and railway;

d. have access to public places or private places and use them temporarily;

e. use land and underground;

f. cross over land or cross over or below buildings built on land or underground; and

g. cut and/or fell obstructing plants.

(2) In the performance of activities as intended by section (1), power supply license holders must exercise them under laws and regulations. Article 28 Power supply license holders must:

a. make available power that meets the prevailing quality and reliability standard;

b. provide as good service as possible to consumers and the public;

c. meet the electricity safety requirements; and

d. give precedence over domestic products and potential.

Part Five Consumer Rights and Obligations Article 29 (1) Consumers shall have the rights to:

a. receive good services;

b. have continuous access to power at good quality and reliability;

c. have access to power as of right at fair price;

d. receive repair services in case of electrical faults; and

e. recover damages in case of outage/failure through operating error and/or neglect by power supply license holders under the terms governed in a power sale/purchase agreement.

(2) Consumers must:

a. secure against any danger that may arise from the use of power;

b. protect the security of consumer power installations;

c. use power within its allocation;

d. pay electric bills; and

e. observe the requirements. electricity technical

(3) Consumers shall be responsible for their negligence resulting in losses to power supply license holders.

(4) Ancillary provisions on consumer responsibility as intended by section (3) shall be governed by Regulation of the Minister.

CHAPTER IX USE OF LAND

Article 30

(1) Use of land by power supply license holders as of right as intended by Article 27 shall require payment of damages with respect to land title or compensation to land, building, and plant title holders under provisions of laws and regulations.

(2) Damages with respect to land title as intended by section (1) shall be paid for land that is directly used by power supply license holders and for buildings and plants related to land.

(3) Compensation as intended by section (1) shall be made for the indirect use of land by power supply license holders who results in depreciation of economic value of land, buildings, and plant over which electricity pylons cross.

(4) Ancillary provisions on calculation of compensation as intended by section (3) shall be governed by Regulation of the Government.

(5) Where power supply license holders use land, the parcels of which are possessed by land title holders or state land users, power supply license holders must prior to commencing any activities solve the land problem under provisions of land laws and regulations.

(6) Where power supply license holders use land, the partial area of which is communal land, any solution thereto shall be achieved under provisions of land laws and regulations subject to the local customary law.

Article 31

Obligations to pay damages with respect to land title or compensation as intended by Article 30 section (1) shall not apply to any person by whom building are erected, plants are planted with willfulness, etc. on land already with a location permit for power supply business and to whom damages or compensation with respect to land title are already paid.

Article 32 (1) Valuation of and procedures for payment of damages with respect to land title or compensation as intended by Article 30 shall be conducted under provisions of laws and regulations.

(2) Damages with respect to land title or compensation as intended by Article 30 shall be for the account of power supply license holders.

CHAPTER X POWER SALES PRICES, ELECTRICAL GRID RENT PRICES, AND TARIFFS

Part One Power Sales Prices and Electrical Grid Rent Prices

Article 33

(1) Power sales prices and electrical grid rent prices shall be set under the principle of sound business.

(2) The competent Government or regional governments shall give approval for power sales prices and electrical grid rent prices.

(3) At no time shall power supply license holders charge power sales prices and electrical grid rent prices without approval of the Government or the regional governments.

Part Two Power Tariffs

Article 34

(1) The competent Government shall set consumer power tariffs upon consent of the House of Representatives of the Republic of Indonesia.

(2) The competent regional governments shall set consumer power tariffs upon consent of the Regional House of Representatives under guidelines adopted by the Government.

(3) Where the regional governments are unable to set power tariffs as intended by section (2), the Government shall set power tariffs for said regions upon consent of the House of Representatives of the Republic of Indonesia.

(4) Consumer power tariffs as intended by section (1), section (2), and section (3) shall be set with due regard to the balanced interests of nation, regions, consumers, and power supply business actors.

(5) Any region within one business area may vary in consumer power tariffs as intended by section (1) and section (2).

Article 35

At no time shall power supply license holders charge consumer power tariffs other than in accordance with tariffs set by the Government or the regional governments as intended by Article 34.

Article 36 Ancillary provisions on procedures for power sale and grid rent pricing, and power tariff as intended by Article 33 and Article 34 shall be governed by Regulation of the Government. Part Three Cross-Border Electricity Trading Article 37 Cross-border electricity trading shall be conducted by power supply license holders under a license by the Government.

Article 38

Cross-border electricity trading may be conducted through power purchase or sale.

Article 39

Cross-border power purchase as intended by Article 38 may be made on the following conditions:

a. local electricity needs are not yet met;

b. only to act as support to meet local electricity needs;

c. not adverse to the state and nation interest with respect to sovereignty, security, and economic development;

d. to improve the quality and reliability of local power supply.

e. to not disregard the development of domestic power supply capability; and

f. to not raise dependence on power procurement from abroad

Article 40

Cross-border power sale as intended by Article 38 shall be made if:

a. electricity needs of the local and surrounding areas have been met;

b. power sale prices are not subsidized

c. not compromise the quality and reliability of local power supply.

Article 41 Ancillary provisions on cross-border electricity trading as intended by Article 37 through Article 40 shall be governed by Regulation of the Government.

CHAPTER XI THE ENVIRONMENT AND ENGINEERING

Part One The Environment

Article 42 Any electricity business activities must meet the provisions as stated in the environmental laws and regulations. Part Two Engineering Article 43 Electricity engineering shall include:

a. electricity safety, and

b. electrical grids utilization in the interests of telecommunications, multimedia, and informatics.

Article 44

(1) Any electricity business activities must meet the electricity safety requirements. (2) The electricity safety requirements as intended by section (1) shall aim at achieving condition that is:

a. reliable and safe for installations;

b. safe from any danger for humans and other living things; and

c. environment-friendly.

(3) The electricity safety requirements as intended by section (1) shall include:

a. to meet power tool and equipment standardization;

b. to secure power installations; and

c. to secure power equipment.

(4) Any power installation in operation must hold a commissioning certificate. (5) Any power tool and equipment must meet the Indonesian national standard requirements. (6) Any technician with electricity business must hold a competency certificate.

(7) Provisions on electricity safety, commissioning certificates, Indonesian national standard, and competency certificates as intended by section (1) through section (6) shall be governed by Regulation of the Government.

Article 45

(1) Electrical grid utilization in the interests of telecommunications, multimedia, and informatics may only be conducted to the extent not affect the viability of power supplies.

(2) Electrical grid utilization as intended by section (1) may only be conducted upon approval of grid owners.

(3) Electrical grid utilization as intended by section (1) shall be conducted upon a grid utilization license that is issued by the competent Government or regional governments. (4) Ancillary provisions on electrical grid utilization as intended by section (1) and section (2) shall be governed by Regulations of the Government.

CHAPTER XII DIRECTION AND SUPERVISION

Article 46 (1) The competent Government or regional governments shall direct and supervise power supply business pertaining to:

a. energy source supply and utilization for power generators;

b. fulfillment of adequate power supplies;

c. fulfillment of technical requirements;

d. fulfillment of environmental protection;

e. preferential use of domestic goods and services;

f. employment of foreign workers;

g. achievement in power supply quality and reliability levels;

h. licensing eligibility;

i. power tariff charge; and

j. achievement of quality of service provided by power support business.

(2) In the exercise of supervision as intended by section (1), the Government and the regional governments may:

a. conduct site inspections;

b. inquire about electricity business reports;

c. conduct research and evaluation of electricity business reports; and

d. impose administrative sanctions against violation of licensing requirements.

(3) In the exercise of engineering supervision as intended by section (1), the Government and the regional governments shall be with the assistance of an electricity inspector(s) and/or a Civil

Servant Investigator(s). (4) Ancillary provisions on direction and supervision shall be governed by Regulation of the Government.

CHAPTER XIII INVESTIGATIONS

Article 47

(1) In addition to Investigators of the State Police of the Republic of Indonesia, specified Civil Service Officials with the scope of duties and responsibilities in the field of electricity shall be granted special authority as Investigators as intended by the Law of Criminal Procedure to conduct criminal investigations in the field of electricity.

(2) Civil Service Investigators as intended by section (1) shall be authorized to:

a. make examinations of the truth of reports or information about criminal acts in electricity business activities;

b. make examinations of any persons that are suspected of perpetrating criminal acts in electricity business activities;

c. call persons to be heard and examined as witnesses or suspects in criminal cases in electricity business activities;

d. search places that are suspected of being used to perpetrate criminal acts in electricity business activities;

e. make examinations of facilities and infrastructure of electricity business activities and cease the use of equipment that is suspected of being used to perpetrate criminal acts;

f. seal and/or seize electricity business activity instruments that are used to perpetrate criminal acts as means of proof;

g. procure experts that are required in connection with examinations of criminal cases in electricity business activities and

h. arrest and detain electricity perpetrators under laws and regulations.

(3) Civil Service Investigators as intended by section (1) shall inform Officials of the State Police of the Republic of Indonesia of the initiation of criminal case investigations under provisions of laws and regulations. (4) Authority as intended by section (2) shall be exercised under provisions of laws and regulations.

CHAPTER XIV ADMINISTRATIVE SANCTIONS

Article 48

(1) Any person in violation of the provisions as intended by Article 16 section (3), Article 17 section (3), Article 27 section (2), Article 28, Article 33 section (3), Article 35, Article 37, Article 42, or Article 45 section (3) shall be imposed administrative sanctions through:

a. written warning:

b. suspension; and/or

c. revocation of license.

(2) Administrative sanctions as intended by section (1) shall be effected by the competent Minister, governors, or regents/mayors.

(3) Ancillary provisions on procedures for imposition of administrative sanctions as intended by section (1) shall be governed by Regulation of the Government.

CHAPTER XV PENAL PROVISIONS

Article 49

(1) Any person who conducts public power supply business without a license as intended by Article 19 section (2) shall be sentenced to imprisonment of at most 3 (three) years and a fine of at most Rp2,000,000,000.00 (two billion rupiah).

(2) Any person who conducts public power supply business without an operating license as intended by Article 22 shall be sentenced to imprisonment of at most 5 (five) years and a fine of at most Rp4,000,000,000.00 (four billion rupiah).

(3) Any person who sells excess power for the benefit of public interest without approval of the Government or the regional governments as intended by Article 23 section (3) shall be sentenced to imprisonment of at most 2 (two) years and a fine of at most Rp2,000,000,000.00 (two billion rupiah).

Article 50

(1) Any person who fails to meet the electricity safety as intended by Article 44 section (1) resulting in death of person from power shall be sentenced to imprisonment of at most 10 (ten) years and a fine of at most Rp500,000,000.00 (five hundred million rupiah).

(2) Power supply license holders or operating license holders who commit acts as intended by section (1) shall be sentenced to imprisonment of at most 10 (ten) years and a fine of at most Rp1,000,000,000.00 (one billion rupiah).

(3) Other than sentences as intended by section (2), power supply license holders or operating license holders are also obligated to compensate the victims in damages.

(4) Requirements and procedures for payment in damages as intended by section (3) shall be implemented under provisions of laws and regulations.

Article 51

(1) Any person who fails to meet the electricity safety as intended by Article 44 section (1) to affect the viability of power supplies shall be sentenced to imprisonment of at most 3 (three) years and a fine of at most Rp500,000,000.00 (five hundred million rupiah).

(2) If an act as intended by section (1) results in power outage/failure to the detriment of the public shall be sentenced to imprisonment of at most 5 (five) years and a fine of at most Rp2,500,000,000.00 (two billion five hundred million rupiah).

(3) Any person who consumes electricity in an unauthorized manner unlawfully shall be sentenced to imprisonment of at most 7 (seven) years and a fine of at most Rp2,500,000,000.00 (two billion five hundred million rupiah).

Article 52

(1) Any person engaged in power supply business not meeting obligations to the land, building, and plant title holders as intended by Article 30 section (1) shall be sentenced to imprisonment of at most 5 (five) years and a fine of at most Rp3,000,000,000.00 (three billion rupiah).. (2) Other than sentences as intended by section (1), additional sanctions of revocation of power supply license or operating license may be imposed.

Article 53

Any person who performs power service business activities in an unlicensed manner as intended by Article 25 section (1) shall be sentenced to imprisonment of at most 5 (five) years and a fine of at most Rp2,000,000,000.00 (two billion rupiah).

Article 54

(1) Any person who operates power installations without a commissioning certificate as intended by Article 44 section (4) shall be sentenced to imprisonment of at most 5 (five) years and a fine of at most Rp500,000,000.00 (five hundred million rupiah). (2) Any person who produces, distributes, or trades in power tools and equipment other than in accordance with the Indonesia national standard as intended by Article 44 section (5) shall be sentenced to imprisonment of at most 5 (five) years and a fine of at most Rp5,000,000,000.00 (five billion rupiah).

Article 55

(1) Where a criminal act as intended by Article 49 through Article 54 is committed by an entity, such an entity and its management shall be imposed sentence. (2) Where sentence as intended by section (1) is imposed on an entity, a maximum fine shall be imposed plus one third of it.

CHAPTER XVI TRANSITIONAL PROVISIONS

Article 56

Upon this Law coming into effect: (1) PT Perusahaan Listrik Negara (Persero) as state- owned entity that is formed under Regulation of the Government Number 23 of 1994 concerning Change in the Form of Public Enterprise into Limited Liability Company shall be deemed to have held a power supply license.

(2) Within a period of not exceeding 2 (two) years, the Government shall already prepare and issue a power supply license to state-owned entity as intended by item 1 under the provisions of this Law.

(3) Public Power Business Licenses, Captive Power Business Licenses, and Power Support Business Licenses that have been issued under Law Number 15 of 1985 concerning Electricity shall remain valid until their expiration.

(4) Within a period of not exceeding 2 (two) years, Public Power Business Licenses, Captive Power Business Licenses, and Power Support Business Licenses that have been issued under Law Number 15 of 1985 concerning Electricity as intended by item 3 are subject to adjustment to the provisions of this Law.

CHAPTER XVII CONCLUDING PROVISIONS

Article 57

(1) Upon this Law coming into effect, Law Number 15 of 1985 concerning Electricity (State Gazette of the Republic of Indonesia Number 74 of 1985, Supplement to State Gazette of the Republic of Indonesia Number 3317) is revoked and declared no longer valid.

(2) Ancillary regulations on electricity already existing under Law Number 15 of 1985 concerning Electricity shall remain valid to the extent not against or not yet replaced under this Law.

(3) Ancillary regulations to this Law must have been issued not longer than 1 (one) year from when this Law is promulgated.

Article 58 This Law shall take effect from the date it is promulgated. In order that every person may know of it, the promulgation of this Law is ordered by placement in the State Gazette of the Republic of Indonesia Stipulated in Jakarta on September 23, 2009

PRESIDENT OF THE REPUBLIC OF INDONESIA, sgd. ANDI MATTALATTA STATE GAZETTE OF THE REPUBLIC OF INDONESIA NUMBER 133 OF 2009. SUPPLEMENT TO STATE GAZETTE OF THE REPUBLIC OF INDONESIA NUMBER 5052

ELUCIDATION ON LAW OF THE REPUBLIC OF INDONESIA NUMBER 30 YEAR 2009 CONCERNING ELECTRICITY

I. General :

Development in electricity sector has aim to improve the public welfare and advance the intellectual life of the people in order to meet the objectives of national development for the creation of just and prosperous society through a fair distribution both in material and spiritual manner under Pancasila and the 1945 Constitution of the Republic of Indonesia.

Electric power as one of the outcomes of natural resource utilization has important role in the realization of attainment of the objectives of national development. Given that electric power is fundamental to the state especially in the realization of public welfare for all sectors and bearing in mind Article 33 section (2) of the 1945 Constitution of the Republic of Indonesia, this Law affirms that electric power supply shall be controlled by the state and used in the best prosperity of the people with its organization being conducted by the Government and the regional governments. The competent Government or regional governments shall make policies, regulation, and supervision and implement electric power supply business.

The Government and the regional governments shall conduct power supply business, of which the implementation shall be undertaken by state-owned entities and region-owned entities. To enhance the capability of the state in the power supply, this Law gives opportunities to private entities, cooperatives, and self-reliant communities to participate in the power supply business. The competent Government or regional governments shall within the regional autonomy principles issue power supply licenses.

Various electricity issues of concerns faced by the nation and state today have been anticipated by this Law that governs, inter alia, division of integrated power supply business areas, application of regional tariffs applicable only for one certain business area, electrical grid utilization in the interest of telecommunications, multimedia, and informatics, as well as cross-border electricity trading former Law Number 15 of 1985 concerning Electricity does not govern. To increase public power supply also requires exercise of legal enforcement powers in the electricity field. The Government and the regional governments shall have the authority to direct and supervise the implementation of electricity, including supervision of engineering. Despite benefit, electric power can also present a danger.

Accordingly, to further guarantee the public safety, occupational safety, installation security, and conservation of environmental functions in the supply electric power and utilization of electric power, electric power installation must use electricity equipment and tools that meet the electricity equipment standards. II. ARTICLE BY ARTICLE : Article 1 : Sufficiently Clear

Article 2 :

Article 2 Section (1) (a): “The principle of benefit” means the results of electricity development should be in beneficial use in the best welfare and prosperity of the people. Article 2 Section (1) (b): “The principle of just efficiency ” means electricity development should be undertaken at a reasonable minimum cost with the results enjoyed in a distributive manner by all of the people.

Article 2 Section (1) (c): “The principle of sustainability” means power supply business should be managed properly to enable sustainable viability.

Article 2 Section (1) (d): “The principle of optimum economics with the beneficial use of energy resources” means energy sources for power generation should be in use with due regard to the availability of energy sources.

Article 2 Section (1) (e): “The principle of self-reliance” means electricity development should be undertaken while setting a value on the domestic capability.

Article 2 Section (1) (f): “The principle of sound business” means electricity business is conducted by serving the principles of transparency, accountability, responsibility, and fairness.

Article 2 Section (1) (g): “The principles of security and safety” means the supply and beneficial use of power should consider installation security, human safety, and the environment around the installations.

Article 2 Section (1) (h): “The principle of conservation of environmental functions” means the provision of power supply should consider the conservation of environmental functions and the surrounding environment.

Article 3 :

Article 3 Section (1): Given power is one of the important and strategic production sectors in national life, power supply business shall be taken control by the state, of which the implementation is dedicated to serving the best interest and prosperity of the people.

Article 4 : Article 4 Section (1):

State-owned entities and region-owned entities shall under this provision be those engaged in the power supply business.

Article 4 Section (2): Participation of private entities, cooperatives, and self-reliant communities is invited to strengthen the power needs. Self-reliant communities may be in legal entity form.

Article 4 Section (3): Sufficiently clear Article 5 : Sufficiently Clear Article 6: Article 6 Section (1): Sufficiently Clear

Article 6 Section (2): New energy and renewable energy sources shall be beneficially used with due regard to their economies.

Article 7 :

Article 7 Section (1): "National energy policy" means national energy policy as intended by Law concerning Energy.

Article 8 : Sufficiently Clear Article 9 : Sufficiently Clear

Article 10 :

Article 10 Section (1): Sufficiently Clear

Article 10 Section (2): Sufficiently Clear

Article 10 Section (3): Sufficiently Clear

Article 10 Section (4): Sufficiently Clear

Article 10 Section (5): A business area shall not be an administrative area of the government.

Article 11 :

Article 11 Section (1): Sufficiently Clear

Article 11 Section (2): Priority to state-owned entities constitutes the embodiment of the control of the state over power supplies.

State-owned entities shall be entities that are solely engaged in the field of power supplies.

Article 12 : Sufficiently Clear

Article 13: "

Captive power supply" means power supply for own consumption, not for trade. "Other institutions/entities" means representatives of foreign institutions or foreign entities.

Article 14: Sufficiently Clear

Article 15: Sufficiently Clear Article 16 :

Article 16 Section (1): Sufficiently Clear

Article 16 Section (2): Sufficiently Clear

Article 16 Section (3): Foreign products and potential may be in use if no domestic products and potential are available.

Article 17 :

Article 17 Section (1): Sufficiently Clear

Article 17 Section (2): Sufficiently Clear

Article 17 Section (3): Foreign products and potential may be in use if no domestic products and potential are available. Article 18 : Sufficiently Clear

Article 19 : Sufficiently Clear Article 20 : Sufficiently Clear

Article 21: In the issue of licenses, the Government or the regional governments shall have due regard to the power supply capability of the power supply license holders in possession of the local business area.

A power supply license shall contain, inter alia, the name and address of the entity, the type of business granted, obligations of the conduct of business, technical requirements, and sanctions.

Article 22 : Sufficiently Clear

Article 23 : Sufficiently Clear

Article 24 : Sufficiently Clear

Article 25 : Sufficiently Clear Article 26 : Sufficiently Clear

Article 27 : Sufficiently Clear

Article 28 : Article 28 (a): Sufficiently Clear

Article 28 (b): Sufficiently Clear Article 28 (c): Sufficiently Clear Article 28 (d): Foreign products and potential may be in use if no domestic products and potential are available. Article 29 : Article 29 Section (1) : Sufficiently Clear

Article 29 Section (2) (a): Sufficiently Clear

Article 29 Section (2) (b): "Consumer power installations" means power installations following the installations of electric meters or load limit controllers.

Article 29 Section (2) (c): Sufficiently Clear

Article 29 Section (2) (d): Sufficiently Clear

Article 29 Section (2) (e): Sufficiently Clear

Article 29 Section (3): Sufficiently Clear

Article 29 Section (4): Sufficiently Clear

Article 30 Section (1): Sufficiently Clear Article 30 Section (2): Damages with respect to land title shall include the remaining land the title holders cannot use as result of partial use of land by power supply license holders. "Directly" means the use of land for the construction of power installations, inter alia, generation, electrical substations, and electricity pylon sites.

Article 30 Section (3): "Indirect use" in this provision includes, inter alia, the use of land over which transmission lines cross. Article 30 Section (4): Sufficiently Clear Article 30 Section (5): Sufficiently Clear

Article 30 Section (6): Sufficiently Clear Article 31 : Sufficiently Clear Article 32: Sufficiently Clear

Article 33 Section (1): "Power sales price" includes all costs in connection with power sale by power generators. "Electrical grid rent price" includes all costs associated with power transmission and/or distribution grid rent.

Article 33 Section (2): In giving approval for power sales prices and electrical grid rent prices, the Government or the regional governments shall have due regard to agreement among entities. Article 33 Section (3): Sufficiently Clear Article 34 Section (1):

Consumer power tariffs include all costs to which consumer power consumption relates, inter alia, fees (Rp/kVA) and billing charges (Rp/kWh), reactive power billing charges (Rp/kVArh), and/or maximum kVA fees paid at subscription rates (Rp/month) within the power limits consumed or other forms.

Article 34 Section (2): Sufficiently Clear

Article 34 Section (3): Sufficiently Clear

Article 34 Section (4): "The interest of regions" includes, inter alia, economic and industrial development in regions.

Article 35: Sufficiently Clear Article 36: Sufficiently Clear Article 37: Sufficiently Clear Article 38: Sufficiently Clear

Article 39 :

Article 39 (a): Sufficiently Clear

Article 39 (b): Sufficiently Clear Article 39 (c): Sufficiently Clear

Article 39 (d): Sufficiently Clear

Article 39 (e): "Development of domestic power supply capability" means development of energy sources, human resources, and technology.

Article 39 (f): Sufficiently Clear Article 40: Sufficiently Clear

Article 41: Sufficiently Clear

Article 42: Sufficiently Clear

Article 43: Sufficiently Clear Article 44: Sufficiently Clear

Article 44:

Sufficiently Clear Article 45: Sufficiently Clear Article 46: Sufficiently Clear Article 47: Sufficiently Clear Article 48: Sufficiently Clear Article 49: Sufficiently Clear Article 50: Sufficiently Clear Article 51: Sufficiently Clear Article 52: Sufficiently Clear Article 53: Sufficiently Clear Article 54: Sufficiently Clear Article 55: Sufficiently Clear Article 56 (1): Sufficiently Clear Article 56 (2): The preparation aims to make state-owned entities more efficient and effective in provision of power supply services to the public. Article 56 (3): Sufficiently Clear

Article 56 (4): Sufficiently Clear

Article 57: Sufficiently Clear

Article 58: Sufficiently Clear

ACT ON NUCLEAR ENERGY[2]

Considering: a. that nuclear energy is related to the life and safety of the people, therefore it must be authorised by the State, and its beneficial use for national development is aimed to the realisation of an equally just and prosperous community, physically and spiritually, based on the Pancasila and the 1945 Constitution;

b. that the development and beneficial use of nuclear energy in the various fields of human life in the world have been so advanced that its beneficial use and development for a sustainable and environmentally insighted national development need to be enhanced and expanded to increase the Nation's welfare and competitiveness;

c. that for the safety, security, peace, health of workers and the public, and environmental protection, the use of nuclear energy shall be implemented correctly and carefully and is aimed to peaceful purposes and for the utmost benefit for the people's welfare and prosperity;

[2] NO. 10, 1997 STATE GAZETTE OF THE REPUBLIC OF INDONESIA OF THE YEAR 1997 NUMBER 23 STATE GAZETTE OF THE REPUBLIC OF INDONESIA SUPPLEMENT NUMBER 3676 NUCLEAR ENERGY CONTROL BOARD JAKARTA 1998

d. that due to the nature of nuclear energy which can not only give advantages but also cause radiation hazards, then every activity that is related to nuclear energy has to be regulated and controlled by the Government;

e. that the Act number 3, 1964 on Basic Stipulations of Atomic Energy does not sufficient to the current developed situation;

f. that based on considerations as stated in paragraph a, b, c, d, and e of this clause, it is necessary to stipulate an Act on Nuclear Energy.

Noting: 1. Article 5 clause (1), Article 21 clause (1), and Article 33 of the 1945 Constitution;

2. Act Number 11, 1967 on Basic Stipulations of Mining (State Gazette of the Republic of Indonesia of the year 1967 number 22, State Gazette of the Republic of Indonesia Supplement number 2831)

With the consent of
THE HOUSE OF THE PEOPLE'S REPRESENTATIVES OF THE REPUBLIC OF INDONESIA
HAS DECIDED

To enact: AN ACT ON NUCLEAR ENERGY

CHAPTER I GENERAL PROVISIONS

Article 1

In the context of this Act,

Nuclear activity is defined as any activity related to the use, the development and the acquisition of nuclear science and technology and the control of any activity related to nuclear energy.

Nuclear energy is defined as any form of energy released in the process of nuclear transformation, including energy from ionising radiation sources.

Ionising radiation is defined as electromagnetic radiation or charged particles that due to their energy are capable of ionising the media of their path.

Utilisation is defined as any activity related to nuclear energy activity that includes research, development, mining, fabrication, manufacturing, production, transportation, storage, transfer, export, import, decommissioning and radioactive waste management to enhance the people's welfare.

Nuclear material is defined as any material that is capable of producing energy by a self-sustaining chain process of nuclear fission or any material that can be changed into such material that is capable of producing energy by a self-sustaining chain process of nuclear fission.

Nuclear ore is defined as raw material used for producing nuclear fuel.

Nuclear fuel is defined as any material that is capable of producing energy by a self-sustaining chain process of nuclear.

Radioactive waste is defined as any radioactive material and any material as well as equipment that has been contaminated by radioactive material or becomes radioactive due to the operation of a nuclear installation and cannot further be used.

Radioactive material is defined as any material that emits ionising radiation with a specific activity greater than 70 kBq/kg (2 nCi/g).

Radioactive waste management is defined as the collection, segregation, process, transportation, storage and/or disposal of radioactive wastes.

Radioisotope is defined as any isotope that is capable of emitting ionising radiation.

Nuclear installation is defined as:

any nuclear reactor;

any facility for the purification, conversion, enrichment of uranium, fabrication of nuclear fuel and/or reprocessing of spent fuel; and/or

any facility which is used for storing nuclear fuel and spent fuel.

Nuclear reactor is defined as any device or installation that is operated using nuclear fuel that is capable of resulting a controlled self-sustaining chain process of nuclear fission and is used to generate power, or for research, and/or for radioisotope production.

Decommissioning is defined as any activity that is aimed to ultimately terminate the operation of a nuclear reactor, i.e., removal of nuclear fuels from reactor core, dismantling of reactor components, decontamination, and final safety process.

Nuclear incident is defined as any occurrence or series of occurrences having the same origin that causes nuclear damage.

Nuclear damage is defined as any loss that can be in the form of loss of life, disablement, personal injury or sickness, or damage to property, contamination and damage to the environment that arises out of radiation or a combination of radiation with toxic, explosive or other hazardous properties as a result of nuclear fuel criticality in a nuclear installation or during transportation, including loss as a result of preventive measures and loss as a result of or measures of reinstatement of impaired environment.

Nuclear installation operator is defined as any individual person or any legal body that is liable in operating a nuclear installation.

Third party is defined as any individual person or any legal body that suffers from nuclear damage, not including the Operator of nuclear installation and workers of nuclear installation, who, according to the organisational structure is under the Operator of nuclear installation.

Article 2

(1) Nuclear materials consist of:

a. nuclear material ore;

b. nuclear fuel, and

c. spent fuel.

(2) Nuclear materials are authorised by the State and their utilisation are regulated and controlled by the Government.

CHAPTER II INSTITUTIONS

Article 3

(1) The Government establishes an Executing Body, under and directly responsible to the President. The executing Body shall have the task to execute the use of nuclear energy.

(2) To accomplish the task under clause (1), the Executing Body conducts research and development, general surveys, explorations and exploitations of nuclear ore, raw material production for manufacturing and fabrication of nuclear fuel, production of radioisotopes for research and development, and radioactive waste management.

Article 4

(1) The Government establishes a Regulatory Body, under and directly responsible to the President. The Regulatory Body shall have the task to control any activity using nuclear energy.

(2) To accomplish the task under clause (1), the Regulatory Body establishes regulations, conduct licensing processes and inspections.

Article 5

The Government establishes a Nuclear Energy Advisory Council, which shall have functions to give advises and considerations on the use of nuclear energy.

Article 6

The status, main tasks, functions, structure and organisation of the Institutions under Article 3, Article 4, and Article 5 shall be further regulated by Presidential Decrees.

Article 7

The Government may establish a State Company that is related to the commercial utilisation of nuclear energy.

CHAPTER III RESEARCH AND DEVELOPMENT

Article 8

(1) Research and development on nuclear energy should be conducted in order to mastering nuclear science and technology for the purpose of safety, security, peace, and welfare of the people.

(2) The research and development under clause (1) is accomplished mainly by and is the responsibility of the Executing Body.

(3) Research and development on nuclear safety need to be notified to reduce negative impacts on the nuclear energy utilisation.

(4) In order to implement research and development under clause (1), the Executing Body may cooperate with other institutions and bodies.

Chapter IV ENTERPRISES

Article 9

(1) General surveys, explorations and exploitations of nuclear material ore shall only be performed by the Executing Body.

(2) The Executing Body under clause (1) may cooperate with State Company, cooperative, private company, and/or other bodies.

Article 10

(1) The production and/or procurement of raw materials for manufacturing nuclear fuel shall only be performed by the Executing Body.

(2) The Executing Body under clause (1) may cooperate with State Company, cooperative, and/or private company.

Article 11

(1) The non-commercial production of nuclear fuel is performed by the Executing Body.

(2) The commercial production of nuclear fuel is performed by State Company, co-operative, and/or private company.

Article 12

(1) The non-commercial production of radioisotopes is performed by the Executing Body.

(2) The commercial production of radioisotopes is performed by State Company, co-operative, and/or private company.

Article 13

(1) The non-commercial construction, operation, and decommissioning of nuclear reactor is performed by the Executing Body.

(2) The Executing Body under clause (1) may cooperate with other government institution and state university.

(3) The commercial construction, operation, and decommissioning of nuclear reactor shall be performed by State Company, co-operative, and/or private company.

(4) The commercial construction of nuclear reactor under clause (3), in the form of nuclear power plant, is established by the Government after a consultation with the House of Representative of the Republic of Indonesia.

CHAPTER V REGULATORY

Article 14

(1) The control on the use of any nuclear energy shall be carried out by the Regulatory Body.

(2) The control under clause (1) should be implemented through regulations, licensing, and inspections.

Article 15

The control under Article 14 is aimed to:

a. assure the welfare, the security and the peace of people;

b. assure the safety and the health of workers and public, and the environmental protection;

c. maintain the legal order in implementing the use of nuclear energy;

d. increase the legal awareness of nuclear energy user to develop a safety culture in nuclear field;

e. prevent the diversion of the purpose of the nuclear material utilisation; and

f. assure for maintaining and increasing the worker discipline on the implementation of nuclear energy utilisation.

Article 16

(1) Any activity related to the utilisation of nuclear energy shall maintain the safety, the security, the peace, the health of the workers and the public, and the environmental protection.

(2) The provisions under clause (1) will be further implemented by Government Regulation.

Article 17

(1) Any use of nuclear energy shall be subjected for licensing, except in certain cases that will be further stipulated by Government Regulation.

(2) The construction and operation of nuclear reactors and other nuclear installations as well as the decommissioning of nuclear reactor shall be subjected for licensing.

(3) The requirements and procedures of licensing under clause (1) and clause (2) will be further stipulated by Government Regulation.

Article 18

(1) Any license pursuant to Article 17 shall be subjected for fee.

(2) The amount of fee under clause (1) is stipulated by the Decree of the Minister of Finance.

Article 19

(1) Any employee who operates a nuclear reactor and any special employee in other nuclear installation and in the installations that use ionising radiation sources shall possess a license.

(2) The requirements to obtain the license under clause (1) will be further stipulated by the Regulatory Body.

Article 20

(1) The inspection to the nuclear installation and installation that uses ionising radiation shall be performed by the Regulatory Body in order to ensure that the licensing requirements and nuclear safety regulations are complied with.

(2) The inspection under clause (1) shall be performed by an Inspector who is appointed and dismissed by the Regulatory Body.

(3) The inspection under clause (1) shall be conducted periodically and at any time.

Article 21

The Regulatory Body provides the guidance and counselling on the implementation of efforts related to the health and safety of workers, and the public, as well as the environmental protection.

CHAPTER VI RADIOACTIVE WASTE MANAGEMENT

Article 22

(1) The radioactive waste management shall be conducted to prevent radiation hazards to workers, the public, and the environment.

(2) The radioactive waste under clause (1) is classified into low level radioactive waste, intermediate level radioactive waste, and high level radioactive waste.

Article 23

(1) The radioactive waste management under Article 22 clause (1) shall be performed by the Executing Body.

(2) In performing radioactive waste management under clause (1), the Executing Body may cooperate with or appoints State Company, co-operative, and/or private company.

Article 24

(1) Low level and intermediate level radioactive waste generator shall collect, segregate, or treat and temporarily store those wastes before being transferred to the Executing Body, pursuant to Article 23.

(2) High level radioactive waste generator shall temporarily store those wastes during the period not less than the life time of nuclear reactor.

Article 25

(1) The Executing Body provides the final repository for high level radioactive wastes.

(2) The siting of final repository under clause (1) shall be stipulated by the Government after having an approval from the House of Representative of the Republic of Indonesia.

Article 26

(1) The radioactive wastes storage pursuant to Article 23 clause (1) and Article 25 clause (1), shall be subjected for fee.

(2) The amount of storage fee under clause (1) will be stipulated by the Decree of the Minister of Finance.

Article 27

(1) The transportation and storage of radioactive wastes shall consider the safety of workers, public, and environment.

(2) The provisions on radioactive waste management, including the waste transportation and disposal under clause (1), shall be further implemented by Government Regulation.

CHAPTER VII LIABILITY FOR NUCLEAR DAMAGE

Article 28

The nuclear installation operator shall be liable for nuclear damage suffered by the third party that is resulted from any nuclear incident occurs in that nuclear installation.

Article 29

(1) In the event of nuclear incident occurs during the transportation of nuclear fuel or spent fuel, the nuclear installation consignor shall be liable for the nuclear damage suffered by third party.

(2) The nuclear installation consignor under clause (1) may transfer the liability to the nuclear installation consignee or the management carrier, if there is a written agreement.

Article 30

(1) Where the liability of nuclear damage under Article 28 engages more than one nuclear installation operator and it is impossible to exactly determine the part of nuclear damage caused by each of the nuclear installation operator, all of the operator shall be jointly liable

(2) The liability of each nuclear installation operator under clause (1) shall not exceed the liability of each operator.

Article 31

If in one site there are several nuclear installations, which are managed by one nuclear installation operator, then the operator shall be liable for every nuclear damage caused by every nuclear installation.

Article 32

No liability shall attach to a nuclear installation operator for nuclear damage caused by a nuclear incident directly due to an act of international or non-international armed conflict, or a grave natural disaster exceeding the limit of safety requirement design established by the Regulatory Body

Article 33

(1) If the nuclear installation operator having paid the compensation under Article 28 is capable of proving that the nuclear damage was resulted from the intent of the third party suffering the damage, then the operator may be relieved wholly or partly from the obligation to pay compensation

(2) The nuclear installation operator under clause (1) shall have a right of recourse of the liability she paid to the third party who act intentionally.

Article 34

(1) The maximum limit of liability of the nuclear installation operators shall be Rp 900.000.000.000,00 nine hundred billion rupiahs for each nuclear incident, either in nuclear installation or in transportation of nuclear fuel or spent fuel.

(2) The limit of liability under clause (1) is established by Presidential Decree.

(3) The liability pursuant to clause (1) and clause (2) are merely for nuclear damage, not including interest and Court fee.

(4) The maximum limit of liability under clause (1) may be reconsidered through the Government Regulation.

Article 35

(1) The nuclear installation operator shall be required to insure this liability to the amount pursuant to Article 34 clause (1) and (2), covered by insurance or other financial security.

(2) able for the nuclear installation consignee management or the carrier management.

(3) If there are more than one nuclear installations in one site managed by one nuclear installation operator, the operator shall insure his/her liability for every installation under his/her management.

Article 36

(1) Should the amount of insurance be reduced due to the compensation for nuclear damage, the nuclear installation operator shall maintain the sum of insurance pursuant to Article 34 clause (1) and (2).

(2) If the insurance policy is terminated or is void due to other reasons, the nuclear installation operator shall renew his/her insurance policy.

(3) Where the nuclear installation operator has not renewed the insurance policy pursuant to clause (2) and a subsequent nuclear incident occurs, the nuclear installation operator shall remain liable for the compensation pursuant to Article 34 and 35.

Article 37

(1) The provisions for insurance coverage under Article 35 shall not be applicable for the government agency that is not a State Company.

(2) The compensation for nuclear loss under clause (1) will be further stipulated by Presidential Decree.

Article 38

(1) The insurance company liable for nuclear damage due to a nuclear incident shall pay the compensation within 7 (seven) days after the date of issuance of nuclear incident statement by the Regulatory Body.

(2) The statement under clause (1) shall be issued within 3 (three) days from the date of nuclear incident.

Article 39

(1) The right to claim the compensation due to nuclear incident shall be extinguished if such claim is not brought within 30 (thirty) years from the date of the statement issuance by the Regulatory Body, pursuant to Article 38.

(2) If a nuclear damage caused by a nuclear incident involving nuclear materials that are stolen, lost, or abandoned, then the period established for applying the claim for compensation pursuant to clause (1) shall be accounted from the date of nuclear incident, provided that the period shall not exceed 40 (forty) years since the nuclear materials was theft, lost, or abandoned.

(3) The right to claim the compensation pursuant to clause (1) and (2) shall be brought within a period of 3 (three) years from the date on which the person suffers the nuclear damage had knowledge or should have had knowledge of the nuclear damage and the nuclear installation operator liable for the damage, provided that the period established pursuant to clause (1) and (2) shall not be exceeded.

Article 40

The Competent Courts of the first instance that have jurisdiction to examine and judge any claims for compensation pursuant to Article 39 are as follows:

a. the Court of the first instance within the place where the nuclear incident occurred; or

b. the Court of the first instance in Central Jakarta in the case that the nuclear incident occurred outside the territory of the Republic of Indonesia during the transportation of nuclear fuel or spent fuel.

CHAPTER VIII PENAL PROVISIONS

Article 41

(1) Any person who is constructing, operating, or decommissioning any nuclear reactor without a license pursuant to Article 17 clause (2) shall be punished by limited penal imprisonment for not more than 15 (fifteen) years and by a fine of not more than Rp. 1.000.000.000,00 (one billion rupiahs).

(2) Any person who commit an act pursuant to clause (1) that cause a nuclear damage shall be punished by imprisonment for life or by limited penal imprisonment for not more than 20 (twenty) years and by a fine of not more than Rp. 1.000.000.000,00 (one billion rupiahs).

(3) In the case that the person is incapable to pay the fine pursuant to clause (1) and clause (2), the person shall be punished by limited penal imprisonment for not more than 1 (one) year.

Article 42

(1) Any person who acts contravene with the provisions pursuant to Article 19 clause (1) shall be punished by limited penal imprisonment for not more than 2 (two) years and/or a fine of not more than Rp. 50.000.000,00 (fifty million rupiahs).

(2) In the case that the person is incapable to pay the fine pursuant to clause (1), the person shall be punished by limited penal imprisonment for not more than 6 (six) months.

Article 43

(1) Any person who acts contravene to the provisions pursuant to Article 17 clause (1) shall be punished by a fine of not more than Rp. 100.000.000,00 (one hundred million rupiahs).

(2) In the case that the person is incapable to pay the fine pursuant to clause (1), the person shall be punished by limited penal imprisonment for not more than 1 (one) year.

Article 44

(1) Any person who acts contravene to the provisions pursuant to Article 24 clause (2) for high level radioactive waste generator shall be punished by limited penal imprisonment for not more than 5 (five) years and by a fine of not more than Rp. 300.000.000,00 (three hundred million rupiahs).

(2) Any person who acts contravene to the provisions pursuant to Article 24 clause (1) for low and intermediate level radioactive waste generator shall be punished by a fine of not more than Rp. 100.000.000,00 (one hundred million rupiahs).

(3) In the case that the person is incapable to pay the fine pursuant to clause (2), the person shall be punished by limited penal imprisonment for not more than 1 (one) year.

CHAPTER IX TRANSITIONAL PROVISIONS

Article 45

At the date of commencement of this act, all the implementation regulations related to atomic energy shall remain in effect, provided that they do not contradict to this act.

Article 46

The National Atomic Energy Agency and any other institutions shall keep performing their function before new institution based on this act is established.

CHAPTER X FINAL PROVISIONS

Article 47

At the date of commencement of this act, the Act Number 31 Year 1964 on Basic Stipulations of Atomic Energy is repealed.

Article 48

This Act shall come into operation on the date of its enactment. In order that all persons be informed, it is ordered that this Act be published in the State Gazette of the Republic of Indonesia.

Sanctioned in Jakarta on April 10, 1997

ELUCIDATION ON ACT NUMBER 10, 1997 ELUCIDATION ON ACT NUMBER 10, 1997 ON NUCLEAR ENERGY

I. GENERAL ELUCIDATION

The purpose of national development is to improve the public welfare, to enhance the intellectual life of the nation, and to increase the quality of Indonesian people in the endeavour to realise an advanced, just, and prosperous society evenly, both physically and spiritually, based on Pancasila and the 1945 Constitution.

Nowadays, in some developed countries, the use of nuclear energy in various fields of human life, such as in research, agriculture, medicine, industries, and energy, has progressed rapidly so that as one of the efforts to participate in the national development and towards the realisation of people's welfare and prosperity as well as achieving the capability to master nuclear technology, then it would be appropriate if the large potency of nuclear energy could be developed and exploited as much as possible for the prosperity of the people. However, besides its large benefit, nuclear energy also has radiation hazard to the workers, the public, and the environment if in the use of nuclear energy, the rules on nuclear safety are neglected and are not controlled appropriately.

The use of nuclear energy in Indonesia has been carried out based on the Act Number 31 Year 1964 on Basic Stipulations of Atomic Energy. Due to the development in times and continuing progress in science and technology in the use of nuclear energy, there are many provisions in this act that have become inappropriate. For example, the authority in executing and controlling the research and utilisation of nuclear energy that fells within one institution so that the function of nuclear energy control has not been optimally achieved. In addition, nuclear materials shall be owned and authorised by the State, whereas the trade of the materials have so far been performed internationally so that the requirement for ownership by the State will impede in the development on the use of nuclear energy. However, the requirement for authorisation by the State shall be maintained. Notwithstanding international free-trading of nuclear material, the government is required to control nuclear material in order to prevent deviation of its use. Therefore, it is necessary to stipulate a new act on nuclear energy to replace the Act Number 31 Year 1964 on Basic Stipulations of Atomic Energy.

In this act, the authority in executing and controlling of nuclear energy are separated into two different institutions to avoid the overlapping of activities on the use and control as well as to optimise the control of nuclear energy in order to improve nuclear safety.

Considering that nuclear energy involves the life and safety of the people, the public role shall be enhanced in the form of an advisory counsel, a non-structural and independent institution consists of experts and public figures that has the task to give advice and considerations on the use of nuclear energy.

To observe the public aspiration on the use of nuclear energy, especially in a construction of a nuclear power plant and in providing a radioactive waste final repository, the Government shall consult with the House of Representative of the Republic of Indonesia before taking any decision.

The utilisation of nuclear energy shall consider the Principles of National Development, the safety, security, peace, health of workers and the public, the protection of the environment, as well as the utmost utilisation for the public prosperity. It shall mean that the use of nuclear energy for public welfare shall be implemented together with the efforts to prevent the radiation hazard to workers, public, and to the environment.

In this relation, it is necessary to consider other regulations related to the Act on Nuclear Energy, such as Act Number 1 Year 1970 on Work Safety, Act Number 8 Year 1978 on the Ratification of Treaty on the Nuclear Proliferation on Nuclear Weapons, Act Number 4 Year 1982 on Basic Stipulations on Environmental Management, Act Number 5 Year 1984 on Industry, Act Number 15 Year 1985 on Electricity, Act Number 3 Year 1992 on the Insurance for Workers, Act Number 23 Year 1992 on Health, Act Number 24 Year 1992 on Landscape, and Act Number 10 Year 1995 on Customs.

Nuclear energy shall mean any form of energy released from a nuclear transformation process, including energy from ionising radiation sources, for example in the form of X-ray. Therefore, this act shall also be applicable for regulating the use of X-ray machine.

The meaning of the utilisation of nuclear energy is so broad that it includes research, development, mining, manufacturing, production, transportation, storage, transfer, export, import, usage, decommissioning, and radioactive waste management. Considering that the utilisation of nuclear energy can enhance the public welfare in many fields, the public, private industry, or Government should be given as much opportunity as possible to the use of nuclear energy under the legal provisions.

The use of nuclear energy shall be controlled cautiously in order to comply with the regulations in nuclear safety, so that the use of nuclear energy will not cause radiation hazard to workers, the public, and the environment. Environment in this sense means the unity of space with all materials, energy, as well as the condition and living things, including human and their behaviour, which influences the survival and the prosperity of human and other living things. The control is conducted by establishing regulations, licensing, and performing inspections. Operators of nuclear reactor and certain workers of other nuclear installations as well as installations that utilise radiation sources are subjected for licensing.

The enhancement and development of the capability of human resources are the absolute requirements in order to support the efforts to use and to control nuclear energy so that the use of nuclear energy shall unquestionably increase the people's welfare with a high level of safety. The enhancement and development are also performed to increase the discipline in operating nuclear installations and to develop the safety culture.

Radioactive material shall means any matter that emits ionising radiation with a specific activity of more than 70 kBq/kg or 2 nCi/g (seventy kilobecquerels per kilogram or two nanocuries per gram). The number of 70 kBq/kg (2 nCi/g) is a reference for any substance, which is called radioactive, established by the International Atomic Energy Agency. However, there exist some materials which have specific activity less than the number, but still be considered as radioactive material. This is necessary because it is impossible to determine an absolute limit for all materials due to the different characteristics of materials.

Radioactive wastes, like any other wastes, are any materials that are not further used and due to their radiation properties, the radioactive wastes have a radiation hazard. Because of their properties, radioactive waste management needs to be regulated and controlled to prevent radiation hazards to workers, the public, and to the environment. The radioactive waste management shall be implemented by the Executing Body, where its implementation may be conducted in co-operation with other parties. Based on their hazard level, radioactive wastes are classified into low, intermediate and high level radioactive wastes.

Low and medium level radioactive wastes are collected, segregated, or treated and stored temporarily by the waste generator before they are sent to the Executing Body for further treatment. Since high level radioactive wastes have high level radiation hazard potential, temporary storage of high level radioactive wastes are accomplished by the waste generator, at least during the lifetime of nuclear reactor, while its final repository becomes the responsibility of the Executing Body.

The nuclear enterprises meant in this act shall mean the commercial business activities in general. In this enterprises, besides State Companies, other parties are also be given some opportunities. However, for the Executing Body, the meaning of authority of enterprises shall be a non-commercial and non-profit enterprises.

Nuclear safety technology, nowadays, has been very advanced and reliable and has minimise the probability of nuclear incident so that it can ensure the safety of workers, the public, and the environment. Nevertheless, in order that the nuclear safety regulations are respected and obeyed, it is necessary to stipulate the compensation of the loss due to nuclear incident suffered by third parties and the environment.

Based on the regulatory system in Indonesia, generally the liability is based on fault. It means that a person who is liable shall pay the compensation after there is a proof that a damage has been caused by his/her fault. If the case is applied in a nuclear incident, the suffering party will have a difficulty in proving the fault. The third party, therefore, shall be protected by a strict liability system. The nuclear installation operator is directly liable to the damage without any proof by the third party regarding whether or not there is any fault of the nuclear installation operator, except in case of nuclear incident occurs as a direct result of international or non-international armed conflict or a grave natural disaster exceeding the established limit of safety requirement design.

On the other hand, in the framework of developing and utilising nuclear industry, protection is also needed for the nuclear installation operator as a liable party in the form of limitation of liability, i.e. either in the amount of compensation or in the period of claim.

Considering the need of the third party and the nuclear installation operator, it is necessary to use a special system in liability for nuclear damage that is stipulated in this act. The system not only gives more assured protection to the third party suffering nuclear damage, but also does not impede the development of nuclear industry, as it has been developed in developed countries or developing countries. The main principles in the system are:

a. strict liability;

b. no person other than the nuclear installation operator shall be liable for nuclear damage;

c. the limitation of liability in amount of compensation and time;

d. the nuclear installation operator shall have the liability coverage in the form of insurance or other financial security.

The scope of liability for nuclear damage caused by nuclear accident in this act is limited only to damage suffered by the third party caused by nuclear incident, in a certain nuclear installation or during the transportation of nuclear fuel or spent fuel, which occurs due to the nuclear fuel criticality. Basically the nuclear installation operator consignor shall be liable for nuclear incident during transportation of nuclear fuel or spent fuel, unless there has been a written agreement. A nuclear installation in this act shall mean:

a. nuclear reactor;

b. any facility used for nuclear fuel purification, conversion, enrichment, fabrication and/or spent fuel reprocessing; and/or

c. any facility used to storage nuclear fuel and spent fuel.

The nuclear fuel criticality is any condition where there is a spontaneous chain reaction of nuclear fuel. The chain reaction results in new neutrons, energy, and radioactive matters. Radioactive matters from the chain reaction in a nuclear incident can cause a nuclear damage. The chain reaction may occur if there is an appropriate combination of mass and dimension of nuclear fuel, in this case the mass and size are specific, which is called critical condition.

Nuclear damage means any damage resulting from the radioactive properties or a combination of radioactive properties with toxic, explosive, or other hazardous properties because of the nuclear fuel criticality.

The third party is any person or body suffering nuclear damage, except the nuclear installation operator and nuclear installation workers that based on organisational structure are under the nuclear installation operator.

The compensation of nuclear damage to the third party in this act is the compensation of damage suffered by the people, such as loss of life, disability, personal injury or sickness, and compensation necessary for the preventive action cost, such as an evacuation done by the competent authority in the location of nuclear installation where the nuclear incident occurred. The compensation to the damage of property shall conform to value of the damage suffered and its rehabilitation cost. Likewise, the damage compensation to the pollution and environment damage shall conform to the damage and the environmental recovery cost.

The damage not caused by the criticality of nuclear fuel is not included in the category of a nuclear damage. Such workers of a nuclear installation or the worker of an installation using radiation shall be entitled for the compensation according to the Act on Social Insurance of Workers or other Workmen's Compensation Insurance.

This act only stipulates principal matters. Therefore, further regulations shall be stipulated in the implementation regulations.

II. ARTICLE BY ARTICLE ELUCIDATION

Article 1

Sufficiently clear.

Article 2

Paragraph (1) Spent fuel is nuclear fuel that has been used as fuel in nuclear reactor. Spent fuel is a high level radioactive waste.

Paragraph (2) Sufficiently clear.

Article 3

Paragraph (1) and paragraph (2) The Executing Body mentioned is a government institution under and directly responsible to the President.

Article 4

Paragraph (1) and paragraph (2) The Regulatory Body mentioned in this act is a government institution under and directly responsible to the President.

Article 5

The Nuclear Energy Advisory Council is an independent and non-structural institution, and its membership consists of experts and public figures. It is established by the Government and has the task to give advice and considerations to the Government.

Article 6

Sufficiently clear.

Article 7 The establishment of State Company is implemented in accordance with the legal provisions.

Article 8

Paragraph (1), paragraph (2), paragraph (3), and paragraph (4) Basically, the research and development can be conducted, either by the Executing Body or any other bodies. However, the responsibility in research and development of nuclear science and technology is given to the Executing Body. The research and development of nuclear technology, especially in the nuclear safety, including nuclear fuel waste treatment to reduce its negative impacts, needs considerations to achieve a breakthrough in technology. Researches producing technological breakthrough should be provided an award by the Government in accordance with the legal regulations. Other bodies in this article shall mean any government institutions or private bodies, either national or foreign bodies.

Article 9

Paragraph (1) and paragraph (2) The Executing Body has an authority for general surveys, non-commercial explorations and exploitations of nuclear ore. In performing the authority, the Executing Body may cooperate with State Companies, co-operatives, private companies, or any other bodies. The form of the co-operation will be further administered by the Government. Other bodies in this article shall mean any foreign government institutions or foreign private bodies.

Article 10

Paragraph (1) and paragraph (2) Because nuclear fuel is a strategic material, the production and/or the supply of raw material for nuclear fuel manufacturing is carried out only by the Executing Body. However, the Executing Body may cooperate with State Companies, co-operatives, and or any private companies.

Article 11

Paragraph (1) Sufficiently clear.

Paragraph (2) Sufficiently clear.

Article 12

Paragraph (1) Sufficiently clear.

Paragraph (2) Sufficiently clear.

Article 13

Paragraph (1) Sufficiently clear.

Paragraph (2) Sufficiently clear.

Paragraph (3) Sufficiently clear.

Paragraph (4) The development of nuclear power plant is established by the Government after consultation with the House of Representative of the Republic of Indonesia. The consultations are conducted for each site where one or more nuclear power plants will be built. In this consultation, the Government shall seriously consider the comments and recommendations from the House of Representative of the Republic Indonesia, and the results of consultations shall be respected and become the guidance by the Government and the House of Representative of the Republic of Indonesia.

Article 14

Paragraph (1) and paragraph (2) The control is necessary to be carried out considering that nuclear energy not only is useful but also has radiation hazards. The purpose of control is to prevent any such radiation hazards from happen. The control is administered by the following :

a. Establishing regulations on the nuclear safety field in order to achieve the goal of nuclear energy control.
b. Carrying-out the licensing to control that the utilisation of nuclear energy is implemented according to the legal provisions. By this licensing, the Regulatory Body is able to know where, by whom, and how the use of nuclear energy is conducted.
c. Conducting inspections periodically and at any time to ensure that the nuclear energy utilisation is according to the legal provisions.

Article 15

Safety culture is a characteristics and attitudes in organisations and individuals that emphasise the importance of safety. Therefore, the safety culture requires that all the obligations related to safety shall be carried out correctly, accurately, and with a full responsibly. One of the objective of controls is to prevent the diversion of the purposes of nuclear material utilisation, i.e. the diversion from peaceful purposes.

Article 16

Paragraph (1)

The safety provisions that need to be further regulated among others, are the provision on working safety against radiation, the provision on safe transportation of radioactive material, the provision on safe mining of nuclear ore, and the provision on reactor safety.

Paragraph (2) Sufficiently clear.

Article 17

Paragraph (1) Specific cases in this paragraph mean the utilisation of materials, tools, substances, equipment or objects that have radiation level and activity less than radiation and activity limit requiring license, i.e., navigation instruments, watches, gas mantles, and smoke detectors.

Paragraph (2) The meaning of construction in this paragraph includes siting and construction of nuclear installation itself.

Paragraph (3) Sufficiently clear.

Article 18

Paragraph (1) Sufficiently clear.

Paragraph (2) The Minister of Finance establishes the amount of licensing fees based on recommendations of the Regulatory Body. The income from the licensing fees are deemed Non Tax State Income and shall be deposited to the State Treasury.

Article 19

Paragraph (1) The status of workers in the operation of a nuclear reactor and the use of radiation source are very important. Due to their role in determining the safety of operation and use of nuclear energy, in obtaining a license, the workers, therefore shall be subjected to a thorough examination to justify their qualifications. The meaning of special workers shall include Industrial radiographers, industrial radiographer assistants, radiation protection officers, dosimetry officers, and maintenance officers.

Paragraph (2) Sufficiently clear.

Article 20

Paragraph (1) Sufficiently clear.

Paragraph (2) Sufficiently clear.

Paragraph (3) The results of inspection by the Regulatory Body are published periodically and open for public.

Article 21

The counselling and guidance are intended to create motivation and awareness of safety in the use of nuclear energy.

Article 22

Paragraph (1) Sufficiently clear.

Paragraph (2) Sufficiently clear.

Article 23

Paragraph (1) Radioactive waste management shall be carried out by the Executing Body based on the safety concern and technical capability of the Executing Body and the facilitation in control implementation. The management is carried out non-commercially.

Paragraph (2) For commercial activities of radioactive waste management, the Executing Body may designate State Companies, co-operatives, and/or any private company in accordance with the legal provisions.

Article 24

Paragraph (1) Low and medium level radioactive waste generator shall manage such wastes within the location of the nuclear installation so that they will not endanger to workers, public and the environment, and enable further management by the Executing Body. The purpose of temporary storage is to reduce the radiation level of short half life radioactive materials before transferring their treatment to the Executing Body.

Paragraph (2)

Sufficiently clear

Article 25

Paragraph (1) and paragraph (2) The determination of final repository for high level radioactive wastes shall be discussed with the House of Representative of the Republic of Indonesia to obtain an approval since it converts the function of a beneficial site to a site that can never be used for anything else. The radioactive wastes from other countries shall not be allowed to be stored in the territory of the Republic of Indonesia.

Article 26

Paragraph (1) Sufficiently clear.

Paragraph (2) The Minister of Finance shall govern the amount of the storage fees based on the Executing Body recommendations. The income from disposal fees of the Executing Body are deemed Non Tax State Incomes and shall be deposited to the State Treasury.

Article 27

Paragraph (1) Sufficiently clear.

Paragraph (2) Sufficiently clear.

Article 28 Basically in case of nuclear incident, the liability shall be burdened to one party only, i.e. the nuclear installation operator. Therefore, there are no other persons except the nuclear installation operator that shall be liable. In the strict liability, to receive the compensation, the third party suffering from nuclear damage shall not be obliged to provide proof whether the nuclear installation operator has made any mistake or not. To avoid the compensation being paid to person or persons having no such right, the third party shall show only legal evidence(s) of the damage originated from the nuclear incident.

Article 29

Paragraph (1) Sufficiently clear.

Paragraph (2) Sufficiently clear.

Article 30

Paragraph (1) "All of the operator shall be jointly liable" shall mean if one of the nuclear installation operator has fulfilled the obligations and other operators shall then be free from such obligation. The operator fulfilling the obligation shall calculate the amount of due liability which shall be burdened by other operators proportional to the type of the nuclear installations and their hazard level. Therefore, the liability that shall be burdened by each operator shall not exceed the amount stipulated in Article 34.

Paragraph (2) If nuclear damage exceeds the amount of the nuclear installation operator liability, the Government shall take any necessary steps to resolve the case.

Article 31

Sufficiently clear.

Article 32 Conflict or International armed conflict shall mean a conflict or an armed conflict involving other countries. Non-international armed conflict or armed conflict shall mean among others rebellions and security disturbances movement. A grave natural disaster is, for example, the types of earthquake which have category levels exceeding S1 (seismic category 1) and S2 (seismic category 2). S1 and S2 are the classification of earthquakes stipulated by the Regulatory Body. S1 is a maximum earthquake that may occur once in the nuclear installation life time, while S2 is a maximum earthquake that may happen in a nuclear installation of more than nuclear installation life time. S1 and S2 are determined based on the maximum earthquake that was happen during a certain time cycle at the site of the nuclear installation, for example 50 (fifty)-year-period for S1 (equivalent to nuclear installation life time) and 1,000 (one thousand)-year-period for S2. A nuclear installation shall be so designed to withstand at the S1 and S2 levels.

Article 33

Paragraph (1) Sufficiently clear.

Paragraph (2) Sufficiently clear.

Article 34

Paragraph (1) Sufficiently clear.

Paragraph (2) Sufficiently clear.

Paragraph (3) To avoid any short payment of compensation to the third party, interest and cost the court fee shall not be included in the liability amount.

Paragraph (4) The purpose of reconsideration to the amount of the operator liability is to adjust it according to the current exchange rate.

Article 35

Paragraph (1) Sufficiently clear.

Paragraph (2) The nuclear installation operator shall means the operator under Article 29

Paragraph (3) Sufficiently clear.

Article 36

Paragraph (1) Sufficiently clear.

Paragraph (2) Sufficiently clear.

Paragraph (3) The provision provides more assured protection to the party suffering the damage.

Article 37

Paragraph (1) The exemption of Government from taking any liability coverage by means of insurance or other financial security does not mean that the Government shall not give any compensation to the third party suffering the damage in case of a nuclear incident, because the Government in principle has the task to protect the public.

Paragraph (2) Sufficiently clear.

Article 38

Paragraph (1) Sufficiently clear.

Paragraph (2) Sufficiently clear.

Article 39

Paragraph (1) The provision on the time period is to give a legal certainty to the parties.

Paragraph (2) Sufficiently clear.

Paragraph (3) Sufficiently clear.

Article 40

Sufficiently clear.

Article 41

Paragraph (1) Sufficiently clear.

Paragraph (2) Sufficiently clear.

Paragraph (3) Sufficiently clear.

Article 42

Paragraph (1) Sufficiently clear.

Paragraph (2) Sufficiently clear.

Article 43

Paragraph (1) Sufficiently clear.

Paragraph (2) Sufficiently clear.

Article 44

Paragraph (1) Sufficiently clear.

Paragraph (2) Sufficiently clear.

Paragraph (3) Sufficiently clear.

Article 45

Sufficiently clear.

Article 46

Sufficiently clear.

Article 47

Sufficiently clear.

Article 48

Sufficiently clear.

REGULATION CONCERNING EXEMPTION OF IMPORT DUTY ON IMPORT OF GOODS FOR OPERATION OF GEOTHERMAL ENERGY BUSINESS[3]

BASED ON CONTRACT PRIOR TO THE VALIDITY OF LAW NO. 27/2003 ON GEOTHERMAL ENERGY

THE MINISTER OF FINANCE,

Considering :

a. that, the provisions ruling about operation of geothermal energy business that have existed prior to the validity of Law No. 27/2003 on Geothermal Energy are ruled by the laws that refer to Law No. 8/1971 on State Petroleum and Natural Gas Mining Companies;

b. that, pursuant to Article 41 of Law No. 27/2003 on Geothermal Energy, all joint operation contracts for Geothermal Energy Business Operation that have existed prior to the validity of Law No. 27/2003 are declared still valid until expiry of the contracts;

c. that, in the contract regarding geothermal energy business operation, as meant in paragraph b, it is stated that import duty imposed in relation to import of operational goods shall be borne by the government so that it is deemed necessary to give exemption of import duty on import of the operational goods;

d. that, based on the considerations, as meant in paragraphs a, b, and c, it is necessary to stipulate Regulation of the Minister of Finance on Exemption of Import Duty on Import of Goods for Geothermal Energy Business Operations based on the Contracts Prior to the Validity of Law No. 27/2003 on Geothermal Energy;

In view of :

1. Law No. 10/1995 on Customs Affairs (Statute Book of 1995 No. 75, Supplement to Statute Book No. 3612);

2. Law No. 27/2003 on Geothermal Energy (Statute Book of 2003 No. 115, Supplement to Statute Book No. 4327);

3. Presidential Decree No. 187/M/2004;

4. Decree of the Minister of Finance No. 547/KMK.01/ 2003 an Designation of Tariff of Import Duty on Import Goods;

DECIDES:

---

[3] No. 78/PMK.010/2005

To stipulate :

REGULATION OF THE MINISTER OF FINANCE ON EXEMPTION OF IMPORT DUTY ON IMPORT OF GOODS FOR OPERATION OF GEOTHERMAL ENERGY BUSINESS BASED ON CONTRACT PRIOR TO THE VALIDITY OF LAW NO. 27/2003 ON GEOTHERMAL ENERGY.

Article 1

Hereinafter referred to as:

1. Geothermal Energy Business Operation is a series of operations in the field of geothermal energy resources consisting of exploration, exploitation, and electricity generation implemented by Joint Operations Contract Contractor (KOB Contractor) based on a joint operation contract signed prior to the validity of Law No. 27/2003 on Geothermal Energy.

2. Joint Operation Contract Contractor (KOB Contractor) is the contractor signing Joint Operation Contract with PERTAMINA prior to the validity of, Law No. 27/2003 on Geothermal Energy, which contract contains clauses on exemption of import duty.

Article 2

Import of goods for purpose of geothermal energy business operations shall be given exemption of import duty.

Article 3

The exemption of import duty, as meant in Article 2, shall be given to Joint Operation Contract Contractor up to the expiry of the Joint Operation Contract.

Article 4

(1) Request for exemption of import duty shall be submitted by the company, as meant in Article 1 paragraph 2, to the Director General of Customs and Excise.

(2) The request, as meant in paragraph (1), shall be completed with Import Schedule (RIB) that will be requested for exemption of import duty for a term of not later than 12 (twelve) months.

(3) Duplicate of RIB, as meant in paragraph (2), shall be submitted to the Director General for Mineral, Coal, and Geothermal Energy of the Department of Energy and Mineral Resources.

(4) RIB, as meant in paragraph (2), shall at least contain the following data:

a. Number and date of RIB;

b. Name of company;

c. Taxpayer's Identification Number (NPWP);

d. Address;

e. Contract;

f. Contract area;

g. Customs and Excise Service Office where goods are introduced;

h. Harmonized Schedule;

i. Description of goods:

j. Quantity and unit of goods;

k. Estimated price/import value;

l. Directors of the company.

Article 5

(1) RIB, as meant in Article 4 paragraph (2), must be signed by the directors/managers/officers of the company authorized to sign RIB.

(2) For the purposes of monitoring of the validity of RIB, the company, as meant in Article 1 paragraph 2, must submit specimen of signature of directors/manager/ officers of the company authorized to sign RIB, to the Directorate General of Customs and Excise.

(3) RIB may only be amended by the directors/managers/officers of the company authorized to sign RIB, as meant in paragraph (1), prior to issuance of Decree of the Minister of Finance on Giving of Exemption of Import Duty.

Article 6

(1) The Director General of Customs and Excise or the officer appointed on behalf of the Minister of Finance shall issue a Decree of the Minister of Finance on Giving of Exemption of Import Duty to the request, as meant in Article 4, based on result of inspection of RIB, as meant in Article 5 paragraph (2).

(2) The decree on the giving of exemption of import duty, as meant in paragraph (1), shall be valid for a term of not later than 12 (twelve) months.

Article 7

The company, as meant in Article 1 paragraph 2, must report realization of import of goods that obtain exemption of import duty to the Directorate General of Customs and Excise with a copy to the Directorate General of Mineral, Coal, and Geothermal Energy of the Department of Energy and Mineral Resources not later than 30 (thirty) days after the end of the period, as meant m Article 6 paragraph (2).

Article 8

The Director General of Customs and Excise shall be instructed to enforce the provisions of this Regulation.

Article 9

When this Regulation starts to be valid, the regulations stipulated by the Minister of Finance on procedure of import of goods in the framework of geothermal energy business operations prior to the validity of Law No. 27/2003, which are in contradiction with this Regulation, shall be declared invalid.

Article 10

This Regulation starts to be valid from the date it is stipulated.

For public cognizance, this Regulation shall be published by placing it in the State Gazette of the Republic of Indonesia.

REGULATION CONCERNING PROCEDURES TO RECOGNIZE CERTIFICATE OF ELECTRIC AND/OR ELECTRONIC EQUIPMENT PRODUCTS FROM CONFORMANCE EVALUATION AGENCIES IN ASEAN COUNTRIES[4]

Article 1

In this Regulation of the Minister of Industry:

1. Electric and electronic equipment products are electric and electronic equipment products produced in ASEAN member countries.

2 ASEAN Conformance Evaluation Agencies (LPK) are agencies performing activities and having the expertise to perform all evaluation process on the conformance of electric and/or electronic equipment products in the form of Product Certification Agency (LSPro) and/or Test Laboratories registered in ASEAN (Listed CABs) based on agreement on the performance of Agreement on ASEAN Harmonize Electrical and Electronic Equipment (EEE) Regulatory Regime (AHEEERR) stipulated by the Joint Sector Committee (JSCEEE).

3. Product Certification Agency, hereinafter abbreviated as LSPro is an agency to perform activities on the certification of products on the use of SNI mark for electric and electronic equipment products.

4. Test Laboratory is a laboratory performing testing activities on sample of goods of electric products and electronic equipment according to testing specification/method of SNI or international standard agreed upon by ASEAN.

[4] No. 51/M-IND/PER/3/2012

5. Certificate of Product is a document issued by a Product Certification Agency (LSPro) registered in ASEAN countries to ensure that an electric product and/or electronic equipment meet SNI requirements.

6. Report on Test Result is a report on the result of test on electric and electronic equipment products issued by a Test Laboratory registered in ASEAN.

7. Certification is process for issuing Product Certificate to Use SNI Sign that is conducted by Product Certification Agency (LSPro).

8. Directorate General of Industry Management is Directorate General of Industry Management within Ministry of Industry environment.

9. BPKIMI is Policy Study Board of Industry Condition and Quality Policy, Ministry of Industry.

10. Minister is Minister that conduct government administration in industrial field.

Article 2

(1) Domestic LSPro shall recognize:

a. Product Certificate of electric and electronic equipment; and

b. Testing Result Report of electric and electronic equipment products, from ASEAN countries origin.

(2) Recognition as intended in paragraph (1) for Product Certificate and Testing Result Report that fulfill following requirements:

a. Product Certificate is issued by registered Product Certification Agency in ASEAN; and/or

b. Testing Result Report that is issued by registered Testing Laboratory in ASEAN.

(3) Recognition for Product Certificate and/or Testing Result Report as intended in paragraph (2) shall be implemented by issues SPPT SNI.

(4) Issuing SPPT SNI as intended in paragraph (3) based on Testing Result Report as intended in paragraph (2) b, by showing fulfillment of SNI requirements or equivalent international standard with SNI requirements and its special conditions.

Article 3

Recognition to Product Certificate and Testing Result Report of electric and/or electronic equipment as intended in Article 2 shall be implemented based on application from:

a. Product Certificate holder for application of Product Certificate recognition; and

b. Owner of Testing Result Report for application for recognition of Testing Result Report;

to LSPro appointed by Minister.

Article 4

(1) Application for recognition for product Certificate and Testing Result Report of electric and/or electronic equipment as intended in Article 3 should be completed with following documents:

a. Status and scope of Conformance evaluation agencies issuing the certificate that is registered in ASEAN;

b. Copy of Testing Result Report and/or Copy of Product Certificate;

c. Copy of certificate of trade mark and/or Letter on Registration of Trade Mark from the Ministry of Law and Human Rights; and

d. Copy of the latest Factory Audit Report.

(2) Copy of Testing Result Report, Product Certificate, and Factory Audit Report as intended in paragraph (1) shall be make in Indonesian Language and English.

Article 5

(1) In the framework of recognition for Product Certificate or Testing Result Report, appointed LSPro as intended in Article 3 can make technical review on the documents based on Attachment D of Presidential Regulation No. 79/2010 concerning Ratification of Agreement on the ASEAN Harmonize Electrical and Electronic Equipment (EEE) Regulatory Regime.

(2) Procedure for technical review as intended in paragraph (1) will be governed further in Regulation of the Director General of Industry Management.

Article 6

(1) Effective period of SPPT SNI as intended in Article 2 is 3 (three) years from the date of issuance.

(2) SPPT SNI issuer LSPro as intended in paragraph (1) should control the effective period of SPPT SNI that is issued.

Article 7

Flow Chard of Procedure for the recognition of Product Certificate and Testing Result Report from Conformance evaluation agencies

Article 8

(1) LSPro as intended in Article 6 paragraph (2) shall report the issuance of SPPT as intended in Article 2 paragraph (3) within 7 (seven) days after issuance SPPT-SNI to the Director General of Industry Management and Head of BPKIMI .

(2) LSPro that publishes SPPT SNI as intended in Article 2 paragraph (1) has responsibility for carrying out surveillance of the use of SPPTSNI that is published.

Article 9

(1) Developing and supervising the implementation of Recognition of Product Certification and Test Report by domestic LSPro as well as the application of SNI on electrical equipment products and/or electronics as intended in Article 2 shall be made by the Minister who is carried out by the Directorate General of Industrial Management.

(2) Supervision as intended in paragraph (1) related to product quality conformance on:

a. endanger the health standards;

b. security and safety; and

c. environmental quality.

(3) In case of inconsistency of product quality to the matters as intended in paragraph (2) the Minister inform the relevant authorities in the origin country of products and the ASEAN Secretariat.

(4) In conducting supervision as intended in paragraph (1) the Directorate General of Industry Management to coordinate with relevant agencies.

(5) BPKIMI implements guidance to the Conformance evaluation agencies in order to implement the SNI electrical appliances and electronic products.

(6) In implementing the guidance as intended in paragraph (5), BPKIMI may give a written warning and sanction to the LSPro which does not submit the report as intended in Article 8 in accordance with the provisions of the regulation.

Article 10

At the time this Ministerial Regulation comes into force, Regulation of the Minister of Industry No. 02/M-IND/PER/1/2011 concerning Procedures for Recognition Certificate of Electrical and Electronic Products from the Conformance Evaluation Agencies in ASEAN Countries shall be revoked and declared null and void.

Article 11

This Ministerial Regulation shall come into force on the date of promulgation.

For public cognizance, this Ministerial Regulation shall be promulgated by placing it in the State Gazette of the Republic of Indonesia.

Stipulated in Jakarta on March 15, 2012

DECREE NUMBER 62 YEAR 2013 REGARDING MANAGING AGENCY FOR THE REDUCTION OF EMISSION FROM DEFORESTATION AND DEGRADATION OF FOREST AND PEATLANDS

Considering:

a. That to reduce global warming that impacts environmental damage, various efforts are needed to reduce the emission of greenhouse gases from deforestation and degradation of forest and peatlands;

b. That the Government of Indonesia made a commitment to reduce emission of greenhouse gases of 26 percent using their own effort and 41 percent with international assistance by the year 2020 from a condition without any action plan (business as usual);

c. That to ensure efforts to reduce emission of greenhouse gases from deforestation and degradation of forest and peatlands to be effective, efficient, just and sustainable there is a need for an agency to manage the reduction of emission of greenhouse gases from deforestation and degradation of forest and peatlands;

d. That in relation to the consideration on letter a, letter b and letter c above, there is a need for a Presidential Decree on a Managing Agency for the Reduction of Emission from Deforestation and Degradation of Forest and Peatlands;

Bearing in Mind:

1. Article 4 point (1) of the 1945 Constitution of the Republic of Indonesia;

2. Law Number 41 of 1999 on Forestry (State Gazette of the Republic of Indonesia 1999 no 167, State Gazette Supplement of the Republic of Indonesia No 3888) amended by Law Number 19 of 2004 on the Enactment of Government Regulation in Lieu of Law Number 1 of 2004 on the Amendment of Law Number 41 of 1999 on Forestry (State Gazette of the Republic of Indonesia Number 86 of 2004, State Gazette Suplement Number 4412);

3. Law Number 17 of 2004 on The Ratification of Kyoto Protocol to The United Nations Framework Convention On Climate Change (State Gazette of Republic of Indonesia Number 72 Year 2004, State Gazette Supplement Number 4403);

4. Law Number 32 of 2009 on Environmental Protection and Management (State Gazette of Republic of Indonesia Number 140 Year 2009, State Gazette Supplement Number 5059);

5. Presidential Decree Number 46 of 2008 on the National Council on Climate Change;

6. Presidential Decree Number 61 of 2011 on the National Action Plan to Reduce Emissions of Greenhouse Gases;

7. Presidential Decree Number 71 of 2011 on the National Greenhouse Gases Inventory System;

HAS DECIDED:

To Enact: PRESIDENTIAL DECREE REGARDING THE MANAGING AGENCY FOR THE REDUCTION OF EMISSION FROM DEFORESTATION AND DEGRADATION OF FOREST AND PEATLANDS.

CHAPTER I GENERAL PROVISIONS

Article 1

In this Presidential Decree what is referred to as:

1. Reducing emissions from deforestation and forest degradation in which from herewith shall be referred to as REDD+ refers to all efforts to reduce emission of greenhouse gases from deforestation and degradation of forest and peatlands carried out on forested land and peatlands in Forest Area and non-Forest Area and carbon maintenance and improvement with the added benefit of the increase of biodiversity, the improvement of local/indigenous people's welfare and the increase of sustainability of other production of ecosystem services.

2. Safeguard framework is a collection of criteria and indicators to ensure the implementation of REDD+ does not deviate from its original objective related to program management and financial accountability, impact to the social relation and position of vulnerable communities, and impacts to the environment.

3. REDD+ Program are efforts implemented through a series of projects and/or activities to achieve REDD+ objective in Indonesia.

4. REDD+ Project is one or more activities managed by a certain administrative unit to achieve REDD+ objective in Indonesia.

5. REDD+ Activity is an activity managed by a certain legal subject to achieve REDD+ objective in Indonesia.

6. Emission of Greenhouse Gases from herewith will be referred to as GHG Emissions is the release of greenhouse gases to the atmosphere in a certain area for a certain period of time.

7. GHG are gases in the atmosphere be it natural or anthropogenic, that absorbs and radiate back infrared radiation.

8. Deforestation is permanent change from forested area to non-forested area.

9. Forest degradation is the decrease in quantity of forest cover and carbon stock in a certain period of time.

10. Forest Area is a certain area designated by the government to be maintained as permanent forest.

11. National Strategy to Reduce Emissions from Deforestation and Degradation of Forest and Peatlands from herewith will be referred to as REDD+ National Strategy is the document containing main strategy guidelines and planning direction as stipulated by the Government that

will be used as reference by all parties concerned in the preparation and implementation of REDD+ activities.

12. Government is the President of the Republic of Indonesia that governs the Republic of Indonesia as stated in the 1945 Constitution of the Republic of Indonesia.

13. Funding Instrument is a fund management instrument formed by the Head of REDD+ Managing Agency to guarantee fund management that is transparent, accountable and effective based on REDD+ funding guidelines and safeguards in accordance to prevailing law.

Article 2

The act to reduce emission of greenhouse gases from deforestation and degradation of forest and peatlands has objectives to:

a. reduce emissions from deforestation;

b. reduce emissions from forest degradation and/or peatlands degradation;

c. maintain and increase carbon stock through forest conservation, sustainable forest management, and/or rehabilitation and restoration of damaged forest area; and

d. provide benefit to the increase of environmental services, biodiversity, and local/indigenous community welfare.

CHAPTER II AGENCY

Part One

REDD+ Managing Agency

Paragraph 1

Formation and Positions

Article 3

(1) To implement objectives as referred to in Article 2, by way of this Presidential Decree will be formed a Managing Agency for the Reduction of Emission from Deforestation and Degradation of Forest and Peatlands that herewith will be referred to as the REDD+ Managing Agency.

(2) REDD+ Managing Agency will be situated in the Capital of the Republic of Indonesia.

(3) REDD+ Managing Agency will be lead by a Head that is directly responsible to the President of the Republic of Indonesia.

Paragraph 2

Tasks and Functions

Article 4

REDD+ Managing Agency is tasked to help the President in coordinating, synchronising, planning, facilitating, managing, monitoring, overseeing and controlling REDD+ in Indonesia.

Article 5

In implementing tasks as referred to in Article 4, REDD+ Managing Agency will carry out the following functions:

a. Forming and developing REDD+ national strategy to implement REDD+ in Indonesia;

b. Forming and developing REDD+ safeguard framework in social, environmental and funding matters;

c. Coordinating the formation and development of REDD+ policies and REDD+ mainstreaming in national development;

d. Preparing and coordinating instruments and mechanisms of REDD+ funding and distribute benefit to parties that are implementing REDD+ programs, projects and/or activities in accordance to prevailing regulations;

e. Managing aid in the form of funds or other legal forms related to REDD+ in accordance to prevailing regulations;

f. Developing standard and methodology to measure GHG emissions and sequestration from REDD+ programs, projects or activities and consolidation and reporting of data on GHG emissions and sequestration from REDD+ programs, projects or activities;

g. Increasing capability and capacity in ministries/agencies, implementing partners and community and quality of implementing instruments in the implementation of REDD+;

h. Preparing recommendation in determining Indonesia's position in international fora;

i. Coordinating law enforcement related to the implementation of REDD+ programs, projects and/or activities;

j. Coordinating and facillitating dispute and conflict resolution related to the implementation of REDD+ programs, projects and/or activities in accordance to prevailing regulations;

k. Monitoring and evaluating implementation of REDD+ programs, projects and/or activities;

l. Carrying out administration of the REDD+ Managing Agency;

m. Carrying out other tasks assigned by the President.

Paragraph 3

Organisational Structure

Article 6

(1) REDD+ Managing Agency consists of:

a. Head;

b. 4 (four) Deputies; and

c. Professional Staffs.

(2) Deputies is directly under and directly responsible to the Head of the REDD+ Managing Agency.

(3) Professional Staffs as referred to point (1) letter c, consist of Expert Assistants, Assistants, Junior Assistants and Skilled Staffs, all totalling not more than 60 (sixty) people.

(4) In carrying out their tasks, function, and authority, the Head of the REDD+ Managing Agency can form special teams or task force to address particular issues.

Article 7

Details of tasks and function of Deputies and Professional Staffs referred to in Article 6, will be determined later by the Head of REDD+ Managing Agency bearing in mind tasks and function of the REDD+ Managing Agency stipulated in this Presidential Decree.

Paragraph 4

Secretariat

Article 8

(1) In providing technical and administrative support to the REDD+ Managing Agency, a REDD+ Managing Agency Secretariat will be formed.

(2) The REDD+ Managing Agency Secretariat as referred to in point (1) is lead by a REDD+ Managing Agency Head Secretary and will be under and directly reponsible to the Head of the REDD+ Managing Agency and administratively coordinated by the Minister-State Secretary.

(3) REDD+ Managing Agency Secretariat is tasked with providing administrative and technical support in implementing tasks and function of the REDD+ Managing Agency.

(4) In implementing their tasks as referred to in point (3), the REDD+ Managing Agency Secretariat will perform the following functions:

(5) REDD+ Managing Agency Secretariat will at most consists of 3 (three) Sections.

(6) Each Parts as referred to in point (5) will at most consists of 2 (two) Sub-Sections.

a. Draft plan for agency program development and budget;

b. Manage finances for the operation of the REDD+ Managing Agency;

c. Manage staffing, general matters and house-keeping;

d. Administration.

Article 9

(1) REDD+ Managing Agency Head Secretary is a structural position of echelon IIa.

(2) Head of Section is a structural position of echelon IIIa.

(3) Head of Sub-Section is a structural position of echelon IVa.

Article 10

Echelon II, echelon III, and echelon IV staffs at the REDD+ Managing Agency Secretariat are appointed and dismissed by the Minister-State Secretary with recommendation of the REDD+ Managing Agency Head.

Paragraph 5

Appointment and Dismissal

Article 11

(1) Head of the REDD+ Managing Agency is appointed and dismissed by the President.

(2) Deputies are appointed and dismissed by the President with recommendation of the Head of the REDD+ Managing Agency.

(3) Professional Staffs, Special Teams and Task Force within the REDD+ Managing Agency are appointed and dismissed by the Head of the REDD+ Managing Agency.

Article 12

Deputies and professional staffs within the REDD+ Managing Agency can be civil servants or non-civil servants.

Article 13

(1) Civil servants appointed as staffs of the REDD+ Managing Agency will be resigned from their organic workstations during their term within the REDD+ Agency without losing their status as civil servant.

(2) Civil servants appointed as staffs of the REDD+ Managing Agency will be granted promotion in accordance to prevailing regulations.

(3) Civil servants as referred to in point (1) bear the status employed.

Article 14

(1) Civil servants that resigned or ended their service as staffs of the REDD+ Managing Agency will be activated again to their previous workstations, when they have not reach their retirement age.

(2) Civil servants that resgined or have reached their retirement age will be dismissed with honour as Civil servants and will be given their rights for their service in accordance to prevailing regulations.

Article 15

Non-civil servant staffs of the REDD+ Managing Agency, after resgining or ended their service period will not be granted pension and/or severance pay.

Paragraph 6

Financial Rights and Facilities

Article 16

Head of the REDD+ Managing Agency will be provided with financial, administrative rights and other facilities at par with Ministers.

Article 17

(1) Deputies will be granted position, financial rights and other facilities at par with structural echelon Ia official.

(2) Professional Staffs, that are appointed as Expert Assistants, will be granted position, financial rights and other facilities at par with structural echelon Ib official.

(3) Professional Staffs, that are appointed as Assistants, will be granted position, financial rights and other facilities at par with structural echelon IIa official.

(4) Professional Staffs, that are appointed as Junior Assistants, will be granted position, financial rights and other facilities at par with structural echelon IIIa official.

(5) Professional Staffs, that are appointed as Skilled Staffs, will be granted position, financial rights and other facilities at par with structural echelon IVa official.

Paragraph 7

Work Procedure

Article 18

Provisions related to work procedure, relations and working coordination mechanisms of the REDD+ Managing Agency with Ministries/Agencies, Provincial Heads and Head of

Regencies/Municipalities and other stakeholders are determined by the Head of the REDD+ Managing Agency.

Article 19

Head of the REDD+ Managing Agency will report at least once every 6 (six) months to the President of the Republic of Indonesia or any time required.

Part Two

Stakeholder Committee

Article 20

In order to institutionalise the involvement of stakeholders, a Stakeholder Committee will be formed and will be under and directly responsible to the REDD+ Managing Agency.

Article 21

The Stakeholder Committee is tasked with providing advice and suggestions to the Head of the REDD+ Managing Agency regarding implementation of REDD+ Programs, Projects and/or Activities based on requests or the Committee's initiative.

Article 22

(1) Stakeholder Committee as referred to in point (1) consists of figures each of which has background and/or expertise in fields including but not limited to:

(2) Members of the Stakeholder Committee are appointed by the Head of the REDD+ Managing Agency for 2 (two) years of service period and can be extended for 1 (one) period.

(3) Members of the Stakeholder Committee are not staff of the REDD+ Managing Agency.

a. Natural environment;

b. Community empowerment including indigenous communities;

c. Women empowerment;

d. Business role in development;

e. Good and clean governance; and

f. Science and technology.

Part Three

REDD+ Agency Instruments

Article 23

(1) To guarantee transparent and accountable fund management, the Head of the REDD+Managing Agency will form a REDD+ funding instrument based on prevailing regulations;

(2) REDD+ funding instrument as referred to in point (1) will adopt REDD+ funding guidelines and safeguard framework;

(3) Further provisions on REDD+ funding instrument as referred to in point (1) will be stipulated by law.

Article 24

(1) Measurement of emission reduction and sequestration [of greenhouse gases] from REDD+ programs, projects or activities will be carried out by each REDD+ program, project or activity managers under coordination of the REDD+ Managing Agency;

(2) Measurement as referred to in point (1) will be conducted based on standards and methodologies of GHG emission and sequestration from REDD+ programs, projects and activities as determined by the REDD+ Managing Agency;

(3) Reporting of GHG emission and sequestration from REDD+ programs, projects and activities will be made by the REDD+ Managing Agency to the minister that governs environmental protection and management for verification;

(4) Verification process of GHG emission and squestration from REDD+ programs, projects and activities will be conducted by the minister that governs environmental protection and management.

CHAPTER III FUNDING

Article 25

All funding needed for the implementation of tasks and function of the REDD+ Managing Agency will be sourced from the National State Budget, Regional State Budget, and other legal and unbinding sources, in which dispensing will be carried out in accordance to prevailing law.

CHAPTER IV OTHER PROVISIONS

Article 26

Head of the REDD+ Managing Agency will be responsible to equip the REDD+ Managing Agency organisation by conducting recruitments, empowerment and training of Professional Staffs and other staffs as needed.

CHAPTER V TRANSITION PROVISIONS

Article 27

(1) Tasks, function and programs carried out by the Task Force for the Preparation of the REDD+ Agency formed under the Presidential Decree Number 25 of 2011 on Task Force for Preparing the Establishment of REDD+ Agency as amended by Presidential Decree Number 5 of 2013 will be continued by the REDD+ Managing Agency.

(2) Before the complete formation of the organisational structure of the REDD+ Managing Agency, tasks and function as referred to in point (1) will be carried out by President's Delivery Unit for Development Monitoring and Oversight .

CHAPTER VI CLOSING PROVISIONS

Article 28

This Presidential Decree comes to effect on the date issued.

To be known by everyone, orders the issuance of this Presidential Decree to be placed in State Gazette of the Republic of Indonesia.

Issued in Jakarta

TRAVEL TO INDONESIA

TRAVEL WARNING (Issued on February 27, 2001): The Department of State urges American citizens to defer nonessential travel to Indonesia and all travel to Aceh, Maluku, Papua, West Timor, Central Kalimantan (Borneo) and Central Sulawesi. Those who must travel to Indonesia, or who are resident there, should exercise extreme caution. Indonesia is experiencing a major political transition, and unrest and violence can erupt with little forewarning anywhere in the country. Bombings of religious, political and business targets have occurred throughout the country. In addition, events in the Middle East have sparked anti-American protests in the past.

Due to the continuing threat of serious violence, all travel should be avoided to the regions of Aceh, Maluku, West Timor, Central Kalimantan and Central Sulawesi. Further, all travel by U.S. and other foreign government officials to Aceh, Papua and the Moluccas (provinces of North Maluku and Maluku) has been restricted by the Indonesian government because of security concerns. Private Americans should adhere to these same restrictions.

A massive bombing campaign struck churches throughout Indonesia on Christmas Eve, leaving at least 16 dead and over one hundred injured. Bombings occurred in the cities of Medan, Pekanbaru, Batam, Bandung, Sukabumi, Bekasi, Jakarta, Mojokerto, Surabaya and Mataram. Bombings have also occurred over the past several months at Indonesian government buildings, foreign diplomatic facilities and business and financial centers, including the Jakarta Stock Exchange. More explosive devices have been discovered in Jakarta since Christmas, and there are indications that the bombings may continue.

On occasion, the U.S. mission in Indonesia may have to suspend services to the public or close because of security concerns. In those situations, the U.S. Embassy will continue to be available by telephone to offer emergency services to American citizens. Some foreign travelers in troubled areas of Indonesia have been subject to arbitrary arrest, detention and deportation and, on at least one occasion, false accusations of espionage. In November, 2000, in the central Java city of Solo, groups opposed to U.S. policy undertook or threatened "sweeps," trying to identify American citizens and order them to depart the country. There also have been a number of acts of intimidation and violence directed at American companies and U.S. diplomatic facilities. Indonesian security officials have sometimes been unwilling or unable to respond.

The Abu Sayyaf terrorist group has been active throughout the islands in the extreme southwest Philippines, near Indonesia, and continues to hold hostage an American citizen kidnapped in the Philippines. American citizens traveling to the border regions in Northern Kalimantan and North Sulawesi, in particular the smaller islands closer to the Philippines, are urged to review their security procedures.

American citizens resident or traveling in Indonesia are advised to exercise caution at all times, be alert to suspicious or unclaimed packages, vary times and routes and other aspects of personal daily routine and keep a low profile.

COUNTRY DESCRIPTION: Indonesia is an independent republic consisting of more than 13,500 islands spread over 3,000 miles. Indonesia's economy is developing, and tourist services are plentiful in the major tourist areas. East Timor voted for independence from Indonesia on August 30, 2004 and is currently under the authority of the United Nations Transitional Administration in East Timor (UNTAET). Please refer to the separate Consular Information Sheet available on East Timor.

ENTRY REQUIREMENTS: A passport valid for six months beyond the intended date of departure from Indonesia is required. A visa is not required for tourist stays up to two months. As of November 2000 the Government of Indonesia has been discussing implementing visa requirements for foreign travelers. Travelers should reconfirm entry requirements before traveling. For additional information about entry requirements for Indonesia, travelers may contact the Embassy of the Republic of Indonesia, 2020 Massachusetts Avenue, N.W., Washington, D.C. 20046, telephone (202) 775-5200, Internet address http://www.kbri.org.

DUAL NATIONALITY: Indonesian law does not recognize dual nationality. Because of this, U.S. citizens who are also documented as Indonesian nationals may experience difficulties with immigration formalities in Indonesia. It may also hamper efforts to provide consular protection when abroad. In addition to being subject to all Indonesian laws affecting U.S. citizens, dual nationals may also be subject to other laws that impose special obligations on Indonesian citizens. For further information, please see the Consular Affairs home page on the Internet at http://travel.state.gov for our Dual Nationality flyer.

SAFETY AND SECURITY: Indonesia is experiencing a major political transition, and unrest and violence can erupt with little forewarning anywhere in the country. Due to the continuing threat of serious violence, American citizens should avoid all travel to the regions of Aceh, Irian Jaya (also known as West Papua), the Moluccas, Central Sulawesi and West Timor. All travel by U.S. and other foreign government officials to Aceh, Irian Jaya and the Moluccas (provinces of North Maluku and Maluku) has been restricted by the Indonesian government because of security concerns. Private Americans should adhere to these same restrictions. Travelers should consult the most recent Public Announcement or Travel Warning for Indonesia for updated information on travel within the country. Public Announcements and Travel Warnings can be found on the Bureau of Consular Affairs' home page at http://travel.state.gov/. American citizens in all parts of Indonesia should exercise prudence, remain vigilant and avoid demonstrations and other situations that could turn violent.

Serious sectarian violence has occurred in the Moluccas, Central Sulawesi and the island of Lombok. Violence broke out on the island of Ambon in January 2004, spreading throughout the Moluccas and eventually leading the Government of Indonesia to declare a "Civil State of Emergency" on June 26, 2000. Violence in Central Sulawesi culminated in the death of some 200 people in May 2000. On January 17, 2000, anti-Christian violence broke out on the resort island of Lombok, leading to looting and the burning of a number of churches.

The cities of Jakarta, Medan, Bandung and Surabaya, among others, have been struck by a number of bombings in recent months. The most recent bombing campaign struck churches across Indonesia on Christmas Eve 2000, leaving at least 16 dead and one hundred injured. The Jakarta Stock Exchange, foreign diplomatic facilities and Indonesian government buildings have also been targeted. There are indications that this wave of bombings may continue.

In November 2000, in the central Java city of Solo, groups opposed to U.S. policy undertook "sweeps," trying to identify American citizens and order them to depart the country. There also have been a number of acts of intimidation and violence directed at American companies and U.S. diplomatic facilities. Security officials have sometimes been unwilling or unable to intervene in instances of unrest. Also, Indonesian authorities arbitrarily detained an American tourist in Irian Jaya without formal charges for two weeks and then deported him after a high ranking Indonesian government official publicly accused him of espionage.

The Abu Sayyaf terrorist group has been active throughout the islands in the extreme southwest Philippines, near Indonesia, and continues to hold an American citizen kidnapped in the Philippines. American citizens traveling to the border regions in Northern Kalimantan and North

Sulawesi, in particular the smaller islands closer to the Philippines, are urged to review their security procedures.

On September 6, 2000, a militia mob in West Timor attacked United Nations offices, killing three UN international staff members. The militias have specifically targeted foreigners. American citizens are strongly advised to avoid all travel to West Timor.

The tourist destination of Bali has been largely free of the disturbances seen in other parts of Indonesia. All tourist facilities are operating normally, and to date foreigners have not been the specific target of any group.

Travelers and residents should ensure that their passports and important personal papers are ready for a quick departure from the country should it become necessary for any reason. Americans traveling in Indonesia should remember that much of the country, including many tourist destinations, can be isolated and difficult to reach by available transportation or communication links.

CRIME: Indonesia has a high crime rate. Credit card fraud is a growing problem. Minor crimes, such as pickpocketing and thefts, occur in popular tourist sites throughout the country. Incidents of robbery are on the increase. One common criminal technique is to puncture automobile tires so that the occupants can be robbed while changing the tire. Thefts and robberies from cars stopped at traffic lights have been reported on occasion. American citizens are advised to keep car doors locked and windows rolled up. Americans in Jakarta are also advised to engage a taxi either from a major hotel queue or by calling a taxi company, rather than hailing one on the street. Poachers and illegal loggers operating in Indonesian parks and nature preserves have threatened researchers, tourists and others in order to discourage foreign presence in those areas.

Claiming to act in the name of religious or moral standards, certain extremist groups have begun attacking nightspots and places of entertainment. Most attacks have been aimed at property destruction rather than injury to individuals. However, in November 2000 similar groups attacked an international HIV/AIDS conference, injuring 25, including two foreigners. These groups have on occasion carried out or threatened hunts for Americans and members of certain religious groups to demand they leave the country.

A number of racially motivated incidents of harassment have recently been reported. Persons of African descent, including American citizens, may be subject to arbitrary stops and questioning by both private and public security officials. There are several credible reports that such incidents have led to harassment and physical abuse. To minimize the risk of an incident with local law enforcement authorities, Americans should carry photocopies of their passports at all times. If stopped and detained, Americans should attempt to comply with all instructions from law enforcement officers, but also make it clear that they are American citizens and that they wish to contact the U.S. Embassy. Any incidents should be reported to the U.S. Embassy immediately.

Maritime piracy is a persistent and growing problem in Indonesian waters, targeting both pleasure and commercial vessels. Pleasure yachters are advised to review the current security situation with their local agent when planning itineraries and to exercise particular care when sailing in the Strait of Malacca between Riau Province and Singapore and in the waters north of Sulawesi and Kalimantan.

The loss or theft abroad of a U.S. passport should be reported immediately to the local police and the nearest U.S. embassy or consulate. U.S. citizens may refer to the Department of State's pamphlet, *A Safe Trip Abroad*, for ways to promote a more trouble-free journey. The pamphlet is available by mail from the Superintendent of Documents, U.S. Government Printing Office, Washington, D.C. 20402, via the Internet at http://www.access.gpo.gov/su_docs, or via the Bureau of Consular Affairs home page at http://travel.state.gov/.

MEDICAL FACILITIES: The general level of sanitation and health care in Indonesia is far below U.S. standards. Some level of routine medical care is available in all major cities, although most expatriates choose to leave the country for serious medical procedures. Serious medical problems requiring hospitalization and/or medical evacuation to Singapore or Australia, the closest locations with acceptable medical care, or the United States, can cost thousands of dollars. Doctors and hospitals often expect immediate cash payment for health services.

MEDICAL INSURANCE: U.S. medical insurance is not always valid outside the United States. U.S. Medicare and Medicaid programs do not provide payment for medical services outside the United States. Doctors and hospitals often expect immediate cash payment for health services. Uninsured travelers who require medical care overseas may face extreme difficulties.

Please check with your own insurance company to confirm whether your policy applies overseas, including provision for medical evacuation, and for adequacy of coverage. Serious medical problems requiring hospitalization and/or medical evacuation to the United States can cost tens of thousands of dollars. Please ascertain whether payment will be made to the overseas hospital or doctor or if you will be reimbursed later for expenses that you incur. Some insurance policies also include coverage for psychiatric treatment and for disposition of remains in the event of death.

Useful information on medical emergencies abroad, including overseas insurance programs, is provided in the Department of State's Bureau of Consular Affairs brochure, Medical Information for Americans Traveling Abroad, available via the Bureau of Consular Affairs home page or autofax: (202) 647-3000.

OTHER HEALTH INFORMATION: Information on vaccinations and other health precautions may be obtained from the Centers for Disease Control and Prevention's hotline for international travelers at 1-877-FYI-TRIP (1-877-394-8747); fax 1-888-CDC-FAXX (1-888-232-3299), or via CDC's Internet site at http://www.cdc.gov/.

TRAFFIC SAFETY AND ROAD CONDITIONS: While in a foreign country, U.S. citizens may encounter road conditions that differ significantly from those in the United States. The information below concerning Indonesia is provided for general reference only, and it may not be totally accurate in a particular location or circumstance.

Safety of Public Transportation: Poor
Urban Road Conditions/Maintenance: Good to Fair
Rural Road Conditions/Maintenance: Fair to Poor
Availability of Roadside Assistance: Poor

In general, traffic in Indonesia is congested and undisciplined. The number and variety of vehicles on the road far exceed the capacity of existing roadways to handle the traffic. Road conditions vary from good (in the case of toll roads and major city roads) to dangerously poor.

There is generally a very low level of road safety awareness in Indonesia, although this situation is improving. Buses and trucks are often dangerously overloaded and tend to travel at high speeds. Most roads outside major urban areas have a single lane of traffic in each direction, making passing dangerous. Most Indonesian drivers do not maintain safe following distance in a manner familiar to U.S. drivers and tend to pass or maneuver with considerably less margin for error than in the United States. Although traffic in Indonesia moves on the left side of the road, drivers tend to pass on both sides and may use the shoulder for this purpose. It is common for drivers to create extra lanes regardless of the lane markings painted on the roads. Throughout the country, motor vehicles share the roads with other forms of transportation such as bicycle pedicabs, horse and ox carts, and pushcarts.

Although Indonesia has a seat belt law requiring the use of seat belts in front seats, most Indonesian automobiles do not have seat belts in the rear passenger seats. The use of infant and child car seats is not common, and it can be very difficult to rent a car seat for temporary use. Helmets are required on motorcycles, but this law is inconsistently enforced. Passengers rarely wear helmets. Accidents on rented motorcycles constitute the largest cause of death and serious accident among foreign visitors to Bali.

The use of professional drivers is common among the expatriate and upper class Indonesian communities. All car rental firms are able to provide drivers for a nominal additional fee.

Driving at night can be extremely dangerous outside of major urban areas. It is common to encounter drivers who refuse to use their lights, and most rural roads are unlit. Residents in rural areas may also sometimes congregate on road surfaces after dark, and at least one American citizen was recently involved in a fatal accident when his car hit a group lying on an unlit stretch of road.

In the event of an accident, Indonesia law requires that both drivers await the arrival of a police officer to report the accident. Although Indonesian law requires third party insurance, most Indonesian drivers are uninsured, and even when a vehicle is insured it is common for insurance companies to refuse to pay damages. If a pedestrian is injured, the driver of the vehicle is normally expected to assist in transporting the injured party to the hospital. (Indonesian ambulance services are unreliable.) In cases of traffic accidents resulting in death, it is not uncommon for bystanders to attack the driver perceived to be responsible. This is more common in rural areas and in accidents involving Indonesian drivers, but expatriate drivers have occasionally been attacked by crowds at the scene of an accident. When an accident occurs outside of a major city, it may be advisable to drive to the nearest police station before stopping.

For additional general information about road safety, including links to foreign government sites, please see the Department of State, Bureau of Consular Affairs home page at http://travel.state.gov/road_safety.html. For specific information concerning Indonesian driving permits, vehicle inspection, road tax and mandatory insurance, please contact the Indonesian national tourist organization via the Internet at http://www.indonesia-tourism.com. Please see also road safety information from the U.S. Embassy home page at http://www.usembassyjakarta.org.

AVIATION OVERSIGHT: In 1997, the U.S. Federal Aviation Administration (FAA) has assessed the government of Indonesia's civil aviation authority as Category 1 - in compliance with international aviation safety standards for oversight of Indonesia's air carrier operations.

For further information, travelers may contact the Department of Transportation within the U.S. at telephone 1-800-322-7873, or visit the FAA's Internet web site at http://www.faa.gov/avr/iasa. The U.S. Department of Defense (DOD) separately assesses some foreign air carriers for suitability as official providers of air services. For information regarding the DOD policy on specific carriers, travelers may contact the DOD at telephone (618) 229-4801.

CUSTOMS REGULATIONS: Indonesia's customs authorities may enforce strict regulations concerning temporary importation into or export from Indonesia of personal and commercial use items such as prescription medicines and Chinese-language material or videotapes. It is advisable to contact the Embassy of Indonesia in Washington, D.C. at telephone (202) 775-5200, or one of Indonesia's consulates in the United States, for specific information regarding customs requirements. Enforcement of customs regulations is not always consistent or transparent.

CRIMINAL PENALTIES: While in a foreign country, a U.S. citizen is subject to that country's laws and regulations, which sometimes differ significantly from those in the United States and do not afford the same protections available to the individual under U.S. law. Penalties for breaking the law can be more severe than in the United States for similar offenses. Persons violating the laws of Indonesia, even unknowingly, may be expelled, arrested, or imprisoned. Penalties for possession, use, or trafficking in illegal drugs in Indonesia are strict, and convicted offenders can expect severe jail sentences and fines. The death sentence can be handed down in cases of drug trafficking, and one U.S. citizen was recently given a life sentence for trafficking.

CONSULAR ACCESS: U.S. citizens are encouraged to carry a copy of their U.S. passports with them at all times, so that, if questioned by local officials, proof of identity and U.S. citizenship are readily available. When U.S. citizens are arrested or detained, formal notification of the arrest is normally provided to the U.S. Embassy in Jakarta in writing, a process that can take several weeks. If detained, U.S. citizens are encouraged to attempt to telephone the nearest U.S. consular office. **SPECIAL CIRCUMSTANCES:** U.S. citizens involved in commercial or property matters should be aware that the business environment is complex. In many cases, trade complaints are difficult to resolve.

DISASTER PREPAREDNESS: Indonesia is located in an area of high seismic activity. Although the probability of a major earthquake occurring during an individual trip is remote, earthquakes can and will continue to happen. General information about natural disaster preparedness is available via the Internet from the U.S. Federal Emergency Management Agency (FEMA) at http://www.fema.gov/.

CHILDREN'S ISSUES: For information on international adoption of children or international parental child abduction, please refer to our Internet site at http://travel.state.gov/children's_issues.html or telephone (202) 736-7000.

REGISTRATION AND EMBASSY LOCATION: Americans living in or visiting Indonesia are encouraged to register at the U.S. Embassy or Consulate where they may obtain updated information on travel and security within the country. The U.S. Embassy is located in Jakarta at Medan Merdeka Selatan 5; telephone:(62)(21)3435-9000; fax (62)(21)3435-9922. The Embassy's web site is located at http://www.usembassyjakarta.org. The consular section can be reached by e-mail at jakconsul@state.gov. The U.S. Consulate General is in Surabaya at Jalan Raya Dr., Sutomo 33; telephone: (62)(31) 567-2287/8; fax (62)(31)567-4492;e-mail consularsuraba@state.gov.

There is a Consular Agency in Bali at Jalan Hayam Wuruk 188, Denpasar, Bali; telephone: (62)(361)233-605; fax (62)(31) 222-426; e-mail amcobali@indo.net.id.

The U.S. Consulate in Medan closed in May 1996. American citizens needing assistance in Sumatra should contact the U.S. Embassy in Jakarta.

SUPPLEMENTS

EMERGING NUCLEAR ENERGY COUNTRIES

Over thirty countries are actively considering embarking upon nuclear power programs.

These range from sophisticated economies to developing nations.

Nuclear power is under serious consideration in over thirty countries which do not currently have it (in a few, not necessarily at government level).

In Europe: Italy, Albania, Portugal, Norway, Poland, Belarus, Estonia, Latvia, Ireland, Turkey. In the Middle East and North Africa: Iran, Gulf states, Yemen, Israel, Syria, Jordan, Egypt, Tunisia, Libya, Algeria, Morocco.

In central and southern Africa: Nigeria, Ghana, Uganda, Namibia.

In South America: Chile, Venezuela.

In central and southern Asia: Azerbaijan, Georgia, Kazakhstan, Mongolia, Bangladesh.

In SE Asia: Indonesia, Philippines, Vietnam, Thailand, Malaysia, Australia, New Zealand.

Despite the large number of these emerging countries, they are not expected to contribute very much to the expansion of nuclear capacity in the foreseeable future - the main growth will come in countries where the technology is already well established.

In all countries governments need to create the environment for investment in nuclear power, including professional regulatory regime, policies on nuclear waste management and decommissioning, and involvement with international non-proliferation and insurance arrangements.

In different countries, institutional arrangements vary. Usually governments are heavily involved in planning, and in developing countries also financing and operation. As emerging nuclear nations lack a strong cadre of nuclear engineers and scientists, construction is often on a turnkey basis, with the reactor vendor assuming all technical and commercial risks in delivering a functioning plant on time and at a particular price. Alternatively the vendor may be set up a consortium to build, own and operate the plant. As the industry becomes more international, new arrangements are likely, including public-private partnerships.

The International Atomic Energy Agency has published a small book Considerations to Launch a Nuclear Power Programme (2007) which addresses the issues involved in a country deciding upon and implementing a nuclear power program. In particular it looks at those considerations before a decision is made, before construction starts and subsequently. It then briefly covers twelve factors for consideration.

In January 2008, the French Nuclear Safety Authority (ASN) indicated that it would pay attention to new nuclear power projects in countries with no experience in this area. It said that the development of nuclear industry in a country needs at least 10 to 15 years in order to build up skills in safety and control and to define a regulatory framework. In a June 2008 position paper the five-member commission of ASN said that building the infrastructure needed to safely operate a nuclear power plant required time and that it would be selective about providing assistance. The commissioners said ASN would give priority to countries using French technologies, that it would apply "geophysical, economic, political, social, and technical" criteria, and require countries to be party to relevant international treaties. ASN said it takes at least five years to set up the legal and regulatory infrastructure for a nuclear power program, two to ten years to license a new plant, and about five years to build a power plant. That means a "minimum lead time of 15 years" before a new nuclear power plant can be started up in a country that does not already have the required infrastructure.

These comments relate to France's creation of Agency France Nuclear International (AFNI) under its Atomic Energy Commission (CEA) to provide a vehicle for international assistance. AFNI will be focused on helping to set up structures and systems to enable the establishment of civil nuclear programs in countries wanting to develop them, and will draw on all of France's expertise in this. It will be guided by a steering committee comprising representatives of all the ministries involved (Energy, Foreign Affairs, Industry, Research, etc) as well as representatives of other major French nuclear institutions including the CEA itself and probably ASN, though this is yet to be confirmed.

ITALY

Electricity consumption in Italy in 2005 was 330 billion kWh, giving per capita consumption of 5640 kWh/yr. In 2006 local production from 81 GWe of plant was 315 billion kWh gross, 50% from gas, 15% from oil, 16% from coal and 14% from hydro. Imports of 50.3 billion kWh (effectively, some 15% of its needs) are required, mostly nuclear power from France . This is equivalent to output from about 7 GWe of capacity at 80%.

Due to the high reliance on oil and gas, as well as imports, Italy's electricity prices are 45% above EU average.

Italy today is now the only G8 country without its own nuclear power, and is the world's largest net importer of electricity.

However, Italy was a pioneer of civil nuclear power and built several reactors which operated 1963-90. But following a referendum in November 1987, provoked by the Chernobyl accident 18 months earlier, work on the nuclear program was largely stopped. In 1988 the government resolved to halt all nuclear construction, shut the remaining reactors and decommission them from 1990. Italy then remained largely inactive in nuclear energy for 15 years.

In 2004 a new Energy Law opened up the possibility of joint venture with foreign companies in relation to nuclear power plants and importing electricity from them. This resulted from a clear change in public opinion, especially among younger people favouring nuclear power for Italy.

In 2005 Electricite de France and ENEL signed a co-operation agreement which gives ENEL some 200 MWe from the new Flamanville-3 EPR nuclear reactor (1700 MWe) in France, and potentially another 1000 MWe or so from the next five such units built. As well as the 12.5% share, Italy's electric utility ENEL will also be involved in design, construction and operation of the plants, which will enhance Italy's power security and improve its economics. A major benefit will be in rebuilding Italy's nuclear skills and competence. ENEL is expected to pay about EUR 350 million for its share in the project.

ENEL has also bought 66% of the Slovak Electric utility which operates six nuclear power reactors, and ENEL's investment plan for SE approved in 2005 by the Slovak government includes EUR 1.6 billion for completion of Mochovce nuclear power plant - 942 MWe gross - by 2011-12.

In May 2008 the new Italian government confirmed that it will commence building new nuclear power plants within five years, to reduce the county's great dependence on oil, gas and imported power. It will work towards having 25% of its electricity from nuclear power by 2030, which will require 8 to 10 large new reactors by then. The government introduced a package of nuclear legislation, including measures to set up a national nuclear research and development entity, to expedite licensing of new reactors at existing nuclear power plant sites, and to facilitate licensing

of new reactor sites. ENEL plans to build new reactors at one of three licensed sites: Garigliano, Latina, or Montalto di Castro. The first two had small early-model reactors operating to 1982 and 1987. At Montalto di Castro two larger reactors were almost complete when the country's November 1987 referendum halted construction.

Several research reactors are operating, including AGN Constanza (since 1960), Uni of Pavia's LENA Triga II (250 kW, since 1965), ENEA's Tapiro (5 kW, since 1971), ENEA's Triga RC-1 (1 MW, since 1960) and a subcritical assembly.

Ansaldo Nucleare, which in conjunction with Canada's AECL, built Cernavoda 2 in Romania, is also involved with international R&D on new reactor systems.

ALBANIA

In 2007 the government proposed construction of a nuclear power plant for both domestic and export supply to Balkans and Italy. A proposed site is at Durres. Italian utility ENEL is looking into the feasibility of a nuclear plant.

PORTUGAL

Portugal's electricity demand is about 46 TWh/yr and this comes one third from coal, a quarter from gas and 20% from hydro. Net imports are about 6.5 TWh/yr from Spain.

Its electricity grid is closely linked with Spain's, so that a large nuclear plant on the Atlantic coast could serve both countries.

In 2004 the government rejected a proposal to introduce nuclear power but this is now being reviewed.

NORWAY

Norway's electricity is almost entirely hydro. In 2006, 121.7 billion kWh gross was generated, from 27.5 GWe of capacity. Exports and imports vary greatly with the season and hydro situation in Scandinavia. In 2004 net imports were 11.5 TWh and in 2005 net exports were 12 TWh, mostly to and from Sweden. Per capita use is about 24,000 kWh/yr.

Leaders of Norwegian industry want nuclear energy to supplement hydro. They believe nuclear power based on thorium, which Norway has plenty of, will prevent future energy crises.

A government-appointed committee reported in February 2008 that building thorium-fuelled power reactors was a possibility, which could be tested by using thorium fuel in the country's Halden research reactor. The committee also said that the country should strengthen its international collaboration in nuclear energy and develop its human resources in nuclear science and engineering so as to keep the thorium option open as complementary to the uranium option. "The potential contribution of nuclear energy to a sustainable energy future should be recognised."

The Norwegian Radiation Protection Agency has licensed an underground repository inside a mountain for radioactive waste from the country's oil and gas industry. It will hold 6000 tonnes of NORM waste, and 400 tonnes has already been placed there.

POLAND

In 2006 Poland produced some 161.7 billion kWh gross from 32 GWe of mostly coal plant, giving per capita consumption of 3700 kWh/yr with very high CO2 emissions. Poland has the largest reserves of coal in the EU (14 billion tonnes), and 93% of electricity comes from coal. Electricity consumption is forecast to grow by 90% to 2025, but the EU has placed stringent restrictions on CO2 emissions. About half of the country's gas supply comes from Russia.

The Polish cabinet decided early in 2005 that for energy diversification and to reduce CO2 and sulfur emissions the country should move immediately to introduce nuclear power, so that an initial plant might be operating soon after 2020. In July 2006 the new Prime Minister reaffirmed the need to build nuclear power plants, and mentioned French technology.

A 2006 feasibility study suggested that 11.5 GWe of nuclear capacity would be optimum for Poland but possibly unaffordable in the medium term, so the figure of 4.5 GWe by 2030 was then targeted. A 2007 draft energy policy proposes 10 MWe of indigenous nuclear capacity by 2030, providing 10% of electricity then, and an interim 7.5% by 2022.

Poland had four 440 MWe Russian units under construction in the 1980s at Zarnowiec, but these were cancelled in 1990 and the components were sold.

In July 2006 Lithuania invited Poland to join with Estonia and Latvia in building a new large reactor in Lithuania, to replace the Ignalina units being shut down at EU insistence. Polish participation would justify a larger and more economical unit such as an EPR. In February 2007 the three Baltic states and Poland agreed to build a new nuclear plant at Ignalina, initially with 3200 MWe. Lithuania as host would have 34% of the project and Poland, Latvia and Estonia 22% each. At least one unit of the project is expected to be operating by 2015. Total cost will be some EUR 6 billion. E.On earlier expressed interest in investing in such a unit. Poland said that unless it has access to at least 1000 MWe of the project, later increased to 1200 MWe, it is not worth building the transmission lines to Poland.

In May 2008 the government formed the Lithuanian Energy Organisation (LEO) to build the new nuclear power plant and also transmission to Sweden and Poland. In July 2008 LEO with energy companies from Latvia, Estonia and Poland (Latvenergo, Eesti Energia and Polska Grupa Energetyczna) established the Visaginas project development company for the new 3200-3400 MWe nuclear power plant. The host country will hold 51% of this and the others 16% each, but the JV will be reconstituted later as a project implementation company with different share split related to long-term equity. Though located close to the Soviet-era Ignalina plant, the new one will be called Visaginas after the nearby town of that name. Lithuania wants at least 34% of the new plant (1090-1160 MWe), Poland wants 1000 MWe, while Latvia and Estonia want 400-600 MWe each.

Meanwhile, and apart from Polish participation in the Baltic states nuclear plant, a high-voltage (400 kV) 1000 MW DC PowerBridge costing EUR 250-300 million to improve transmission capacity between Lithuania and Poland is to be built by 2015. It will be half funded by the EC. This follows inauguration of an interconnector between Estonia and Finland - Estlink, a 150 kV, 350 MW DC cable costing EUR 110 million and also supported by EC funding.

A further major transmission link, of 700 to 1000 MWe is proposed undersea between Sweden and Lithuania, to allow power from the new joint reactor to be exported to Scandinavia.

A public opinion poll in December 2006 carried out for the National Atomic Energy Agency showed that 60% supported construction of nuclear power plants to reduce the country's dependence on natural gas and to reduce CO2 emissions. In contrast to NIMBY attitudes elsewhere, 48% said they would favour such a plant being built in their neighbourhood because of its immediate local benefits including lower power cost.

Poland joined the Global Nuclear Energy Partnership (GNEP) in September 2007.

BELARUS

Belarus produces only 31 billion kWh/yr from 7 GWe of plant, mostly gas-fired, giving per capita consumption of 3330 kWh/yr.

The country imports 90% of its gas from Russia - much of it for electricity, and overall aims for 2530% energy independence, compared with half that now. A single nuclear plant would be expected to reduce gas imports by US$ 200-400 million per year. There have been studies on both a domestic plant using Russian technology, and Belarus participation in a new nuclear unit at Smolensk or Kursk in Russia.

Plans to build a new coal-fired plant were shelved in 2005 because no coal supply could be found, but a 600 MWe coal-fired plant is now under consideration.

In mid 2006 the government approved a plan for the construction of an initial 2000 MWe PWR nuclear power plant in the in the Mogilev region of eastern Belarus. This is expected to provide electricity at half the cost of that from Russian gas (5 billion cubic metres per year for same capacity) and to provide some 30% of the electricity by 2020 at a cost of about EUR 4 billion (January 2008 estimate) on a turnkey basis.

After expressions of interest from international reactor vendors were invited, the energy ministry announced in August 2008 that proposals had been received from Atomstroyexport, Westinghouse-Toshiba and Areva. Anything from USA would need several years for an intergovernmental agreement, and Areva's EPR was noted as being too big for the first plant. In addition, the energy ministry received a proposal from the Guangdong Nuclear Power Corporation. Russia is considered as the most likely supplier for the 2 x 1000 MWe plant. Operation of the first unit is envisaged for 2016 and the second in 2018. Two further units are proposed for operation by 2025. In June 2007 Russia's Eximbank offered a US$ 2 billion credit line to enable purchase of equipment from Russia's Power Machines company as a major part of the overall cost.

In November 2007 a presidential decree defined the organizations responsible for preparing for the construction of the country's first nuclear power plant and budgeted money for engineering and site selection, to be finalized at the end of 2008. The candidate sites are Krasnopolyansk and Kukshinovsk (both in the Mogilev region) and Ostrovetsk in the Grodno region, which is favoured. Site works are expected to begin early in 2009 and construction possibly in 2010. Ownership of the plant could be partly or wholly private, and the Bulgarian precedent is being watched with interest.

The decree also aims to ensure that nuclear and radiation safety is in line with the recommendations of the International Atomic Energy Agency (IAEA). A Directorate for the Construction of a Nuclear Power Plant will be established under the Ministry of Energy. A Nuclear & Radiation Safety Department will also be set up as part of the Emergencies Ministry to act as the state nuclear regulator and licensing authority. The state-run Belnipienergoprom enterprise has been designated as the general designer of the plant and will be responsible for negotiating and signing contracts, carrying out feasibility studies and preparing tender documents.

In January 2008 the Security Council confirmed that the country is to embark upon building a nuclear power plant, to provide some 30% of the electricity by 2020. The cost is expected to be about EUR 4 billion.

(A VVER-1000 unit was earlier being built near Minsk but construction was abandoned in 1988 after the Chernobyl accident.)

IRELAND

Ireland produces about 28 billion kWh/yr gross from 6 GWe of plant for its population of 4.1 million, giving per capita consumption of 6300 kWh/yr. About half of its electricity is generated by gas, but relative to the rest of Europe it is heavily dependent on oil for its electricity - 10%. Coal provides 28%. Wind provides around 6% from about 800 MWe of capacity. Ireland has a target of 3000 MWe wind capacity by 2020 In 1981 the government considered building a 650 MWe nuclear power plant (PWR) at Carnsore Point, but the plan was dropped as energy demand flattened. It would have required a link across the Irish Sea to the UK to be viable, due to its large size relative to the Irish grid.

A government-commissioned report by Forfas in April 2006 pointed to the need for Ireland again to consider nuclear power in order "to secure its long-run energy security". Relatively small-scale nuclear plants were envisaged. The report also suggested accelerating plans for greater east-west interconnection with the UK, which would draw on its nuclear capacity and also provide an export channel for any Irish nuclear power development.

In 2007 Ireland's Electricity Supply Board made it known that it would consider a joint venture with a major EU energy company to build nuclear capacity. In April 2008 the Irish Energy Regulator proposed a nationwide debate on the issue of nuclear power to address the country's pending energy crisis. It referred to the need to find an alternative to meet future energy needs since neither wind power or any other renewable energy sources could satisfy demand.

ESTONIA AND LATVIA

These countries have not been planning to build any nuclear capacity themselves and are participants in a plan to replace the Ignalina reactor in Lithuania with much larger capacity which will serve those three Baltic states and Poland. SeeLithuania paper, and Poland section above.

However, in 2008 Estonia took steps to identify sites for a possible nuclear power plant of its own, and investigate possible involvement in a sixth Finnish plant. Estonia has recently completed a 350 MW DC cable interconnector with Finland - Estlink, costing EUR 110 million and supported by EC funding.

Estonia has two small Soviet naval reactors originally used for submarine training. They date from 1968 and 1983 and were closed down in 1989. They are in Safestor mode and will be dismantled after fifty years. The fuel has been returned to Russia.

TURKEY

In 2006 Turkey produced 176 billion kWh/yr gross from 35 GWe of plant, giving per capita consumption of almost 2000 kWh/yr. 44% of electricity comes from gas (two thirds of this from Russia, most of the rest from Iran), 26% from coal and 25% from hydro. Demand growth is 8% pa.

Several nuclear power projects have been proposed: In 1970 a feasibility study concerned a 300 MWe plant, in 1973 the electricity authority decided to build a 80 MWe demonstration plant but didn't, then in 1976 the Akkuyu site on the Mediterranean coast near the port of Mersin was licensed for a nuclear plant. In 1980 an attempt to build several plants failed for lack of government financial guarantee, in 1993 a nuclear plant was included in the country's investment program following a request for preliminary proposals in 1992. Then revised tender specifications were not released until December 1996. Bids for a 2000 MWe plant at Akkuyu were received from Westinghouse + Mitsubishi, AECL and Framatome + Siemens. Following the final bid deadline in October 1997, the government delayed its decision no less than eight times between June 1998 and April 2000, when plans were abandoned due to economic circumstances.

Early in 2006 the province of the port city of Sinop on the Black Sea was chosen to host a commercial nuclear power plant. This has the advantage of cooling water temperatures about 5 degrees below those at Akkuyu, allowing about 1% greater power output from any thermal unit. A 100 MWe demonstration plant was to be built there first, then 5000 MWe of further plants to come into service from 2012. Some kind of public-private partnership is envisaged for construction and operation.

In August 2006 the government said it planned to have three nuclear power plants total 4500 MWe operating by 2012-15, a US$ 10.5 billion investment. Discussions have been under way with Atomic Energy of Canada Ltd re two 750 MWe CANDU units as an initial investment. These and the PWR type are apparently preferred. The first units of some 5000 MWe total will be built at Akkuyu, since the site is already licensed, but licensing is proceeding for Sinop.

In 2007 a new bill concerning construction and operation of nuclear power plants and sale of their electricity was passed by parliament and subsequently approved by the President. The bill provides for the Turkish Atomic Energy Authority (TAEK) to set the criteria for building and operating the plant. The Turkish Electricity Trade & Contract Corporation (TETAS) will buy all the power under 15-year contracts. The bill also provides for public institutions to build the plants if other offers are not satisfactory. It also addresses waste management and decommissioning, providing for a National Radioactive Waste Account (URAH) and a Decommissioning Account (ICH) which generators will pay into progressively.

TETAS called for tenders in March 2008, inviting bids for the first nuclear power plant at Akkuyu. TAEK issued specifications, allowing for PWR, BWR or PHWR types of at least 600 MWe and with 40 year service life. Design certification in country of origin will be acceptable, allowing TAEK to concentrate on site-specific aspects of the project. In the event, only one bid was received, from Atomstroyexport, for an AES-2006 power plant with twin 1200 MWe reactors. TAEK has found that it does not meet technical criteria however. Following technical advice from TAEK and commercial advice from TETAS, a government decision on plant type and construction arrangements is expected by the end of 2008. Operation is now expected in 2014. A mid February 2008 announcement said that preparatory work is under way to build a second nuclear plant at Sinop, along with a EUR 1.7 billion nuclear technology centre. There are proposals to build a further 10 to 12 reactors by 2020. Near Istanbul, eight Organised Industrial Parks comprising 70,000 firms and using 1.5 billion kWh per year have set up a joint venture - IOSBB - to construct the country's first nuclear power plant(s) of 1500 MWe each. Likely sites mentioned include Sinop on Black Sea and Gokova on Mediterranean.

In May 2008 a civil nuclear cooperation agreement with the USA entered into force.

IRAN

Iran produces some 180 billion kWh/yr from 31 GWe of plant, giving per capita consumption of 1943 kWh/yr. 73% of electricity comes from gas, 18% from oil.

In the mid 1970s construction of two 1,200 MW(e) PWR units was started at Bushehr by Siemens KWU. In 1979 this was suspended. The Islamic Republic of Iran revived the nuclear power program in 1991 with a bilateral agreement with China for the supply of two 300 MW(e) PWR units of Chinese design, similar to the Qinshan power plant, but nothing eventuated.

In 1994, Russia's Minatom and the Atomic Energy Organization of Iran (AEOI) agreed to complete unit 1 of Bushehr nuclear power plant with a VVER-1000 unit, using mostly the infrastructure already in place. This long-awaited 915 MWe plant, being constructed by Atomstroyexport, is nearing completion and is expected to start up late in 2008, with commercial operation mid 2009. A second reactor is planned at the site.

After two years delay due to Iran's reluctance to return spent fuel to Russia without being paid for it, two agreements were signed early in 2005 covering both supply of fresh fuel for the new Bushehr nuclear reactor and its return to Russia after use. Supply of the fuel was originally contingent upon Iran's signing the Additional Protocol to its safeguards agreement with the IAEA. It has done this but not ratified it. The Russian agreement means that Iran's nuclear fuel supply is secured for the foreseeable future, removing any justification for enrichment locally. It also means that the anticipated 6-7 TWh/yr from the new reactor will free up about 1.6 million tonnes of oil or 1800 million cubic metres of gas per year which can be exported for hard currency.

Russia's Atomstroyexport in December 2007 delivered the first of 163 fuel assemblies for the initial core of Bushehr. The fuel is enriched to 3.62% or less and is under full international safeguards. The Russian government had withheld supply as negotiations over Iran's uranium enrichment activities proceeded. The AEOI has announced that a new indigenous 360 MWe nuclear power plant is to be built at Darkhovin in Khuzestan province in the southwest, at the head of the Gulf, where two Framatome 900 MWe plants were about to be constructed in 1970s. It has also invited bids for two units of up to 1600 MWe to be built near Bushehr and come on line about 2016.

Iran also has a major project developing uranium enrichment capability and in November 2007 it announced that the initial target of 3000 centrifuges had been reached - evidently 18 cascades operating. This program is heavily censured by the UN, since no commercial purpose is evident.

GULF STATES, UAE

In December 2006 the six member states of the Gulf Cooperation Council - Kuwait, Saudi Arabia, Bahrain, the United Arab Emirates (UAE), Qatar and Oman - announced that the Council was commissioning a study on the peaceful use of nuclear energy. France agreed to work with them on this, and Iran pledged assistance with nuclear technology.

Together they produce 273 billion kWh per year, all from fossil fuels (2003) and 5-7% annual demand growth. They have total installed capacity of about 80 GWe, with a common grid. There is also a large demand for desalination, currently fuelled by oil and gas. UAE itself has some 18 GWe installed and 85% of its power is from gas, the rest from oil.

In February 2007 the six states agreed with the IAEA to cooperate on a feasibility study for a regional nuclear power and desalination program. Saudi Arabia was leading the investigation and thought that a program might emerge about 2009.

The six nations are all signatories of the NPT and the UAE ratified a safeguards agreement with IAEA in 2003. In mid 2008 it appointed an ambassador to IAEA.

In January 2008 three French companies Areva, Suez and Total signed a partnership agreement to propose to UAE the construction of two EPR units there. Suez and Total would each invest up to 25% of the project with Abu Dhabi entities providing at least 50%. Suez would be operator, Areva would supply the plant and manage the fuel. Total and Suez are well established in the region and together operate a power and desalination plant for Abu Dhabi, 100 km west of Dubai. The consortium's first EPR would not be operating before about 2017.

In April 2008 the UAE independently published a comprehensive policy on nuclear energy. This projects escalating electricity demand from 15.5 GWe in 2008 to over 40 GWe in 2020, with natural gas supplies sufficient for only half of this. Imported coal is dismissed as an option due to environmental and energy security implications. Nuclear power "emerged as a proven, environmentally promising and commercially competitive option which could make a significant base-load contribution to the UAE's economy and future energy security." Hence 20 GWe nuclear is envisaged from about 14 plants, with nearly one quarter of this operating by 2020. Two reactors are envisaged for a site between Abu Dhabi and Ruwais, and a third possibly at Al Fujayrah on the Indian Ocean coast.

Accordingly, and as recommended by the IAEA, the UAE established a Nuclear Energy Program Implementation Organization which has set up the Emirates Nuclear Energy Corporation (ENEC) as a public entity, initially funded with $100 million, to evaluate and implement nuclear power plans within UAE.

Secondly, the UAE "will also draft a comprehensive national nuclear law" which establishes a fully independent Federal Authority for Nuclear Regulation (FANR), to be headed by a senior US regulator. Thirdly, it will "offer joint-venture arrangements to foreign investors for the construction and operation of future nuclear power plants" similar to existing Independent Water and Power Producer structures which have 60% owned by the government and 40% by the JV partner(s). ENEC is seeking bids from international companies to manage its nuclear power program. The UAE is setting up a model of managing its nuclear power program based on contractor services rather than more slowly establishing indigenous expertise.

"In lieu of domestic enrichment and reprocessing, the UAE would seek to conclude long-term arrangements .... for the secure supply of nuclear fuel, as well as the safe and secure transportation and, if available, the disposal of spent fuel via fuel leasing or other emerging fuel supply arrangements."

The UAE is reported to have invited expressions of interest from nine short-listed companies for construction of its first nuclear power plant. By 2020 it hopes to have three 1500 MWe nuclear plants running and producing electricity at a quarter the cost of that from gas. ENEC has appointed the global full-service program management, engineering, construction and operations firm C2HM Hill to manage the UAE's plans for bringing nuclear power to the country.

The USA has bilateral nuclear energy cooperation agreements with Saudi Arabia, Bahrain and UAE. In May 2008 the UK signed a Memorandum of Understanding on nuclear energy cooperation with UAE. France has a nuclear cooperation agreement with UAE and has discussed nuclear energy development with Saudi Arabia, offering Atomic Energy Commission (CAE) assistance. Yemen

It was reported in September 2007 that Yemen had signed an agreement with Texas-based PowerEd Corporation to build 5000 MWe of nuclear power capacity by 2017. However, with 2005 production of 4.7 billion kWh (corresponding to about 670 MWe of base-load capacity) this did not seem plausible, and the government apparently cancelled the deal.

ISRAEL

Israel produces 50 billion kWh per year, about 75% from coal and 25% from imported gas. It has little reserve capacity. Net exports are 1.6 billion kWh/yr.

In the 1980s the state-owned Israel Electric Corporation (IEC) set aside a site in the southern Negev at Shivta for a nuclear power plant, and discussions were held with France regarding equipment. The question was raised again in 2007 by the National Infrastuctures Ministry and Atomic Energy Commission. A twin reactor nuclear plant of 1200-1500 MWe under IAEA safeguards is envisaged for the site. However, Nuclear Suppliers Group guidelines would constrain any supply of nuclear plant to Israel.

Israel has a 5 MWt research reactor at Nahal Soreq near Tel Aviv under IAEA safeguards and another 70 MWt French-built heavy water reactor at Dimona in the Negev, which is understood to have been used for military plutonium production.

Israel is one of three significant countries which have never been part of the Nuclear Non-Proliferation Treaty (NPT), so any supply of nuclear equipment or fuel from outside the country would be severely constrained. Unlike India and Pakistan, Israel has had no civil nuclear power program.

SYRIA

Syria produced 35 billion kWh in 2005, 90% of this from fossil fuels, the balance from hydro.

Syria had plans in the 1980s to build a VVER-440 reactor but abandoned these after the Chernobyl accident and due to the collapse of Soviet Union. With escalating oil and gas prices, nuclear power is now being considered again.

Meanwhile over 2001-07 Syria built at a remote location what appeared to be a gas-cooled reactor similar to the plutonium production unit at Yongbyong in North Korea. This was destroyed by an Israeli air strike in 2007 and the remains then demolished. The project was clandestine and in breach of Syria's obligations under the NPT.

JORDAN

Jordan imports about 95% of its energy needs. It generated and imported 10.4 billion kWh of electricity in 2005 for its six million people. It has 2030 MWe of generating capacity and expects to need an additional 1200 MWe by 2015. Also it has a "water deficit" of about 500 million cubic metres per year.

The energy minister has said that the country expects to have a nuclear power plant operating by 2015, for electricity and desalination. Jordan's Committee for Nuclear Strategy has set out a program for nuclear power to provide 30% of electricity by 2030, and to provide for exports.

In mid 2008 an agreement between the Jordan Atomic Energy Commission (JAEC) and Atomic Energy of Canada Ltd (AECL) with SNC-Lavalin is to conduct a 3-year feasibility study on building an AECL 740 MWe Enhanced Candu-6 reactor using natural uranium fuel, for power and desalination. Site selection is planned for 2009, though options appear to be limited to 30 kilometres of Red Sea coast near Aqaba. In August 2008 it was reported that the government intended to sign up for an Areva reactor, and discussions with Areva in November pointed to an 1100 MWe unit, presumably from Atmea, the Areva-Mitsubishi joint venture which is developing such a unit for countries embarking upon nuclear power programs.

The country has low-cost uranium resources of 140,000 tU plus another 59,000 tU in phosphate deposits, and plans to mine these have been announced by the government. A feasibility study on recovering uranium as a by-product of phosphate production is under way. In October 2008 a joint venture between JAEC and Areva was established to define uranium resources in central Jordan and mine them. Mining at the rate of 2000 t/yr is envisaged from 2012. Areva anticipates that the agreement will lead to further involvement with JAEC.

Jordan has signed nuclear cooperation agreements with the USA, France and UK, in respect to both power and desalination, and is seeking help from the IAEA. It has signed a nuclear cooperation agreement with China, covering uranium mining in Jordan and nuclear power. Jordan joined the Global Nuclear Energy Partnership (GNEP) in 2007.

EGYPT

Egypt produced 109 billion kWh in 2005 from 18 GWe of plant, giving per capita consumption of 1350 kWh/yr. 84% of electricity comes from gas, 16% from hydro. Demand growth is about 7% pa.

In 1964 a 150 MWe nuclear plant with 20,000 m^3/day desalination was proposed then in 1974 a 600 MWe plant was proposed. The government's Nuclear Power Plants Authority (NPPA) was then established in 1976, and in 1983 the El Dabaa site lon the Mediterranean coast was selected for a nuclear plant. This plan was aborted following the Chernobyl accident. More recently the NPPA carried out a feasibility study for a cogeneration plant for electricity and desalination, updating it in 2003.

A new agreement on peaceful uses of atomic energy was signed with Russia at the end of 2004, and a further one in March 2008, reviving Egypt's plans for a nuclear power and desalination plant there, supported by Rosatom. . In 2006 a nuclear cooperation agreement was reached with China.

Egypt already has a 1961-vintage 2 MW Russian research reactor serviced by Russia, and a 22 MW Argentinian research reactor partly supported by Russia and which started up in 1997.

On the basis of the feasibility study for a cogeneration plant for electricity and potable water at El-Dabaa, in October 2006 the Minister for Energy announced that a 1000 MWe reactor would be built there by 2015. The US$ 1.5 to 2 billion project would be open to foreign participation.

TUNISIA

Tunisia produced 13.7 billion kWh in 2005, almost all of this from fossil fuels.

The government is reported to be evaluating the possible construction of a 600 MWe nuclear plant costing US$ 1.14 billion.

In December 2006 a nuclear cooperation agreement was signed with France, focused on nuclear power and desalination, and in April 2008 this was amplified.

LIBYA

In 2005 Libya produced 22.5 billion kWh of electricity.

Early in 2007 it was reported that Libya was seeking an agreement for US assistance in building a nuclear power plant for electricity and desalination. In 2006 an agreement with France was signed for peaceful uses of atomic energy and in mid 2007 a memorandum of understanding related to building a mid-sized nuclear plant for seawater desalination. Areva TA would supply this. In 2008 Libya signed a civil nuclear cooperation agreement with Russia.

In 2003 Libya had halted a clandestine program developing uranium enrichment capability, and fully opened itself to IAEA inspections.

Libya has a Russian 10 MW research reactor which is under IAEA safeguards.

ALGERIA

Algeria produced 34 billion kWh of electricity in 2005 from natural gas and is a major gas exporter.

In January 2007 Russia signed an agreement to investigate the establishment of nuclear power there. The government has also made positive statements on the matter, but referring to a 20year time frame. A further nuclear energy cooperation agreement, with the USA, was to be signed in June 2007 and one with France was signed in December 2007, coupled with strong commercial interest from Areva. An accord with China on developing peaceful nuclear power was signed in March 2008.

The country has operated two research reactors since 1995, at Draria and Ain Ouessara. One was built by INVAP of Argentina, the other by China.

MOROCCO

Morocco has growing electricity demand and produced 22.6 billion kWh in 2005. It also has requirements for desalination. Most electricity is supplied by coal, some by oil.

The government has plans for building an initial nuclear power plant in 2016-17 at Sidi Boulbra, and Atomstroyexport is assisting with feasibility studies for this.

Morocco has a 2 MW Triga research reactor under construction.

For desalination, it has completed a pre-project study with China, at Tan-Tan on the Atlantic coast, using a 10 MWt heating reactor which produces 8000 m^3/day of potable water by distillation.

In October 2007 a partnership with France to develop a nuclear power plant near Marrakesh was foreshadowed and a nuclear energy cooperation agreement was signed.

NIGERIA

Nigeria produced 23.5 billion kWh in 2005 from about 6 GWe of plant and had final consumption of 17 TWh, giving per capita consumption of only 113 kWh/yr.

To address rapidly increasing base-load electricity demand, Nigeria has sought the support of the International Atomic Energy Agency to develop plans for up to 4000 MWe of nuclear capacity by 2025. Nigeria is Africa's most populous country and its power demand is expected to reach 10,000 MWe by 2007 - current grid-supplied capacity is 2600 MWe.

Early in 2008 the Minister of Science and Technology said that the government has reaffirmed its determination to initiate its nuclear energy program by approving a technical framework for it. This is to proceed through manpower and infrastructure development, power reactor design certification, regulatory and licensing approvals, construction and start-up. In mid 2008 the target was moved forward to having up to 5000 MWe of nuclear capacity by 2017.

Nigeria's first research reactor was commissioned at Ahmadu Bello University in 2004. It is a 30 kW Chinese Miniature Neutron Source Reactor similar to other Chinese units operating in Ghana, Iran, Syria and China. The IAEA assisted the Nigerian government with the project, to "reinforce and widen the human resource base to sustain nuclear technology" in relation to medical technology, geochemistry, mineral and petrochemical analysis and exploration.

GHANA

Ghana produced 6.8 billion kWh in 2005. In April 2007 the government announced that it planned to introduce nuclear power on energy security grounds. In May 2008 the government said it planned to have 400 MWe of nuclear capacity by 2018. Ghana joined the Global Nuclear Energy Partnership (GNEP) in September 2007.

Ghana has a small Chinese research reactor, operating since 1994.

Uganda Uganda's Atomic Energy Bill came into effect in 2008, to regulate the use of ionising radiation and provide a framework to develop nuclear power generation. The government has signed an agreement with IAEA to initiate moves in that direction.

NAMIBIA

Namibia's electricity supply of 3.2 billion kWh in 2005 was almost half supplied by South Africa, which faces supply constraints itself. A coal-fired plant is planned for Walvis Bay.

Namibia holds about 7% of the world's uranium reserves, which are mined to fuel nuclear power stations around the world. Now the government has committed to a policy position of supplying its own electricity from nuclear power. The country faces severe challenges in power supply.

AZERBAIJAN

In 2005 the country produced 21 billion kWh of electricity. The government is planning construction of a 1000-1500 MWe nuclear power reactor, possibly starting 2010 in the Avai region in the south of the country, supporting proposed industrialisation there.

In June 2008, the International Atomic Energy Agency (IAEA) issued a preliminary agreement to support a 10-15 megawatt research reactor outside of Baku. The $119-million reactor will be operated by the National Academy of Sciences Institute for Radiation Problems, which specializes in nuclear energy research. Construction is expected to begin in 2012.

GEORGIA

Georgia generated 7.3 billion kWh in 2005, largely from natural gas, and imported another 1.5 billion kWh.

It is heavily dependent on Russia for energy supplies and there is some discussion about building a nuclear power plant to assist its energy independence. This could be in collaboration with Azerbaijan or Armenia. In November 2006 Russia threatened to double the price of gas to Georgia. In August 2008 it invaded Georgia.

KAZAKHSTAN

Kazakhstan produced 68 billion kWh in 2005 from about 17 GWe of plant, mostly now privatised, and had net exports of 12.5 billion kWh. Per capita consumption is 3460 kWh/yr. 52% of electricity comes from coal, 26% from gas and 17% from oil.

It has no national electricity grid, but a northern grid links to Russia and a southern one links to Kyrgystan and Uzbekistan.

The BN-350 fast reactor at Aktau (formerly Shevchenko), on the shore of the Caspian Sea, successfully produced up to 135 MWe of electricity and 80,000 m^3/day of potable water over some 27 years until it was closed down in mid 1999. About 60% of its power was used for heat and desalination. It was operated by the Mangyshlak Power Generation Co. (MAEK).

The Russian plant, built under Minatom supervision, was designed as 1000 MWt but never operated at more than 750 MWt and was most recently quoted at 520 MWt, but it established the feasibility and reliability of such cogeneration plants. (In fact, oil/gas boilers were used in conjunction with it, and total desalination capacity through ten multi-effect distillation (MED) units was 120,000 m^3/day.)

There are proposals for a new nuclear power plant near Lake Balkhash in the south of the country near Almaty.

A July 2006 joint venture with Russia's Atomstroyexport envisages development and marketing of innovative small and medium-sized reactors, starting with OKBM's VBER-300 as baseline for Kazakh units. Atomstroyexport expects to build the initial one.

In April 2007 two agreements with Japan relate to assistance in building nuclear power plants, one between Japan Atomic Power Co and three Kazakh entities, the other between Toshiba Corp and Kazatomprom.

At Kurchatov (aka Semipalatinsk-21) on the former Semipalatinsk nuclear test site three research reactors owned by the National Nuclear Centre are operated by the Institute of Atomic Energy. A fourth is at Almaty. The three larger ones are tank-type units of 6, 10 and 60 MW, the newest is a 400 kW high-temperature gas reactor. All were supplied by Russia and use high-enriched fuel.

The internationally-significant Ulba Metallurgical Plant at Ust Kamenogorsk in the east of the country was commissioned in 1949. It has a variety of functions relevant to uranium, the most basic of which since 1997 is to refine most Kazakh mine output of U3O8. (It also produces beryllium, niobium and tanatalum.)

Since 1973 Ulba has produced nuclear fuel pellets from Russian-enriched uranium which are used in Russian and Ukrainian VVER and RBMK reactors. Some of this product incorporates gadolinium and erbium burnable poisons. Other exports are to the USA and in 2007 it plans to market to Asia. It briefly produced fuel for submarines (from 1968) and satellite reactors. Since 1985 it has been able to handle reprocessed uranium.

Ulba is majority owned by Kazatomprom and 34% by Russia's TVEL and has major new investment under way. It has secured both ISO 9001 and ISO 14001 accreditation. In 2007 a technological assistance agreement was signed with Japan apparently in line with government announcements that it would move towards selling its uranium as fabricated fuel or at least fuel pellets rather than just raw material. Ulba's web site speaks of establishing uranium conversion capacity (it has produced HF since 1952), which would fit in with Russian JV enrichment arrangements. Kazakhstan joined the Global Nuclear Energy Partnership (GNEP) in September 2007.

MONGOLIA

Russia is reported to be examining the feasibility of building nuclear power plants in Mongolia, and in April 2008 Russia and Mongolia signed a high-level agreement to cooperate in identifying and developing Mongolia's uranium resources.

CHILE

Chile imports over 70% of its energy, mostly as hydrocarbons. It produced 50 billion kWh in 2005 from some 11 GWe of plant and imported 2 billion kWh. There is a need to build a further 5 - 7 GWe by 2020 to secure a measure of energy security as Argentina cuts back gas supplies. Per capita consumption is 2760 kWh/yr. 40-50% of electricity comes from hydro, depending on how much rain it has and about 35% comes from imported natural gas. Up to 20% comes from coal and there are proposals for new coal-fired plants.

In February 2007 the Energy Ministry announced that it was beginning technical studies into the development of nuclear power. A major business group has already had discussions with Areva about building a nuclear power plant to connect Chile's northern and central power grids. In November 2007 the President asked the Energy Minister to prepare new studies regarding the country's nuclear energy options for the next administration.

VENEZUELA

Venezuela produced 101 billion kWh in 2005, 34% of this from fossil fuels, 66% from hydro.

The National Assembly is working on legislation which includes nuclear power as an option. The President announced in November 2007 that the country will pursue a nuclear power program, inspired by Brazil and Argentina. In September 2008 he announced that this would be with Russian help.

BANGLADESH

Bangladesh produced 22.6 billion kWh in 2005 from some 4 GWe of plant, giving per capita consumption of 114 kWh/yr. 89% of electricity comes from natural gas. Electricity demand is now rising rapidly.

Building a nuclear power plant in the west of the country was proposed in 1961. Since then a number of feasibility reports have affirmed the technical and economic feasibility. The Rooppur site in Pabna district was selected in 1963 and land was acquired. The government gave formal approval for a succession of plant proposals, then after independence a 125 MWe nuclear power plant proposal was approved in 1980 but not built.

With growth in demand and grid capacity since then, a much larger plant looked feasible, and the government in 1999 expressed its firm commitment to build this Rooppur plant. In 2001 it adopted a national Nuclear Power Action Plan and in 2005 it signed a nuclear cooperation agreement with China. In 2007 the Bangladesh Atomic Energy Commission proposed two 500 MWe nuclear reactors for Rooppur by 2015, quoting likely costs of US$ 0.9-1.2 billion for a 600 MWe unit and US$ 1.5-2.0 billion for 1000 MWe. In April 2008 the government reiterated its intention to work with China in building the Rooppur plant and China offered funding for the project.

Russia and South Korea had earlier offered financial and technical help to establish nuclear power. The country has had a Triga 3 MW research reactor operational since 1986.

INDONESIA

Indonesia's population of 242 million is served by power generation capacity of only 21.4 GWe, producing 127 billion kWh/yr (2005). This gives per capita electricity consumption: 475 kWh/yr.

With an industrial production growth rate of 10.5%, electricity demand is estimated to reach 175 TWh in 2013 and 450 billion kWh in 2026. At present a low reserve margin with poor power plant availability results in frequent blackouts.

About 45% of Indonesia's electricity is generated by oil and gas, so as well as catering for growth in demand in its most populous region, the move to nuclear power will free up oil for export. Remaining electricity supply is from coal (36%), hydropower (12%) and geothermal energy (7%).

Three research reactors are operated by the National Atomic Energy Agency (BATAN), the third of them being intended to support the introduction of nuclear power to the country. It is a 30 MW (thermal) unit at the Serpong Nuclear Facility near Jakarta, and started up in 1987.

Following earlier tentative proposals, in 1989 the government initiated a study focused on the Muria Peninsula in central Java and carried out by the National Atomic Energy Agency (BATAN). It led to a comprehensive feasibility study for a 7000 MWe plant, completed in 1996, with Ujung

Lemahabang as the specific site, selected for its tectonic stability. Plans for the initial plant on the Muria Peninsula in central Java were then deferred indefinitely early in 1997.

Then a 2001 power generation strategy showed that introduction of a nuclear plant on the Java-Bali grid would be possible in 2016 for 2 GWe rising to 6-7 GWe in 2025, using proven 1000 MWe technology with 85% capacity factor and investment cost $2000/kWe. The Java-Bali interconnected system accounts for more than three quarters of Indonesia's electricity demand.

Late in 2003 BATAN was reported to have narrowed the choice of plant to two: a South Korean 1000 MWe pressurised water reactor or a Canadian 700 MWe pressurised heavy water reactor - probably the KSNP+ (OPR-1000) and ACR-700 respectively. Subsequent reports point to the Korean OPR-1000 option and suggest an increasing sense of urgency due to power shortages.

Under the 2006 Law on Nuclear Reactors the project may be given to an Independent Power Producer to build and operate, on one of three sites on the central north coast of Java. Plans are to call tenders in 2008 for two 1000 MWe units, Muria 1 & 2, leading to decision in 2010 with construction starting soon after and commercial operation from 2016 and 2017. Fuel services will be purchased from abroad and fuel would preferably be leased. Used fuel would be stored centrally in the medium term. Tenders for Muria units 3 & 4 are expected to be called for in 2016, for operation from 2023.

The government has said that it has $8 billion earmarked for four nuclear plants of total 6 GWe to be in operation by 2025. Under current plans it aims to meet 2% of power demand from nuclear by 2017. It is anticipated that nuclear generation cost would be about 4 cents/kWh (US) compared with 7 c/kWh for oil and gas.

In July 2007 Korea Electric Power Corp. and Korea Hydro & Nuclear Power Co. (KHNP) signed a memorandum of understanding with Indonesia's PT Medco Energi Internasional to progress a feasibility study on building two 1000 MWe OPR-1000 units from KHNP at a cost of US$ 3 billion. This was part of a wider energy collaboration.

In addition, Batan has undertaken a pre-feasibility study for a small Korean SMART reactor for power and desalination on Madura. However, this awaits the building of a reference plant in Korea. Also the province of Gorontalo on Sulawesi is reported to be considering a floating nuclear power plant from Russia, and late in 2007 a cooperation agreement with Japan was signed, envisaging its help in building and operating nuclear power plants.

The Japanese and Indonesian governments signed a cooperation agreement in November 2007 relating to assistance to be provided for the preparation, planning, and promotion of Indonesia's nuclear power development and assistance for public relations activities.

The IAEA is reviewing the safety aspects of both Muria and Madura proposals, with Indonesia's Nuclear Technology Supervisory Agency.

During the 1980s Indonesia trained many technical people in anticipation of nuclear power development then, many of these are still available for the new project.

Indonesia has several nuclear facilities in operation. In addition to three research reactors, it has front-end capabilities in ore processing, conversion and fuel fabrication, all at a laboratory scale. There have been no experiments in reprocessing, but there is a radwaste program for spent fuel from the research reactors.

There are some uranium resources in Kalimantan.

PHILIPPINES

The Philippines produced 56.5 billion kWh of electricity in 2005.

In response to the 1973 oil crisis, the Philippines decided to build the two-unit Bataan Nuclear Power Plant (BNPP). Construction of Bataan 1 - a 621 MWe Westinghouse pressurized water reactor - began in 1976 and it was completed in 1984 at a cost of $460 million. However, due to financial issues and safety concerns, the plant was never loaded with fuel or operated. In April 2007, the Philippine government made the final payment for the plant. The government was considering converting it into a natural gas-fired power plant, but this was impractical, and the plant has simply been maintained.

In 2007 the Philippines Department of Energy (DOE) set up a project to study the development of nuclear energy, in the context of an overall energy plan for the country. Nuclear energy would be considered in order to reduce the country's dependency on imported oil and coal. In its 2008 update of the national energy plan, 600 MWe was projected on line in 2025, with further 600 MWe increments in 2027, 2030 and 2034 to give 2400 MWe.

In 2008 an IAEA mission commissioned by the government advised that Bataan-1 could be refurbished and economically and safely be operated for 30 years. Refurbishment is estimated to cost $800 million. The IAEA was also to recommend a policy framework for nuclear power development in the country.

The government is establishing a working group with a view to proceeding along the same lines as Thailand, retaining engineering consultants to guide progress. The Department of Energy is considering how to rebuild local skills in nuclear sciences and engineering. The state-owned National Power Corporation originally had 710 nuclear engineers who were trained by Westinghouse and Ebasco Overseas Corp. in the 1980s, but this has declined to about one hundred, many of whom are due to retire in the next five to ten years.

VIETNAM

Vietnam produced 53.5 billion kWh in 2005 from 11.4 GWe of plant, giving per capita consumption of 445 kWh/yr. 60% of electricity comes from hydro and demand is growing rapidly. In mid 2008 capacity was about 12.5 GWe and demand significantly higher than this, resulting in rationing. According to the government at the end of 2006, electricity demand is expected to grow 15% pa to 2010 and it plans to increase generating capacity to 25 GWe.

In the early 1980s two preliminary nuclear power studies were undertaken, followed by another which reported in 1995 that: "Around the year 2015, when electricity demand reaches more than 100 billion kWh, nuclear power should be introduced for satisfying the continuous growth in the country's electricity demand in that time and beyond".

More recently, a national energy plan approved by Vietnam's National Assembly included at least 2000 MWe of nuclear power capacity to be commenced by 2010. This followed a feasibility study in 2002, and establishment of nuclear cooperation agreements with Russia, South Korea and the USA, the first related principally to Vietnam's 500 kW Da Lat research reactor.

In February 2006 the government announced that a 2000 MWe nuclear power plant would be on line by 2020. A feasibility study for this due to be completed in mid 2009 and formal approval will then be required to open a bidding process with a view to construction start in 2011 and commissioning in 2017. This general target was confirmed in a nuclear power development plan approved by the government in August 2007, with the target being raised to a total of 8000 MWe nuclear by 2025. In October 2008 it was reported that two reactors total 2000 MWe was planned for the southern Ninh Thuan province to be constructed from 2014 and come into operation from about 2018, followed by another 2000 MWe at Vinh Hai. These plants would be followed by a further 6000 MWe by 2030. South Korea has expressed interest in bidding for the project, and its Ministry of Education, Science & Technology in 2006 committed to long-term cooperation on nuclear energy. In May 2008 a nuclear assistance agreement was signed with Japan. There are also agreements with France, China and Canada. In June 2008 the US Nuclear Regulatory Commission and the Vietnam Agency for Radiation and Nuclear Safety & Control officially signed a cooperation agreement to share technical information on nuclear energy. The accord calls for an exchange of information about regulations, environmental impact and safety of nuclear sites. The Vietnam Atomic Energy Commission was established in 1976 and is under the Ministry of Science & Technology. Vietnam Electricity will be the company responsible for building and operating the plants.

THAILAND

Peak demand is about 20 GWe and some 132 billion kWh per year is supplied (2005). About 70% is from natural gas. . Installed capacity is about 28 GWe, half of it gas-fired. Forecast peak demand in 2021 is almost 50 GWe. Thailand has the potential to be regional electricity hub for ASEAN countries.

Tentative plans to embark on a nuclear power program have been revived by a forecast growth in electricity demand of 7 per cent per year for the next twenty years. Capacity requirement in 2016 is forecast at 48 GWe. As gas prices rise, the Atomic Energy Commission and its Office of Atoms for Peace (OAP) however are assessing the feasibility of nuclear power, and any initial plants would probably be built by the Electricity Generating Authority of Thailand (EGAT). Independent power producers have also expressed interest. The Ministry of Science & Technology is responsible for the issue.

Thailand's National Energy Policy Council commissioned a feasibility study for a nuclear power plant in the country, and among the options in the draft power development plan for 2007-2021 was the construction of 5000 MWe of nuclear generating capacity, starting up in 2020-21.

In June 2007 the Energy Minister announced that EGAT will proceed with plans to build a 4000 MWe nuclear power plant, and has budgeted some US$ 53 million between 2008 and 2011 on preparatory work, half of it coming from oil revenues. Construction will commence in 2014, with four reactors to operate from 2020. The capital cost is expected to be US$ 6 billion and electricity cost about USD 6 cents/kWh, slightly less than from coal. The government plans to establish safety and regulatory infrastructure by 2014 and commissioned a formal 3-year feasibility study early in 2008. Then in October 2008 the engineering firm Burns & Roe was commissioned to undertake a 20-month study to recommend siting, technology and reactor size for the firs plant. The project will then go out to tender with a view to starting construction in 2014. The government plans to have 2000 MWe nuclear on line in 2020 and double that two years later.

Thailand has had an operating research reactor since 1977 and a larger one is under construction.

MALAYSIA

Malaysia produced 87 billion kWh in 2005, 93% of this from fossil fuels - mostly natural gas, the balance from hydro. It has about 24 GWe of capacity, a 55% increase over two years to 2005. Government policy is to reduce reliance on natural gas by building coal-fired capacity.

A comprehensive energy policy study including consideration of nuclear power will be completed before 2010. The main state-owned utility Tenaga Nasional Bhd (TNB) is tentatively in favour of nuclear power and in August 2006 the Malaysian Nuclear Licensing Board said that plans for nuclear power after 2020 should be brought forward and two reactors built much sooner. This intention has since been reiterated from the Ministry of Science, Technology & Innovation. In July 2008 the government directed TNB to set up a task force to look at the feasibility of nuclear power, the study likely to take two years. In September the government announced that it had no option but to commission nuclear power due to high fossil fuel prices, and set 2023 as target date. It then sent a draft energy policy blueprint back to the Energy Commission as it was not comprehensive enough.

http://www.malaysiaenergy.com

The Malaysian Institute for Nuclear Technology Research has operated a 1 MW Triga research reactor since 1982. In April 2007 MINT was renamed the Malaysian Nuclear Agency (or Nuclear Malaysia) to reflect its role in promoting the peaceful uses of atomic energy.

AUSTRALIA

Australia produced 255 billion kWh from 46 GWe of capacity in 2006, with 23 billion kWh/yr being embedded in aluminium exports. Final consumption was 187 billion kWh, hence per capita consumption (net of Al exports) is 9100 kWh/yr. Low-cost power is a competitive advantage of the country. 80% of electricity comes from coal-fired plant, 12% from gas and 7% from hydro. This gives it a high output of CO_2, which is the main reason that discussion has started on possible nuclear generation in the future. Australia joined the Global Nuclear Energy Partnership (GNEP) in September 2007.

Australia has operated a research reactor since 1956 and has now commissioned its 20 MWt replacement.

About 1970 the Australian government sought tenders for building a nuclear power reactor at Jervis Bay, NSW. Designs from UK, USA, Germany and Canada were short listed, but a change in leadership led to the project being cancelled in 1972. However, until 1983 there were various plans and proposals for building an enrichment plant.

At the end of 2006 the report of the Prime Minster's expert taskforce considering nuclear power was released. It said nuclear power would be 20-50% more expensive than coal-fired power and (with renewables) it would only be competitive if "low to moderate" costs are imposed on carbon emissions (A$ 15-40 - US$ 12-30 - per tonne CO_2). "Nuclear power is the least-cost low-emission technology that can provide base-load power" and has low life cycle impacts environmentally. The first nuclear plants could be running in 15 years, and looking beyond that, 25 reactors at coastal sites might be supplying one third of Australia's (doubled) electricity demand by 2050. Certainly "the challenge to contain and reduce greenhouse gas emissions would be considerably eased by investment in nuclear plants." "Emission reductions from nuclear power could reach 8 to 18% of national emissions in 2050".

NEW ZEALAND

New Zealand produces some 41 billion kWh/yr from 8.4 GWe of plant. For 4.04 million people, average per capita consumption is thus about 10,000 kWh/yr or 9000 kWh if aluminium smelting is treated as largely an electricity export. 75% of electricity comes from hydro.

New Zealand has depended primarily on hydro-electric power for its electricity for many years, but scope for expansion is limited and even the reliability of present capacity depends on capricious rainfall, as in 2001 and 2003.

The last major hydro scheme was raising the level of Lake Manapouri to provide low-cost power for NZAS aluminium smelter in the South Island.

Hydro output has not increased over the last 15 years, and that growth in demand since 1990 has been mostly met by gas-fired plant, at least until the 1000 MWe state-owned Huntly plant shifted to using coal for 80% of its energy. However, its operation is severely constrained in hot weather.

Of 41 billion kWh of electricity generated in NZ in 2003, 58% was hydro, 24.5% gas, 7.6% coal, 6.7% geothermal, 2% wind and 1.3% biomass. The power is produced from 8.4 GWe capacity, including 5.25 GWe hydro, 1.38 GWe gas-fired, 1.0 GWe coal-fired and 0.42 GWe geothermal - mainly run as base-load (2002 data). In 2004 there was 0.17 GWe of wind capacity installed and 0.07 GWe committed.

In 1968 the national power plan first identified the likely need for nuclear power in NZ a decade or more ahead, since readily-developed hydro-electric sites had been utilised. Plans were made and a site at Oyster Point on the Kaipara harbour near Auckland was reserved for the first plant. Four 250 MWe reactors were envisaged, to supply 80% of Auckland's needs by 1990. But then the Maui gas field was discovered, along with coal reserves near Huntly, and the project was abandoned by 1972.

In 1976 a Royal Commission was set up to enquire further into the question. Its 1978 report said that there was no immediate need for NZ to embark upon a nuclear power program, but suggested that early in 21st century "a significant nuclear programme should be economically possible."

IMPORTANT CONTACTS

Bank Negara Indonesia (BNI) – www.bni.co.id

Center for Research on Energy Policy, ITB – www.crep.itb.ac.id

Directorate General of New Renewable Energy and Energy Conservation (DJEBTKE) – www.ebtke.esdm.go.id

Directorate General of Electricity (DJK) – www.djlpe.esdm.go.id

Indonesian Geothermal Association (API) – www.inaga.org

Indonesia Renewable Energy Society (METI) – www.meti.or.id

Ministry of Energy and Mineral Resources (KESDM) – www.esdm.go.id

National Energy Council (DEN) – www.den.go.id

PT PLN (Persero) – www.pln.co.id

PT Navigat Organic Energy Indonesia – www.noei.co.id

Research and Development Center of Technology for Electricity, Renewable Energy and Energy Conservation – www.p3tkebt.esdm.go.id

The Indonesian Solar Module Manufacturer Association (APAMSI) – www.apamsi.org

US AND COUNTRY CONTACTS

1. CHAMBERS OF COMMERCE

Based in New York City, the membership of the American-Indonesian Chamber of Commerce is comprised of over 150 companies and individuals doing business in Indonesia. It has an active program of monthly luncheons, featuring speakers knowledgeable about Indonesia and briefing programs for newly appointed American and Indonesian Government officials. The Chamber also publishes "OUTLOOK/ INDONESIA," a quarterly publication containing interpretations of new Indonesian policies, sectoral reviews, summaries of recent Chamber activities, and an Executive Director's column. In addition, the American-Indonesian Chamber of Commerce provides business translations, either from English to Indonesian or Indonesian to English. Their contact information is as follows:

The American-Indonesian Chamber of Commerce
317 Madison Avenue, Suite 520New York, NY 10017
Tel.: (212) 687-4505
Fax : (212)687-5844
Email: aiccny@bigplanet.com
Internet: http://www.aicc.globalnetlink.com
Contact: Wayne Forrest, Executive Director

The American Chamber of Commerce in Indonesia, known as AMCHAM, is affiliated with the Asia-Pacific Council of American Chambers of Commerce (APCAC). Its 600 members include leading U.S. firms with offices in Indonesia, associates, individuals, and special members. The chamber prepares a number of useful guides to doing business in Indonesia and a membership directory. AMCHAM assists U.S. firms in assessing business opportunities by staging briefing breakfasts at the requestor's expense. Their contact information is as follows:

American Chamber of Commerce (AMCHAM)
World Trade Center, 11th Floor,
Jl. Jendral Sudirman Kav. 29 - 31
Jakarta 12920, Indonesia
Tel: (62-21) 526-2860
Fax: (62-21) 526-2861
Email: amcham_indonesia@ibm.net
Internet: http//www.amcham.or.id
Contact: James W. Castle, President

2. RELATED BUSINESS COUNCILS AND ASSOCIATIONS IN THE U.S.A.:

California-Asia Business Council
1946 Embarcadero, Suite 200
Oakland, CA 94606
Tel.: (510) 536-1967
Fax : (510) 261-9598
E-mail: shanahan@calasia.org
Internet: www.CalAsia.org
Contact: Mr. Shanahan Lim, Program Manager, Ms. Jeremy W. Potash, Executive Director

Indonesian American Business Association
P.O. Box 420398 Houston, TX 77242-0398
10900 Richmond Avenue, Houston, TX 77242
Tel.: (713) 789-0702
Fax : (713) 789-0559
E-mail: iaba@onramp.net
Internet: www.iaba-usa.org
Contact: Engeline Tan, Executive Director

Pacific Basin Economic Council
PBEC International Secretariats
900 Fort St., suite 1080
Honolulu, Hawaii, USA 96813
Tel.: (808) 521-9044
Fax : (808) 521-8530
E-mail: info@pbec.org
Internet: www.pbec.org
Contact: Robert G. Lees, Secretary General

U.S.-ASEAN Business Council
Support Services/MIS
1400 L Street NW #375
Washington, DC 20005
Tel.: (202) 289-1911
Fax : (202) 289-0519
E-mail: mail@usasean.org
Internet: www.us-asean.org
Contact: Jackie Darden

3. INDONESIAN TRADE ASSOCIATIONS

The major trade association in Indonesia is the Indonesian Chamber of Commerce and Industry (KADIN). Members include representatives from private industry, cooperatives, public corporations, utilities, as well as state-owned enterprises. In addition, there are numerous other specialized and professional organizations that represent the interests of various other sectors and trades in the economy. Contact information for KADIN is as follows:

Indonesian Chamber of Commerce & Industry (KADIN)
Menara Kadin, 29th Floor
Jl. H.R. Rasuna Said X-5 Kav. 2-3
Jakarta 12940
Tel.: (62-21) 916-5535, 916-5538, 527-4485
Fax : (62-21) 527-4486
Associations of importers and exporters, most of whom are organized on a commodity basis,

include the Importers Association of Indonesia (GINSI) and the Indonesian Association of Exporters (GPEI). Both organizations have head offices in Jakarta. Contact information is as follows:

Importers Association of Indonesia (GINSI)
Gedung Bank Niaga, 1st Floor
Jl. Raya Pasar Minggu #45
Jakarta Selatan
Tel.: (62-21) 7919-8986, 7919-8987, 7919-8988
Fax : (62-21) 790-0808
Contact: Mr. H. Amirudin Saud, Chairman

Indonesian Exporters Association (GPEI)
21 Jl. Wolter Monginsidi No. 88A
Jakarta 12170
Tel.: 725-2612, 725-2613
Fax : 725-2615
Contact: Dipl. Ing. H. Benny Soetrisno

The Indonesian Government established the National Agency for Export Development within the Ministry of Trade to promote the export of less renowned products. These products include handicrafts (ie, jewelry, batik, handwoven fabric, and wood carvings), agricultural and cottage industry products, and new manufactured products. The agency will also assist foreign buyers and importers in establishing contacts with Indonesian companies. Contact information is as follows:

National Agency for Export Development
Jl. Gajah Mada No. 8
Jakarta 10130
Tel.: (62-21) 634-1082
Fax : (62-21) 633-8360

4. TRADE AND PROJECT FINANCING

U.S. Trade and Development Agency
American Business Center
25th Flr., Ayala Life-FGU Center
6811 Ayala Ave., Makati City
Philippines
Tel: (632) 887-1196/-1192
Fax: (632) 887-1225 fax
E-mail: tdasia@skyinet.net

Commercial Liaison to the Asian Development Bank
U.S. and Foreign Commercial Service
2nd Floor, GSO Building Seafront Compound
Roxas Boulevard, Pasar City
Manila, Philippines
Tel.: (632) 804-0460
Fax : (632) 804-0357
E-mail: Stewart.Ballard@mail.doc.gov

Internet: http://www.ita.doc.gov/mdbo
Contact: Stewart Ballard, Senior Commercial Officer

U.S. Trade and Development Agency
1621 North Kent Street, Suite 300
Arlington, VA 22209-2131
Tel: (703) 875-4357
Fax: (703) 875-4009
Internet: www.tda.gov

TDA Regional Office – Manila
John Herrman, Deputy Regional Manager
2/F, RSC Bldg. Seafront Compound
Roxas Blvd.
Manila, Philippines
Tel: (632) 804-0381 or –0363
Fax: (632) 804-0367
E-mail: tdasia@skyinet.net

Overseas Private Investment Corporation (OPIC)
1100 New York Avenue, N.W.
Washington, D.C. 20527
Tel.: (202) 336-8799
Fax : (202) 408-9859
Internet: http://www.opic.gov

Export-Import Bank of the United States (EX-IM Bank)
811 Vermont Ave., NW
Washington, D.C. 20571
Tel.: (800) 565-EXIM(3946), (202) 565-3946
Fax : (202)565-3380
Asia hotline : (800) 565-3946 ext.3905
Internet: http://www.exim.gov

Islamic Development Bank
P.O. Box 5925
Jeddah, 21432
Kingdom of Saudi Arabia
Tel.: (966-2) 636-1400 (10 lines)
Fax : (966-2) 636-6871
Tlx: 601 137 ISDB SJ
E-mail idb.comm@mail.oicisnet.org or idb.archives@mail.oicisnet.org
Internet: http://www.isdb.org

The World Bank
Office of the U.S. Executive Director
Commercial Service Liaison Staff
Jill Wilkins or Kimberly Versak
1818 H Street, NW
Washington, DC 20433
Tel.: (202) 477-1234
Fax : (202) 522-3405

Internet: http://www-esd.worldbank.org/
E-mail: kversak@worldbank.org or jwilkins@worldbank.org

In Indonesia:
Mark Baird, Indonesia Country Manager
Jakarta Stock Exchange Building
Tower 2, 12th Floor
Jl. Jendral Sudirman, Kav. 52-53
Jakarta 12190 Indonesia
Tel.: (62-21) 529-93000
Fax : (62-21) 529-93111

5. INDONESIAN GOVERNMENT CONTACTS

The following is a listing of President Abdurraham Wahid's cabinet as of March 2000. Because of the uncertainties surrounding the Indonesian government at the time, business persons who wish to get in touch with these cabinet members can contact the U.S. Commercial Service in Jakarta for up-to-date contact information or to forward correspondence.

6. Coordinating Ministers:

Political and Security: Lt. Gen (Ret.) Suryadi Sudirja
Economic, Finance, and Industry: Drs. Kwik Kian Gie
People's Welfare and Poverty Alleviation: Basri Hasanuddin, MA

Ministry:
Agriculture: Dr. Ir. Muhammad Prakosa, Ph.D
Defense/Armed Forces Commander: Prof. DR Juwono Sudarsono
Education: DR Yahya A. Muhaimin
Finance: DR. Bambang Sudibyo, MBA
Foreign Affairs: DR Alwi Shihab
Forestry and Plantation: DR Nurmahmudi Ismail
Health: Dr. Achmad Suyudi, MHA
Home Affairs: Lt. Gen. (Ret.) Suryadi Sudirja
Industry and Trade: Lt. Gen. Luhut B. Panjaitan
Law and Legislation: DR Yusril Ihza Mahendra, SH, MSC
Manpower: DR Bomer Rasaribu, SH
Maritime Exploration: Ir. Sarwono Kusumaatmadja
Mining and Energy: Lt. Gen. (Ret.) Susilo Bambang Yudhoyono
Religious Affairs: Drs. K.H. Muhammad Tolchah Hasan
Residential and Area Development: Ir. Erna Witoelar, M. Si
Transportation and Communications: Lt. Gen. Agum Gumelar

State Ministers:
State Secretary: Dr. Ali Rahman, M. Sc.
Cooperatives, Small and Medium Enterprises: Drs. Zarkasih Noer
Environment: DR Soni Keraf
Human Rights: Dr. Hasballah M. Saad
Investment/Head of BKPM: H.M. Rozy Munir, S.E., MSC
Public Affairs: DR Anak Agung Gde Agung
Public Works: Dr. Ir. Rozik Boedioro Soetjipto
Reform of State Apparatus: Rear Adm. Freddy Numberi

Regional Autonomy: DR Ryaas Rasyid
Research and Technology: Dr. A.S. Hikam
Tourism, Art and Culture: Drs. Hidayat Djaelani
Transmigration and Population: Ir. Al-Hilal Hamdi
Women's Affairs: Dra. Khofifah Indar Parawansa (Ms)
Youth and Sports: Drs. Mahadi Simambela

Officials of Ministerial Rank:
Attorney General: Marzuki Darusman, SH
Governor of the Bank of Indonesia: DR. Syahril Sabrin
Indonesian Armed Forces (TNI) Commander-in-Chief: Admiral Widodo A.S.

*In Indonesia DR refers to a male with a Doctorate degree; Dr. refers to a male with a medical degree; Drs. refers to a male with a bachelor degree; Dra. refers to a female with a bachelor degree; SH. refers to a male or female with a law degree; MSC is a male or female with a Master of Science

6. U.S. GOVERNMENT CONTACTS

U.S. Embassy, Jakarta

Mailing Address from U.S.: International Mail:
American Embassy - Jakarta American Embassy-Jakarta
Box 1 Unit 8129 Jl. Medan Merdeka Selatan #5
FPO, AP 96520 Jakarta 10110, Indonesia

U.S. Ambassador Robert S. Gelbard Deputy Chief of Mission Stephen Mull
1st Aide to Ambassador Janice G. Smith Sec. to DCM Brenda Dudley

Internet: http://www.usembassyjakarta.org

U.S. Commercial Center
Alice Davenport, Counselor for Commercial Affairs
Judy R. Reinke, Commercial Attaché
Michael Carroll, Commercial Officer
Wisma Metropolitan II, 3rd Floor
Jl. Jendral Sudirman Kav. 29-31
Jakarta 12910, Indonesia
(From the U.S., use Embassy's FPO mailing address – see above)
Tel.: (62-21) 526-2850
Fax : (62-21) 526-2855
Email : jakarta.office.box@mail.doc.gov
Internet : http://www.jakarta.uscc.org

U.S. Asian Environmental Partnership
Gerald Sanders, Technical Representative
Wisma Metropolitan II, 3rd Floor
Jl. Jendral Sudirman Kav. 29-31
Jakarta 12910, Indonesia
Tel.: (62-21) 526-2844, 526-2848
Fax : (62-21) 526-2846, 526-2849

Email: usaepdir@rad.net.id
Internet : http://www.usaep.org

U.S. Asian Environmental Partnership
Office of Environmental Infrastructure
Tel.: (62-21) 344-2211 Ext. 2480
Fax : (62-21) 380-6694

U.S. Agricultural Trade Office (ATO)
Dennis Voboril, Agricultural Attaché
Wisma Metropolitan II, 3rd Floor
Jl. Jendral Sudirman Kav. 29-31
Jakarta 12910, Indonesia
Tel.: (62-21) 526-2850
Fax : (62-21) 571-1251

The following U.S. Embassy Offices are located at:
Jl. Medan Merdeka Selatan #5
Jakarta 10110, Indonesia

Foreign Agricultural Service
Kent D. Sission, Counselor for Agricultural Affairs
Deanna M. Johnson, Agricultural Attaché
Tel.: (62-21) 344-2211 (ext. 2161)
Fax : (62-21) 380-1363

Economic Section
Judith Fergin, Economic Counselor
Michael Uyehara, Resources Officer
Robin McClellan, Science and Technology Officer
David DiGiovanna, Trade Officer
William Heidt, Finance Officer
Vicky Alvarado, Economic Officer
Tel.: (62-21) 344-2211 (ext. 2073)
Fax : (62-21) 386-2259

Public Affairs Section
Greta Morris, Public Affairs Counselor
Karl Fritz, Press Attaché
Tel.: (62-21) 344-2211 (ext. 2500)
Fax : (62-21) 381-0243
E-mail: gmorris@pd.state.gov/kfritz@pd.state.gov

Informational Resource Center (IRC)
Melling Simanjuntak, Library Director
Tel.: (62-21) 350-8467
Fax : (62-21) 350-8466
E-mail: irc@usembassyjakarta.org

Agency for International Development (U.S. AID)
Terry Myers, Director
Paul Deuster, Chief, Private Sector Development Office

Tel.: (62-21) 344-2211 (ext. 2308)
Fax : (62-21) 380-6694

Defense Attaché Office (DAO)
Col. Joseph H. Daves, Defense Attaché
Tel.: (62-21) 344-2211 (ext. 2191)
Fax : (62-21) 386-2259

Office of the Military Attaché for Defense Programs
Col. James Tompkins, Military Attaché
Tel.: (62-21) 344-2211 (ext. 2601)
Fax : (62-21) 386-2259

U.S. Consulate General – Surabaya
Robert Pollard, Consul General
Joaquin F. Monserrate, Pol/Econ/Commercial Officer
Lina Jusuf, Commercial Specialist
American Consulate General-
Jl. Raya Dr. Sutomo #33
Surabaya, Indonesia 60264
Tel.: (62-31) 567-6880, 568-2287, 568-2288
Fax : (62-31) 567-4492 Telex : 34331
E-mail: amconsby@rad.net.id

U.S. Consulate Medan (to be opened sometime in the year 2000)
Ronald Deutch
Jl. Walikota No. 13
Medan, Sumatra Utara
Tel.: (62-61) 415-2200, 453-1251
Fax: (62-61) 451-8711

U.S. Consular Agent Bali
Andrew Toth, Executive Office
Jl. Hayam Wuruk 188, Denpasar, Bali, Indonesia 80235
Tel.: (62-361) 233-605, 222-424
Fax : (62-361) 222-426
E-mail : tabuh@denpasar.wasantara.net.id

7. U.S. GOVERNMENT – WASHINGTON D.C.

Trade Information Center
U.S. Department of Commerce
Ronald Reagan Building, Mezzanine Level
Washington, D.C. 20230
Tel.: 1-800-USA-TRAD(E)8723 or (202) 482-0543
Fax : (202) 482-4473
Internet: http://www.ita.doc.gov/tic or http://tradeinfo.doc.gov//
Flashfax: 1-800-872-8723 (Use flashfax for rapid access to the most frequently requested information on doing business throughout Asia.)

Elena Mikalis / Farah Press
Desk Officers for Indonesia (Policy Issues)

U.S. Department of Commerce
Room 2324
14th and Constitution Ave., N.W.
Washington, D.C. 20230
Tel.: (202) 482-3894
Fax : (202) 482-3316
E-mail: Elena_Mikalis@ita.doc.gov, Farah_Press@ita.doc.gov
Internet: http:\\www.ita.doc.gov

U.S. Department of State Office of Business Affairs
Commercial Coordinators
22201 C St., N.W., Room 2318
Washington D.C., 20520
Tel.: (202) 647-1625
Fax : (202) 647-3951
E-mail: cbaweb@state.gov
Internet: www.state.gov/www/about_state/business/index.html

U.S. Department of Agriculture, Foreign Agricultural Service, Trade Assistance and Promotion Office
1400 Independence Ave., S.W., Room 6958-8
Washington D.C. 20250
Tel.: (202) 720-7349
Contact: Scott Cook, Contracting & Acquisition Management Division

United States-Asia Environmental Partnership (US-AEP)
Asia/Near East Bureau
U.S. Agency for International Development
1300 Pennsylvania Ave, NW
Washington, D.C. 20523-4101
Tel.: (202) 712-0270
Fax : (202) 216-3379
Internet : http://www.usaep.org

8. INDONESIAN EMBASSY AND CONSULATES IN THE UNITED STATES:

Embassy of the Republic of Indonesia
2020 Massachusetts Avenue, N.W.
Washington, D.C. 20046
Tel.: (202) 775-5200
Fax : (202) 775-5365
Internet: http://www.kbri.org

Consulate General Office-New York
5 East 68th St.
New York, NY 10021
Tel.: (212) 879-0600
Fax : (212) 570-6206

Consulate General Office-Los Angeles
3457 Wilshire Boulevard
Los Angeles, CA 90010

Tel.: (213) 383-5126
Fax : (213) 487-3971
E-mail : kabidpen@kjri-la.com
Internet : http://www.kjri-la.com

Consulate General Office-Houston
10900 Richmond Ave.
Houston, TX 77042
Tel.: (713) 785-1691
Fax : (713) 780-9644

Consulate General Office-Chicago
Illinois Center Suite 1422
233 North Michigan Ave
Chicago, IL 60601
Tel.: (312) 938-0101
Fax : (312) 938-3148

Consulate General Office-San Francisco
1111 Columbus Ave.
San Francisco, CA 94133
Tel.: (415) 474-9571
Fax : (415) 441-4320

9. MARKET RESEARCH, CONSULTANTS, AND LAWYERS

Performing market research in Indonesia is difficult because detailed statistics on production and consumption are often not available through published sources. External trade statistics, however, are fairly detailed and additional data can be obtained for a fee from the Central Bureau of Statistics (Biro Pusat Statistik):

BPS
Ir. Sutomo No. 8
P.0. Box 3
Jakarta Pusat
Tel.: (62-21) 384-1195, 384-2508, 381-0281-4
Fax : (62-21) 385-7046
E-mail : bpshq@bps.go.id
Internet : http://www.bps.go.id

Unrecorded trade may distort import statistics and trends. For example, BPS figures tend to understate import values, as these figures exclude duty-free imports, including duty-free imports for investment and certain other transactions. Although there is a growing number of Indonesian organizations active in market research, the number remains small and expertise varies. Branches of American banks will often conduct market surveys for their customers, and several U.S. consulting firms now have affiliates in Jakarta. A growing number of foreign law firms, including some from the United States, are also entering the Indonesian business community as business consultants. Members of INKINDO, the Association of Indonesian Consultants, are able to perform a wide range of research and consulting services. INKINDO was established by Indonesian consultants based in Jakarta. The contact information for this association is as follows:

Association of Indonesian Consultants (INKINDO)
Jl. Bendungan Hilir Raya, No. 29.
Jakarta 10210
Tel.: (62-21) 573-8577/78
Fax : (62-21) 573-3474.
E-mail : inkindo@pu.go.id
Contact: Ir. H. Muhayat, Chairman

PT. Data Consult publishes a biweekly newsletter entitled "Indonesian Commercial Newsletter." The newsletter contains a sectoral survey in each issue and other market information. The addresses of several market research firms in Jakarta are listed in the "Sources of Information" section of the Overseas Business Report. Information regarding firms that are capable of conducting market surveys, including the aforementioned ones, may be obtained from the Foreign Commercial Service, U.S. Embassy in Jakarta, or in the following list of consultants and market research firms.

10. CONSULTANTS AND MARKET RESEARCH FIRMS

AC Nielsen Indonesia
15/F Wisma Bank Dharmala
Jl. Jendral Sudirman Kav. 28
Jakarta 12920
Tel.: (62-21) 521-2200
Fax : (62-21) 521-2203/2204
E-mail: acn@acnielsen.co.id
Internet : www.acnielsen.com
Expertise: Public relations consulting, media relations, crisis management, public relations planning, government consulting.
Contact: Mariana M.G. Warokka

Business Advisory Indonesia (PT. Laksana Tata Indonesia)
Wisma Bank Dharmala, 11th floor, Suite 1103
Jl. Jendral Sudirman Kav. 28
Jakarta 12920
Tel.: (62-21) 522-8613
Fax : (62-21) 522-8612
E-mail: bai@prima.net.id
Expertise: Management consulting, government and corporate research, official sworn-in English/Indonesia language translations of documentation's.
Contact: Mariana Warokka

Castle Group
Menara Batavia, 8th floor
Jl. K.H. Mas Mansyur Kav. 126
Jakarta 10220
Tel.: (62-21) 572-7321
Fax : (62-21) 572-7329
E-mail: castle@castleasia.com
Internet: http://www.castleasia.com
Expertise: Market entry strategy consulting, customized market and industry analysis, customized profiles of local business groups and potential, local regulations and business practices, syndicated sector reports, and business briefing program for senior managers.
Contact: James Castle

CIC Consulting Group
Jl. Raden Saleh No. 46
Jakarta 10330
Tel.: (62-21) 310-1081, 314-7433
Fax : (62-21) 310-1505
Cable: CISIRAYA-JAKARTA
E-mail: cisi-cic@idola.net.id
Expertise: Market and feasibility Studies (including pharmaceuticals, cosmetics, medical supplies, health equipment, food and beverages, hotels, golf course, resorts, and recreation facilities), periodical/business Reports, credit information services, partner seeking services, project reports, consumer research, business to business research, social research, and agricultural research.

Citra Duta Artistry (CDA International)
Niaga Tower, 25th Floor
Jl. Jendral Sudirman Kav.58
Jakarta 12190
Tel.: (62-21) 250-5346
Fax : (62-21) 250-5347
E-mail: cda@idola.net.id
Expertise: Office renovation, project management, architectural design, and interior design.

Consensus MBL
Division of the MBL Group Plc., an NFO Worldwide Company
Jl. Bangka Raya 18, Pela Mampang
Jakarta 12720
or
P.O. Box 4334
Jakarta 12043
Tel.: (62-21) 719-1858/9, 719-1861, 719-5355
Fax : (62-21) 799-0152, 719-9550
E-mail: cmblindo@dnet.net.id
Internet: www.mblasia.com
Contact Name: Gordon Guscott, Sjafril Djalal
Expertise: Market research and strategic planning and consultancy.

PT. Data Consult, Inc.
Maya Indah Building II
Jl. Kramat Raya No.5-L
Jakarta 13210
Tel.: (62-21) 390-4711, 390-4712, 390-1879
Fax : (62-21) 475-3227
Or Jl. Pulomas Raya 31
Jakarta 13210
Tel.: (62-21) 475-3302, 475-3304, 475-3226
Fax : (62-21) 475-3227
E-mail : datacon@idola.net.id
Internet : http://www.datacon.co.id
Expertise: Market Research.

Ganesha Aggies Jaya
Jl. Pertogogan II No. 22 A
Kebayron Baru
Jakarta 12160

Tel.: (62-21) 7279-3904
Fax : (62-21) 739-5049
E-mail: ganesha1@cbn.net.id
Expertise: Indonesian labor laws and practices, expatriate staffing, residence and work permits, company establishment, corporate and individual document processing, and human resource related activities and publishing Indonesian Labor Law Library CD-ROM.

Grant Thornton
Wisma Grant Thornton
Jl. Sisingamangaraja No. 3C
Jakarta Selantan 12120
Tel.: (62-21) 720-2605
Fax : (62-21) 720-2606
E-mail: gti@grant-thornton.co.id
Internet: http://www.grant-thornton.co.id
Expertise: Accounting and audit services, tax advice and compliance, government advisory, municipal finance, corporate recovery and reconstruction, merger and acquisition advisory, information technology, hospitality and leisure consulting, business establishment and maintenance, inbound investment services, corporate finance, valuation and business planning, employee compensation and benefit planning.

Harvest International Inc.
Wisma Metropolitan I, 10th floor
Jl. Jendral Sudirman, Kav. 29
Jakarta 12920
Tel.: (62-21) 525-1641, 570-1491
Fax : (62-21) 520-7789
E-mail: hgoldstein@harvest-international.com or
Expertise: Consulting in business development, investment, project development, and trade.

PT. IBIS Dharma Nusa
Fatmawati Mas Blok 1/113
Jl. R.S. Fatmawati No. 20
Jakarta 12430
Tel.: (62-21) 769-9757
Fax : (62-21) 765-4906
E-mail : henri@ibisworld.com
Expertise: Market research on industrial sectors.

Indonesia Executive Search (I.E.S.)
Graha Irama, 9th floor
Jl. H.R. Rasuna Said Kav. X-1 No. 1-2
Jakarta 12950
Tel.: (62-21) 526-1250
Fax : (62-21) 525-5529
E-mail: dansearch@attglobal.net
Contact Names: Dan Goldsmith, Managing Director
Expertise: Executive search services, human resource and management consultancy.

Penelitian Hukum Indonesia (PHI)
(PT Terataimas Indocitra)
Wisma Bank Dharmala, 11th floor, Suite 11-03
Jl. Jendral Sudirman Kav. 28

Jakarta 12920
Tel.: (62-21) 522-8613
Fax : (62-21) 522-8612
E-mail: hukum@rad.net.id
Expertise: CD-ROMs containing full text of Indonesian law in English and Indonesian, development of full text databases for business and industry, information systems consulting.

Plansearch Associates
Jl. Limau 1/28 A
Kebayoran Baru
Jakarta 12130
P.O. Box 83 JKSBA
Tel.: (62-21) 739-5017, 722-8052, 727-94117
Fax : (62-21) 739-5017
E-mail: plans@indo.net.id
Contact: Gunter J. Schwarze, Guido B. Schwarze, Ronald A. Schwarze.
Expertise: Intellectual Property, partner searches, financial investigations for money collections, market research, and environmental studies.

SOFRES FSA Jakarta (Taylor Nelson Sofres Group)
Menara Thamrin Suite 1401
Jl. M.H. Thamrin Kav. 3
Jakarta 10250
Tel.: (62-21) 230-2788
Fax : (62-21) 230-2794
E-mail: JakartaOffice@id.tnsofres.com
Internet: www.tnsofres.com
Contact Name: Hans Lang and Yanti Zen
Expertise: Quantitative and qualitative market research, customer satisfaction and awareness studies, advertising testing, test marketing, concept testing, brand tracking, trade, economic and industrial research.

12. LAW FIRMS AND ATTORNEYS

Ali Budiardjo, Nugroho, Reksodiputro
Graha Niaga, 24th Floor
Jl. Jendral Sudirman Kav. 58,
Jakarta 12190
Tel.: (62-21) 250-5125(11 lines), 250-5136
Fax : (62-21) 250-5121, 250-5122, 250-5392, 250-5001
E-mail: abnrco@abnr.co.id
Singapore Office:
50 Raffles Place
37-00 Singapore Land Tower
Singapore 048623
Tel.: (65) 320-8708
Fax : (65) 320-8709
Expertise: Commercial, corporate, and financing matters, including joint venture arrangements, licensing, banking, energy, mining, forestry, project finance, stock & bond issues, construction & engineering projects, maritime law, labor issues, patents, trademark, and other intellectual property matters.
Contact: T.M. Zahirsjah, Ferry P. Madian, Gregory Churchill

PT. CB Indonesia
16th Fl. Central Plaza
Jl. Jendral Sudirman Kav.47
Jakarta 12930
Tel.: (62-21) 525-1985, 570-1425, 525-3340
Fax : (62-21) 525-0734
E-mail: cbindon@attglobal.net.
Expertise: Joint ventures, insolvency and corporate reorganizations, finance and capital markets, infrastructure and property development, and commercial transactions and disputes.
Contact: Michael Horn

Dewi Soeharto & Rekan
Plaza Exim, 24th Fl.
Jl. Jendral Gatot Soebroto Kav.36-38
Jakarta 12190
Tel.: (62-21) 526-3473
Fax : (62-21) 526-3474
E-mail: dsrekan@indosat.net.id
Contact: Dewi Kamaratih Soeharto, Garry W. Christian
Expertise: Aviation, banking and finance, commercial, construction, corporate, energy and resources, entertainment industry contracts, environmental, foreign investment and joint ventures, government, intellectual property protection and enforcement, international trade, labor and employment, maritime, media and board casting, real property, securities, tax, telecommunications.

Frans Winarta & Partners
Kelapa Gading Boulevard TB 2/24
Kelapa Gading Permai
Jakarta Utara 14240
Tel.: (62-21) 453-3236, 453-2143
Fax : (62-21) 452-0083, 451-6605, 452-0933
E-mail: frans_winarta@hotmail.com
Contact: Frans H. Winarta
Expertise: Corporate law (cross-border acquisitions and mergers, investments, business, and matters regarding intellectual property), commercial laws and banking regulations, general international trade laws, employment and land laws, civil and administrative law, litigation and international commercial arbitration.

George Widjojo & Partners
Patents, Trademarks & Copyrights
Advocates & Solicitors
Jl. Kali Besar Barat No. 5
Jakarta 11230
Indonesia
P.O. Box 2102 / JKT
Jakarta 10021, Indonesia
Tel.: (62-21) 691-2226, 692-4533, 690-1707
Fax: (62-21) 692-3648, 693-0712
E-mail: gwidjojo@indosat.net.id
Contact person: Mr. George Widjojo
Expertise: Patents, Trademarks, Copyrights and Designs

Hadiputranto, Hadinoto & Partners
Bursa Efek Jakarta Tower II, 21st Floor
Jl. Jendral Sudirman Kav. 52-53
Jakarta Selatan
Tel.: (62-21) 515-5090, 515-5091, 515-5092
Fax : (62-21) 515-4840, 515-4845
Contact: Ms. Sri Indrastuti Hadiputranto
Expertise: International and general practice, corporate, commercial, foreign investment, tax: labor and employment law, banking and finance (including project finance), capital markets, intellectual property, construction, natural resources telecommunications, insurance, mergers and acquisitions, and environmental law.

Jusuf Indradewa & Partners-Legal Consultants
Bank Artha Graha Tower, 15th Floor
Lot 25, Sudirman Central Business District
Jl. Jendral Sudirman Kav. 52-53
Jakarta 12190
Tel.: (62-21) 515-2122
Fax : (62-21) 515-2382
E-mail: jusufind@uninet.net.id
Contact: Cecilia Sianawati (Managing Partner), Jusuf Indradewa (Senior Partner), Indra D. Santosa (Technical Advisor), Maya Pradjono (Partner)
Expertise: Legal matters relating to establishment of businesses, opening representative and branch offices, joint ventures, direct foreign investment companies, intellectual property rights, mergers and acquisitions, initial public offerings.

Karim Sani Law Firm
Wisma Danamon Aetna Life, 11th Floor
Jl. Jendral Sudirman Kav. 45-46
Jakarta 12930
Tel.: (62-21) 577-1177
Fax : (62-21) 577-1947, 577-1587
E-mail: kksm@indosat.net.id
Expertise: Investment, tax, energy, mining, information technology law, and arbitration/mediation.
Contact: Karen Mills (International Legal Advisor), Kitty Sugondo-Kramadibrata (Project Legal Structure & Commercial Citigation), Iswahjudi A. Karim (Banking & Finance), Asrul Sani (intellectual Property Rights & Commercial Litigation), Tio M. Manihuruk (Corp. Finance/Capital Markets & Property Investment)

Kusnandar & Co.
Aetna Danamon Tower II, 24th Fl.
Jl. Jendral Sudirman Kav. 45
Jakarta 12930
P.O. Box 2057, JKT 10001
Tel.: (62-21) 577-1435
Fax : (62-21) 577 1436
E-mail: kusnalaw@ub.net.id
Internet: http://www.kusnandar.co.id/
Expertise: Foreign and domestic investment, banking, multi finance, the capital market, securities, tax, natural resources, plantations civil, commercial & criminal trademark, patent & copyright litigation & enforcement trade secret, industrial design, liquidators & bankruptcy lawyers, merger, acquisition, consolidation & integrated circuit official receivers, administrators project finance, licensing, counterfeit domestic & overseas arbitration, admiralty, insurance trade, anti-monopoly

company establishment in land, construction & conveyance, labor, unions & CLA, Indonesia or offshore tax review, maintenance accounting service, sworn translation.
Contact: Winita E. Kusnandar, John S. Beaumont

Lontoh & Kailimang
Jl. H.O.S. Cokroaminoto No. 47
Jakarta 10350
Tel.: (62-21) 392-3355
Fax : (62-21) 392-3366
E-mail: lonka@indo.net.id
Contact: Harry Ponto
Expertise: Civil law, including adoption and child custody, wills, collections, contracts, corporations, government relations, foreign claims, insurance, labor, direct foreign investment, trademarks/copyrights, auto accidents, immigration, capital market, narcotics, customs, and criminal defense.

Lubis, Ganie, Surowidjojo
Menara Imperium, 30th Fl.
Jl. H.R. Rasuna Said Kav. 1
Jakarta 12480
Tel.: (62-21) 831-5005, 831-5025
Fax : (62-21) 831-5015, 831-5035
E-mail: lgslaw@rad.net.id
Contact: Lubis, Ganie, Surowidjojo
Expertise: Representation of foreign and Indonesian clients on commercial and corporate law matters, including acquisition, corporate reorganization, mergers, agency and distributorship, arbitration, banking, capital market, commercial litigation, construction, corporate finance and secured transactions, energy, foreign, and domestic investment, commercial litigation and arbitration, international trade, lease financing, maritime and shipping, mining, real estate transactions, tax, telecommunications, intellectual proprietary rights including trademark, copyrights and patents, venture capital, insurance and labor.

Makarim & Taira S
Summitmas I, 17th Fl. & 18th Fl.
Jl. Jendral Sudirman Kav. 61-62
Jakarta 12069
Tel.: (62-21) 252-1272, 252-2460
Fax : (62-21) 252-2750, 252-2751, 252-2234
E-mail: m_ts@pacific.net.id
Expertise: International and general practice. foreign investment, commercial, capital markets, banking and finance, constructions, energy and natural resources, intellectual property, commercial litigation, tax, mergers and acquisitions, property development.

Makes & Partner (in association with SKADDEN, ARPS, SLATE, MEAGHER & FLOM)
Menara Batavia, 7th Floor
Jl. K.H. Mas Mansyur Kav. 126
Jakarta 10220
Tel.: (62-21) 574-7181
Fax : (62-21) 574-7180
E-mail: makes@indosat.net.id
Contact: Iwan Setiawan
Expertise: Corporate and commercial law, restructuring and reorganizations, mergers and acquisitions (including tender offers), foreign and domestic capital investment (including joint

ventures), banking and corporate finance (including bank financing, issuance of debt instruments and establishment of venture capital companies and local branches of international investment and commercial banks), capital markets (including debts and equity offerings by Indonesian companies in the Indonesian and International capital markets as well as listing on the Jakarta, Surabaya, Singapore, New York, London, and Luxemburg Stock Exchanges and Nasdaq), telecommunications, energy and infrastructure (including oil and gas contracts and project finance) and real property law.

Mochtar, Karuwin & Komar
Wisma Metropolitan II, 14th fl.
Jl. Jendral Sudirman, Kav 31
Jakarta 12920
Tel.: (62-21) 571-1130
Fax : (62-21) 571-1162, 570-1686
E-mail: mail@mkk.co.id
Contact: Emir Kusumaatmadja
Expertise: General corporate and commercial law, specializing in the legal aspects of international finance, foreign investment, infrastructure, and privatization.

Prof. Dr. Sudargo Gautama
Jl. Merdeka Timur 9
Jakarta 10110
Tel.: (62-21) 345-6529, 384-1358, 384-7165, 380-9488
Fax : (62-21) 384-6180, 381-3166
E-mail: sgautama@rad.net.id
Contact: Prof. Dr. Sudargo Gautama
Expertise: General, civil and criminal law, patents and trademarks, private international law, consulting, and litigation.

Rosetini Ibrahim & Associates
Wisma GKBI 33rd. Fl., Suite 3309
Jl. Jendral Sudirman #28
Jakarta 10210
Tel.: (62-21) 574-1225
Fax : (62-21) 574-1226
E-mail: rosetini@rosetini.com
Internet: www.rosetini.com
Contact: Rosetini Ibrahim, Richard Weiss, Dennie Rahmanto
Expertise: Corporate law, finance and foreign investments.

Soewito, Suhardiman, Eddymurthy & Kardono
Wisma Bank Dharmala, 14th Fl., Suite 1403
Jl. Jendral Sudirman Kav.28
Jakarta 12920
Tel.: (62-21) 521-2038
Fax : (62-21) 521-2039
E-mail: ssek@ssek.com
Internet: http://www.ssek.com
Contact: Darrell R. Johnson, Michael D. Twomey
Expertise: Corporate, commercial and financial law practice, energy and natural resources, maritime, banking, capital markets and securities law, insurance, intellectual property, investment, labor, franchising, real estate, construction and engineering, mergers and acquisitions, tax law,

arbitration, hotel and tourist development, environmental law, international trade, government contracts, immigration, oil and gas law.

Taira & Suyudono
Menara Dea Suite 1104
Kawasan Mega Kuningan
Jakarta 12950, Indonesia
Tel.: (62-21) 576-3030 Fax: (62-21) 576-3033 E-mail: finlaw@taira.co.id
Contact: Frank Taira Supit, Benny Suyudono
Expertise: Corporate law, corporate finance, capital markets, infrastructure, financial litigation, financial transactions, complicated business

BASIC TITLES ON INDONESIA

IMPORTANT!
All publications are updated annually!
Please contact IBP, Inc. at ibpusa3@gmail.com for the latest ISBNs and additional information

TITLE
Indonesia A "Spy" Guide - Strategic Information and Developments
Indonesia A Spy" Guide"
Indonesia Air Force Handbook
Indonesia Air Force Handbook
Indonesia Art Handicrafts, Children's Toys, and Paintings Export-Import and Business Opportunities Handbook
Indonesia Art Handicrafts, Children's Toys, and Paintings Export-Import and Business Opportunities Handbook
Indonesia Banking & Financial Market Handbook
Indonesia Banking & Financial Market Handbook
Indonesia Banks, Financial Companies Export-Import and Business Opportunities Handbook
Indonesia Banks, Financial Companies Export-Import and Business Opportunities Handbook
Indonesia Business and Investment Opportunities Yearbook
Indonesia Business and Investment Opportunities Yearbook
Indonesia Business and Investment Opportunities Yearbook Volume 1 Strategic Information and Opportunities
Indonesia Business and Investment Opportunities Yearbook Volume 2 Leading Export-Import, Business, Investment Opportunities and Projects
Indonesia Business Intelligence Report - Practical Information, Opportunities, Contacts
Indonesia Business Intelligence Report - Practical Information, Opportunities, Contacts
Indonesia Business Law Handbook - Strategic Information and Basic Laws
Indonesia Business Law Handbook - Strategic Information and Basic Laws
Indonesia Business Law Handbook - Strategic Information and Basic Laws
Indonesia Business Law Handbook - Strategic Information and Basic Laws
Indonesia Business Law Handbook Volume 2 Investment and Trade Laws Volume 2 Investment and Trade Laws
Indonesia Business Success Guide - Basic Practical Information and Contacts
Indonesia Chemical Industry Export-Import and Business Opportunities Handbook
Indonesia Chemical Industry Export-Import and Business Opportunities Handbook
Indonesia Clothing & Textile Industry Handbook

TITLE
Indonesia Clothing & Textile Industry Handbook
Indonesia Clothing Manufacturing Export-Import and Business Opportunities Handbook
Indonesia Clothing Manufacturing Export-Import and Business Opportunities Handbook
Indonesia Company Laws and Regulations Handbook
Indonesia Constitution and Citizenship Laws Handbook - Strategic Information and Basic Laws
Indonesia Construction Industry Export-Import and Business Opportunities Handbook
Indonesia Construction Industry Export-Import and Business Opportunities Handbook
Indonesia Country Study Guide - Strategic Information and Developments
Indonesia Country Study Guide - Strategic Information and Developments
Indonesia Country Study Guide - Strategic Information and Developments Volume 1 Strategic Information and Developments
Indonesia Criminal Laws, Regulations and Procedures Handbook - Strategic Information, Regulations, Procedures
Indonesia Customs, Trade Regulations and Procedures Handbook
Indonesia Customs, Trade Regulations and Procedures Handbook
Indonesia Diplomatic Handbook - Strategic Information and Developments
Indonesia Diplomatic Handbook - Strategic Information and Developments
Indonesia Ecology & Nature Protection Handbook
Indonesia Ecology & Nature Protection Handbook
Indonesia Ecology & Nature Protection Laws and Regulation Handbook
Indonesia Economic & Development Strategy Handbook
Indonesia Economic & Development Strategy Handbook
Indonesia Education System and Policy Handbook
Indonesia Electoral, Political Parties Laws and Regulations Handbook - Strategic Information, Regulations, Procedures
Indonesia Electrical Machinery, Equipment, Apparatus Export-Import and Business Opportunities Handbook
Indonesia Electrical Machinery, Equipment, Apparatus Export-Import and Business Opportunities Handbook
Indonesia Electronic, Electrical, Communication Export-Import and Business Opportunities Handbook
Indonesia Electronic, Electrical, Communication Export-Import and Business Opportunities Handbook
Indonesia Energy Policy, Laws and Regulation Handbook
Indonesia Energy Policy, Laws and Regulations Handbook
Indonesia Energy Policy, Laws and Regulations Handbook
Indonesia Export-Import Trade and Business Directory
Indonesia Export-Import Trade and Business Directory
Indonesia Fishing and Aquaculture Industry Handbook - Strategic Information, Regulations, Opportunities
Indonesia Food, Beverage and Tobacco Export-Import and Business Opportunities Handbook
Indonesia Food, Beverage and Tobacco Export-Import and Business Opportunities Handbook
Indonesia Footwear Manufacturing Export-Import and Business Opportunities Handbook
Indonesia Footwear Manufacturing Export-Import and Business Opportunities Handbook
Indonesia Foreign Policy and Government Guide

TITLE
Indonesia Foreign Policy and Government Guide
Indonesia Government and Business Contacts Handbook
Indonesia Government and Business Contacts Handbook
Indonesia Health Care, Cosmetics, Pharmaceuticals Producers Handbook
Indonesia Health Care, Cosmetics, Pharmaceuticals Producers Handbook
Indonesia Healthcare Sector Organization, Management and Payment Systems Handbook - Strategic Information, Programs and Regulations
Indonesia Immigration Laws and Regulations Handbook - Strategic Information and Basic Laws
Indonesia Industrial and Business Directory
Indonesia Insolvency (Bankruptcy) Laws and Regulations Handbook - Strategic Information and Basic Laws
Indonesia Internet and E-Commerce Investment and Business Guide - Strategic and Practical Information: Regulations and Opportunities
Indonesia Internet and E-Commerce Investment and Business Guide - Strategic and Practical Information: Regulations and Opportunities
Indonesia Investment and Business Guide - Strategic and Practical Information
Indonesia Investment and Business Guide - Strategic and Practical Information
Indonesia Investment and Business Guide - Strategic and Practical Information
Indonesia Investment and Business Guide - Strategic and Practical Information
Indonesia Investment and Business Guide Volume 2 Business, Investment Opportunities and Incentives
Indonesia Investment and Business Profile - Basic Information and Contacts for Succesful investment and Business Activity
Indonesia Investment and Trade Laws and Regulations Handbook
Indonesia Investment, Trade Strategy and Agreements Handbook - Strategic Information and Basic Agreements
Indonesia Justice System and National Police Handbook
Indonesia Justice System and National Police Handbook
Indonesia Labor Laws and Regulations Handbook - Strategic Information and Basic Laws
Indonesia Land Ownership and Agriculture Laws Handbook
Indonesia Leather and Leather Product Manufacturing Export-Import and Business Opportunities Handbook
Indonesia Leather and Leather Product Manufacturing Export-Import and Business Opportunities Handbook
Indonesia Machinery & Equipment Export-Import and Business Opportunities Handbook
Indonesia Machinery & Equipment Export-Import and Business Opportunities Handbook
Indonesia Medical & Pharmaceutical Industry Handbook
Indonesia Medical & Pharmaceutical Industry Handbook
Indonesia Metals and Metal Products Export-Import and Business Opportunities Handbook
Indonesia Metals and Metal Products Export-Import and Business Opportunities Handbook
Indonesia Mineral & Mining Sector Investment and Business Guide - Strategic and Practical Information
Indonesia Mineral & Mining Sector Investment and Business Guide - Strategic and Practical Information
Indonesia Mining Laws and Regulations Handbook
Indonesia Mining, Oil & Gas Industry Export-Import and Business Opportunities Handbook

TITLE
Indonesia Mining, Oil & Gas Industry Export-Import and Business Opportunities Handbook
Indonesia Oil & Gas Sector Business & Investment Opportunities Yearbook
Indonesia Oil & Gas Sector Business & Investment Opportunities Yearbook
Indonesia Oil and Gas Exploration Laws and Regulation Handbook
Indonesia Oil and Gas Industry Handbook
Indonesia Oil and Gas Industry Handbook
Indonesia Paper and Paper Product, Printing and Publishing Export-Import and Business Opportunities Handbook
Indonesia Paper and Paper Product, Printing and Publishing Export-Import and Business Opportunities Handbook
Indonesia Plastic Product Manufacturing Export-Import and Business Opportunities Handbook
Indonesia Plastic Product Manufacturing Export-Import and Business Opportunities Handbook
Indonesia Porcelain and Glass Manufacturing Export-Import and Business Opportunities Handbook
Indonesia Porcelain and Glass Manufacturing Export-Import and Business Opportunities Handbook
Indonesia Privatization Programs and Regulations Handbook
Indonesia Privatization Programs and Regulations Handbook
Indonesia Real Estate Companies Export-Import and Business Opportunities Handbook
Indonesia Real Estate Companies Export-Import and Business Opportunities Handbook
Indonesia Recent Economic and Political Developments Yearbook
Indonesia Recent Economic and Political Developments Yearbook
Indonesia Recent Economic and Political Developments Yearbook
Indonesia Research & Development Policy Handbook
Indonesia Research & Development Policy Handbook
Indonesia Research & Development Policy Handbook
Indonesia Rubber and Rubber Product Manufacturing Export-Import and Business Opportunities Handbook
Indonesia Rubber and Rubber Product Manufacturing Export-Import and Business Opportunities Handbook
Indonesia Starting Business (Incorporating) in Indonesia Guide
Indonesia Starting Business (Incorporating) in....Guide
Indonesia Stone, Marble, and Asbestos Manufacturing Export-Import and Business Opportunities Handbook
Indonesia Stones, Marble, Granite and Asbestos Manufacturing Export-Import and Business Opportunities Handbook
Indonesia Tax Guide Volume 1 Strategic Information and Basic Regulations
Indonesia Taxation Laws and Regulations Handbook
Indonesia Technical Assistance Proposals 2006-2009 Vol 1 Project Assistance Proposals
Indonesia Technical Assistance Proposals 2006-2009 Vol 2 Technical Assistance Proposals
Indonesia Telecom Laws and Regulations Handbook
Indonesia Telecommunication Industry Business Opportunities Handbook
Indonesia Telecommunication Industry Business Opportunities Handbook
Indonesia Textile, Clothing Products Export-Import and Business Opportunities Handbook
Indonesia Textile, Clothing Products Export-Import and Business Opportunities Handbook

TITLE
Indonesia Transport Equipment and Ships Producers Handbook
Indonesia Transport Equipment and Ships Producers Handbook
Indonesia Transportation Policy and Regulations Handbook
Indonesia Wood, Bamboo, Grass, Household Export-Import and Business Opportunities Handbook
Indonesia Wood, Bamboo, Grass, Household Export-Import and Business Opportunities Handbook
Indonesia: How to Invest, Start and Run Profitable Business in Indonesia Guide - Practical Information, Opportunities, Contacts

WORLD ENERGY POLICY, BUSINESS & INVESTMENT OPPORTUNITIES LIBRARY

Price: $149.95 Each

Title
Albania Energy Policy, Laws and Regulations Handbook
Algeria Energy Policy, Laws and Regulations Handbook
Angola Energy Policy, Laws and Regulations Handbook
Australia Energy Policy, Laws and Regulations Handbook
Azerbaijan Energy Policy, Laws and Regulations Handbook
Bahrain Energy Policy, Laws and Regulations Handbook
Bangladesh Energy Policy, Laws and Regulations Handbook
Benin Energy Policy, Laws and Regulations Handbook
Brazil Energy Policy, Laws and Regulations Handbook
Brunei Energy Policy, Laws and Regulations Handbook
Cameroon Energy Policy, Laws and Regulations Handbook
Canada Energy Policy, Laws and Regulations Handbook
China Energy Policy, Laws and Regulations Handbook
Colombia Oil & Gas Sector Energy Policy, Laws and Regulations Handbook
Congo Energy Policy, Laws and Regulations Handbook
Congo, Dem. Republic Energy Policy, Laws and Regulations Handbook
Cote d'Ivoire Energy Policy, Laws and Regulations Handbook
Dubai Energy Policy, Laws and Regulations Handbook
Ecuador Energy Policy, Laws and Regulations Handbook
Egypt Energy Policy, Laws and Regulations Handbook
Equatorial Guinea Energy Policy, Laws and Regulations Handbook
Gabon Energy Policy, Laws and Regulations Handbook
India Energy Policy, Laws and Regulations Handbook
Indonesia Energy Policy, Laws and Regulations Handbook
Iran Energy Policy, Laws and Regulations Handbook
Iraq Energy Policy, Laws and Regulations Handbook
Kazakhstan Energy Policy, Laws and Regulations Handbook

Title
Kuwait Energy Policy, Laws and Regulations Handbook
Libya Energy Policy, Laws and Regulations Handbook
Malaysia Energy Policy, Laws and Regulations Handbook
Mexico Energy Policy, Laws and Regulations Handbook
Morocco Energy Policy, Laws and Regulations Handbook
Namibia Energy Policy, Laws and Regulations Handbook
New Zealand Energy Policy, Laws and Regulations Handbook
Nigeria Energy Policy, Laws and Regulations Handbook
Norway Energy Policy, Laws and Regulations Handbook
Oman Energy Policy, Laws and Regulations Handbook
Pakistan Energy Policy, Laws and Regulations Handbook
Papua New Guinea Energy Policy, Laws and Regulations Handbook
Papua New Guinea Oil & Gas Sector Energy Policy, Laws and Regulations Handbook
Peru Energy Policy, Laws and Regulations Handbook
Philippines Energy Policy, Laws and Regulations Handbook
Qatar Energy Policy, Laws and Regulations Handbook
Sri Lanka Energy Policy, Laws and Regulations Handbook
Romania Energy Policy, Laws and Regulations Handbook
Russia Energy Policy, Laws and Regulations Handbook
Saudi Arabia Energy Policy, Laws and Regulations Handbook
Sudan Energy Policy, Laws and Regulations Handbook
Suriname Energy Policy, Laws and Regulations Handbook
Syria Energy Policy, Laws and Regulations Handbook
Trinidad and Tobago Oil & Gas Sector Energy Policy, Laws and Regulations Handbook
Tunisia Energy Policy, Laws and Regulations Handbook
Turkey Energy Policy, Laws and Regulations Handbook
Turkmenistan Energy Policy, Laws and Regulations Handbook
Ukraine Oil & Gas Sector Energy Policy, Laws and Regulations Handbook
United Arab Emirates Energy Policy, Laws and Regulations Handbook
United Kingdom Energy Policy, Laws and Regulations Handbook
United Kingdom Oil & Gas Sector Energy Policy, Laws and Regulations Handbook
United States Energy Policy, Laws and Regulations Handbook
Venezuela Energy Policy, Laws and Regulations Handbook
Vietnam Energy Policy, Laws and Regulations Handbook
Vietnam Energy Policy, Laws and Regulations Handbook
Yemen Energy Policy, Laws and Regulations Handbook

Made in the USA
Monee, IL
13 September 2026

62976392R00164